ADVANCES IN CHEMICAL ENGINEERING
Volume 25

ADVANCES IN CHEMICAL ENGINEERING

Editor-in-Chief
JAMES WEI
School of Engineering and Applied Science
Princeton University
Princeton, New Jersey

Editors

KENNETH B. BISCHOFF
Department of Chemical Engineering
University of Delaware
Newark, Delaware

MORTON M. DENN
College of Chemistry
University of California at Berkeley
Berkeley, California

JOHN H. SEINFELD
Department of Chemical Engineering
California Institute of Technology
Pasadena, California

GEORGE STEPHANOPOULOS
Department of Chemical Engineering
Massachusetts Institute of Technology
Cambridge, Massachusetts

Volume 25

ACADEMIC PRESS
San Diego   London   Boston
New York   Sydney   Tokyo   Toronto

TP
145
.D7
v.25
1999

This book is printed on acid-free paper.

Copyright © 2000 by ACADEMIC PRESS

All Rights Reserved.
No part of this publication may be reproduced or transmitted in any form or by any means, electronic or mechanical, including photocopy, recording, or any information storage and retrieval system, without permission in writing from the Publisher.

The appearance of the code at the bottom of the first page of a chapter in this book indicates the Publisher's consent that copies of the chapter may be made for personal or internal use of specific clients. This consent is given on the condition, however, that the copier pay the stated per copy fee through the Copyright Clearance Center, Inc. (222 Rosewood Drive, Danvers, Massachusetts 01923), for copying beyond that permitted by Sections 107 or 108 of the U.S. Copyright Law. This consent does not extend to other kinds of copying, such as copying for general distribution, for advertising or promotional purposes, for creating new collective works, or for resale. Copy fees for pre-2000 chapters are as shown on the title pages. If no fee code appears on the title page, the copy fee is the same as for current chapters.
0065-2377/00 $30.00

Explicit permission from Academic Press is not required to reproduce a maximum of two figures or tables from an Academic Press chapter in another scientific or research publication provided that the material has not been credited to another source and that full credit to the Academic Press chapter is given.

Academic Press
A Harcourt Science and Technology Company
525 B Street, Suite 1900, San Diego, California 92101-4495, U.S.A.
http://www.apnet.com

Academic Press
24-28 Oval Road, London NW1 7DX, UK
http://www.hbuk.co.uk/ap/

International Standard Book Number: 0-12-008525-9

PRINTED IN THE UNITED STATES OF AMERICA
99 00 01 02 03 04 QW 9 8 7 6 5 4 3 2 1

CONTENTS

Contributors vii
Preface ix

Process Data Analysis and Interpretation

J. F. Davis, M. J. Piovoso, K. A. Hoo, and B. R. Bakshi

I. Introduction 2
   A. Overview 2
   B. Data Analysis 3
   C. Data Interpretation 5
   D. Challenges 7
   E. Overview of Discussion 8
II. Data Analysis: A Unifying Framework 10
III. Input Analysis Methods 13
   A. Methods Based on Linear Projection 14
   B. Methods Based on Nonlinear Projection 27
IV. Input-Output Analysis 32
   A. Methods Based on Linear Projection 33
   B. Methods Based on Nonlinear Projection 40
   C. Partition-Based Methods 41
V. Data Interpretation 43
   A. Nonlocal Interpretation Methods 47
   B. Local Interpretation Methods 55
VI. Symbolic-Symbolic Interpretation: Knowledge-Based System Interpreters . . 64
   A. Families of KBS Approaches 67
VII. A. Managing Scale and Scope of Large-Scale Process Operations 72
   A. Degradation of Interpretation Performance 73
   B. Hierarchical Modularization 79
VIII. Comprehensive Examples 82
   A. Comprehensive Example 1: Detection of Abnormal Situations 82
   B. Comprehensive Example 2: Data Analysis of Batch Operation Variability . 86
   C. Comprehensive Example 3: Diagnosis of Operating Problems in a Batch Polymer Reactor 90
IX. Summary 96
   References 97

Mixing and Dispersion of Viscous Liquids and Powdered Solids

J. M. Ottino, P. DeRoussel, S. Hansen, and D. V. Khakhar

I. Preliminaries 105

II. Mixing and Dispersion of Immiscible Fluids	124
A. Breakup	130
B. Coalescence	151
C. Breakup and Coalescence in Complex Flows	155
III. Fragmentation and Aggregation of Solids	159
A. Fragmentation	163
B. Aggregation	180
IV. Concluding Remarks	194
References	198

Application of Periodic Operation to Sulfur Dioxide Oxidation

Peter L. Silveston, Li Chengyue, and Yuan Wei-Kang

I. Introduction	206
II. Air Blowing of the Final Stage of a Multistage SO_2 Converter	208
A. Experimental Studies	208
B. Mechanistic Model	215
C. Application to the Final Stage of SO_2 Converter with Composition Forcing	216
D. Application to an Isothermal Back-Mixed Reactor	217
III. SO_2 Converters Based on Periodic Reversal of the Flow Direction	223
A. Industrial Applications	225
B. Experimental Results	230
C. Modeling and Simulation	234
D. Overview	248
IV. Conversion of SO_2 in Trickle-Bed Catalytic Scrubbers Using Periodic Flow Interruption	248
A. Experimental Studies	249
B. Modeling and Simulation	256
C. Application to Stack-Gas Scrubbing	261
D. Physical Explanation	269
V. The Future of Periodic Operations for SO_2 Conversion	272
Nomenclature	273
References	278

Index	283
Contents of Volumes in This Serial	291

CONTRIBUTORS

Numbers in parentheses indicate the pages on which the author's contributions begin.

B. R. BAKSHI, *Department of Chemical Engineering, Ohio State University, Columbus, Ohio 43210* (1)

LI CHENGYUE, *Department of Chemical Engineering, Beijing University of Chemical Technology, Beijing, 100029, China* (205)

J. F. DAVIS, *Department of Chemical Engineering, Ohio State University, Columbus, Ohio 43210* (1)

P. DEROUSSEL, *Department of Chemical Engineering, Northwestern University, Evanston, Illinois 60208* (105)

S. HANSEN, *Department of Chemical Engineering, Northwestern University, Evanston, Illinois 60208* (105)

K. A. HOO, *Department of Chemical Engineering, Swearingen Engineering Center, University of South Carolina, Columbia, South Carolina 29208* (1)

D. V. KHAKHAR, *Department of Chemical Engineering, Northwestern University, Evanston, Illinois 60208* (105)

J. M. OTTINO, *Department of Chemical Engineering, Northwestern University, Evanston, Illinois 60208* (105)

M. J. PIOVOSO, *Dupont, Wilmington, Delaware 19880* (1)

PETER L. SILVESTON, *Department of Chemical Engineering, University of Waterloo, Waterloo, Ontario, Canada* (205)

YUAN WEI-KANG, *Department of Chemical Engineering, East China University of Chemical Technology, Shanghai 201107, China* (205)

PREFACE

This is the Silver Anniversary for *Advances in Chemical Engineering,* founded by Thomas Drew and John Hoopes in 1956. The first volume contained review papers by legendary luminaries such as Westwater, Metzner, Bird, Sage, Treybal, Schrage, and Henley. Let me quote from the first preface:

> The end *Advances in Chemical Engineering* seeks to serve is to provide the engineer with critical running summaries of recent work; some that bring standard topics up to date; others that gather and examine the results of new or newly utilized techniques and methods of seeming promise in the field .... The editors salute and thank the authors who have braved the vicissitudes of this new venture, and commend their efforts to their colleagues: the engineers of chemical industry.

This Silver Anniversary volume contains review papers by contemporary luminaries such as Ottino, Davis, Silveston, Li, Yuan. *Advances* is still dedicated to providing engineers with the timely information to update and to explore new fields. However, the chemical industry is no longer our only concern, as we have enlarged our frontier to include many new industries in which chemistry plays a significant role. Chemical engineers are now making major contributions in such fields as biochemical processing, biomedical devices, electronic and optical material and devices, advanced materials, and safety and environmental protection. Chemical engineers are also discovering new scientific concepts and tools in the fields of mathematical and computation analysis, computer-assisted process and control, and surface sciences.

We are looking forward to the Golden Anniversary in the year 2024, and we cannot wait to see how the world of chemical engineering will shape up.

JAMES WEI

PROCESS DATA ANALYSIS AND INTERPRETATION

J. F. Davis

Department of Chemical Engineering, Ohio State University, Columbus, Ohio 43210

M. J. Piovoso

DuPont, Wilmington, Delaware 19880

K. A. Hoo

Department of Chemical Engineering, University of South Carolina, Columbia, South Carolina 29208

B. R. Bakshi

Department of Chemical Engineering, Ohio State University, Columbus, Ohio 43210

I. Introduction	2
A. Overview	2
B. Data Analysis	3
C. Data Interpretation	5
D. Challenges	7
E. Overview of Discussion	8
II. Data Analysis: A Unifying Framework	10
III. Input Analysis Methods	13
A. Methods Based on Linear Projection	14
B. Methods Based on Nonlinear Projection	27
IV. Input–Output Analysis	32
A. Methods Based on Linear Projection	33
B. Methods Based on Nonlinear Projection	40
C. Partition-Based Methods	41
V. Data Interpretation	43
A. Nonlocal Interpretation Methods	47
B. Local Interpretation Methods	55
VI. Symbolic–Symbolic Interpretation: Knowledge-Based System Interpreters	64
A. Families of KBS Approaches	67
VII. Managing Scale and Scope of Large-Scale Process Operations	72
A. Degradation of Interpretation Performance	73

B. Hierarchical Modularization   79
VIII. Comprehensive Examples   82
  A. Comprehensive Example 1: Detection of Abnormal
     Situations   82
  B. Comprehensive Example 2: Data Analysis of Batch
     Operation Variability   86
  C. Comprehensive Example 3: Diagnosis of Operating
     Problems in a Batch Polymer Reactor   90
 IX. Summary   96
     References   97

I. Introduction

A. Overview

The recent, unprecedented ease with which process data can be collected and recorded is the use of data to support increasingly sophisticated process management activities such as increasing process efficiency, improving product quality, identifying hazardous conditions, and managing abnormal situations. To support these activities, the raw process data must be transformed into meaningful descriptions of process conditions. As shown in Fig. 1, the data analysis and interpretation techniques that affect the required transformation can be viewed as a classical pattern recognition problem with two primary tasks: *data analysis* (or feature extraction), which consists of numerically processing the data to produce numerical features of interest, and *data interpretation* (or feature mapping), which consists of assigning symbolic interpretations (i.e., labels) to numerical features.

The general pattern recognition problem can be described as follows. The input data, or pattern X, is defined by a number of specific data measurements, x_i, defined at a particular point in time:

$$X = (x_1, x_2, \ldots, x_i, \ldots, x_d). \tag{1}$$

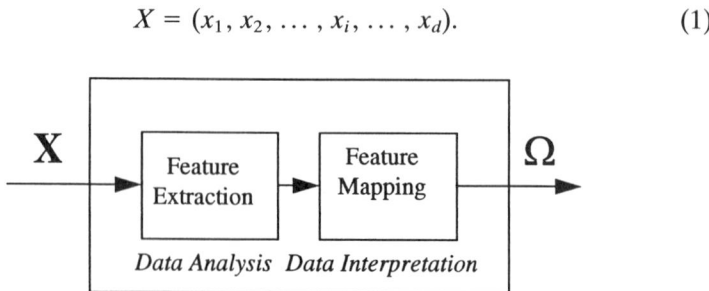

FIG. 1. Pattern recognition view of data analysis and interpretation.

The representation space, Ξ, consists of the set of all possible patterns, $X \in \Xi$. The set of all possible class labels ω_j forms the interpretation space Ω:

$$\Omega = (\omega_1, \omega_2, \ldots, \omega_j, \ldots, \omega_p). \tag{2}$$

Pattern recognition then corresponds to determining the numeric-symbolic mapping, ζ, from the representation space to the interpretation space:

$$\zeta: \Xi \to \Omega \quad \text{or} \quad \zeta: (x_1, x_2, \ldots, x_d) \to \omega_j. \tag{3}$$

Typically, determining ζ is easier after extracting relevant features, Z, from X and thereby using a mapping of the form

$$\zeta: \{\zeta': (x_1, x_2, \ldots, x_d) \to (z_1, z_2, \ldots, z_f)\} \to \omega_j. \tag{4}$$

Using this notation, X corresponds to any time series of data with x_i being a sampled value, and Z represents the processed forms of the data (i.e., a pattern). The z_i are the pattern features, ω_j is the appropriate label or interpretation, ζ' is the feature extraction or data analysis transformation, and ζ is the mapping or interpretation that must be developed.

The primary purpose of pattern recognition is to determine class membership for a set of numeric input data. The performance of any given approach is ultimately driven by how well an appropriate discriminant can be defined to resolve the numeric data into a label of interest. Because of both the importance of the problem and its many challenges, significant research has been applied to this area, resulting in a large number of techniques and approaches. With this publication, we seek to provide a common framework to discuss the application of these approaches.

B. Data Analysis

The objective of data analysis (or feature extraction) is to transform numeric inputs in such a way as to reject irrelevant information that can confuse the information of interest and to accentuate information that supports the feature mapping. This usually is accomplished by some form of *numeric–numeric* transformation in which the numeric input data are transformed into a set of numeric features. The numeric–numeric transformation makes use of a process model to map between the input and the output.

Models are generally built from either fundamental knowledge about a system or empirical data collected from a system. Models based on fundamental knowledge attempt to directly predict actual plant behavior. Therefore, they can be especially useful for those operating situations that have not been previously observed. However, accurate fundamental models are difficult to build for many chemical processes because the underlying phenomena are often not well understood. Thus, empirical models are an extremely popular approach to map between input and output variables. In terms of feature extraction, a model captures the relationship between data features and labels of process behaviors.

As with any empirical technique, the development of a feature mapping scheme is an inherently inductive process that requires a good knowledge source to produce discriminants relevant to the numeric features to be interpreted and the labels of interest. Determination of the mapping requires a set of reference patterns with correct class labels. This *training set* consists of multiple *pattern exemplars*. The distinguishing features used in generating the empirical model can be expressed either explicitly or implicitly, depending on the underlying technique.

Data analysis techniques can be viewed as either *input mapping* or *input–output mapping,* as shown in Fig. 2. The transformations are illustrated in Fig. 3. *Input mapping* refers to manipulation of input data, X, without consideration of any output variables or features. This type of mapping is generally used to transform the input data to a more convenient representation, Z', while retaining all the essential information of the original data. *Input–output mapping* extracts important features, Z, by relating input variables, X, output variables, Y, or features of interest, Z, together. This type of mapping may include an implicit input transformation step for reducing input dimensionality.

Input mapping methods can be divided into *univariate, multivariate,* and *probabalistic* methods. *Univariate* methods analyze the inputs by extracting the relationship between the *measurements*. These methods include various types of single-scale and multiscale filtering such as exponential smoothing, wavelet thresholding, and median filtering. *Multivariate* methods analyze

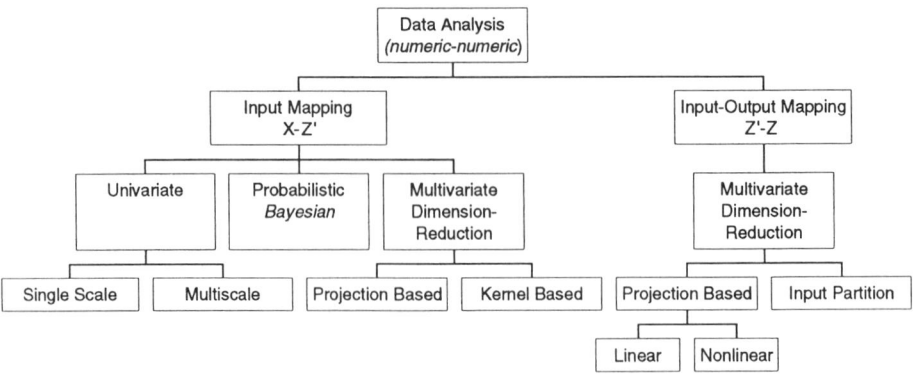

FIG. 2. A classification of data analysis methods.

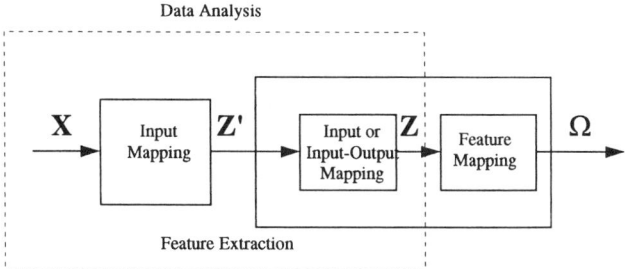

FIG. 3. Data analysis applied to data interpretation.

the inputs by extracting the relationship between the *variables*. These methods reduce dimensionality and include both projection and kernel based approaches, such as principal component analysis and clustering. All share the characteristic of transforming a given matrix of data without information on any output variables. As will be discussed, these methods either produce or support the extraction of features that can be used to assign labels. As shown in Fig. 3, it is often useful to apply an input mapping approach to preprocess data to produce a more useful representation of the input data. Input–output methods map input numeric data to a transformed set of numeric features *with* knowledge of the output variables. These methods can be classified based on the method used to transform the input space (linear projection, nonlinear projection, or partition). Methods based on *linear projection* combine the inputs as a linear weighted sum. These methods include popular empirical modeling methods such as partial least squares (PLS) regression (both linear and nonlinear) and back propagation networks (BPNs) with a single hidden layer. Methods based on *nonlinear projection* can be divided further into *local* and *nonlocal* methods. Nonlocal methods have at least one infinite dimension in a projection surface. Local methods include such approaches as radial basis function networks (RBFNs) or Adaptive Resonance Theory (ART), whereas nonlocal methods include approaches like BPNs with multiple hidden layers. *Subset selection* or *partition-based* methods include classification and regression trees (CARTs) and multivariate adaptive regression splines (MARSs).

C. Data Interpretation

In this publication, the purpose of data analysis is to drive toward data interpretation that consists of assigning various types of labels. These label types include *state descriptions* (e.g., normal or high), *trends* (e.g., increasing or pulsing), *landmarks* (e.g., process change), *shape descriptions* (e.g., skewed, tail), and *fault descriptions* (e.g., flooding or contamination). Assignment of labels requires that feature extraction be extended by establishing decision discriminants based on labeled (known and observed) situations of sufficiently similar characteristics (i.e., *pattern exemplars*). The combination of feature extraction and feature mapping specifically for label assignment is called *numeric–symbolic mapping*, as shown in Fig. 4. A number of approaches integrate both the feature extraction and feature mapping aspects of numeric–symbolic mapping. Although both aspects are occurring in these methods, the individual steps cannot be distinguished. The resultant symbolic output of a numeric–symbolic interpreter may itself need further interpretation. This requires a *symbolic–symbolic* mapper such as a knowledge-based system (KBS) that can further refine symbolic interpretations. Figure 4 indicates this as a mapping from Ω' to Ω.

Feature mapping (i.e., numeric–symbolic mapping) requires decision mechanisms that can distinguish between possible label classes. As shown in Fig. 5, widely used decision mechanisms include linear discriminant surfaces, local data cluster criteria, and simple decision limits. Depending on the nature of the features and the feature extraction approaches, one or more of these decision mechanisms can be selected to assign labels.

Labels are distinguished based on whether they are context dependent or context-free. *Context-dependent* labels require simultaneous consideration of time records from more than one process variable; *context-free* labels do not. Thus, generating context-free trend, landmark, and fault descriptions is considerably more simple than generating context-dependent descriptions. Context-free situations can take advantage of numerous methods for common, yet useful, interpretations. Context-dependent situations, however, require the application of considerable process knowledge to get a useful interpretation. In these situations, performance is dependent on the availability, coverage, and distribution of labeled process data from

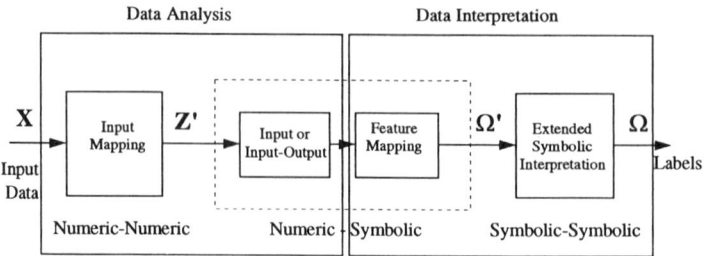

FIG. 4. The numeric–symbolic and symbolic–symbolic views of data interpretation.

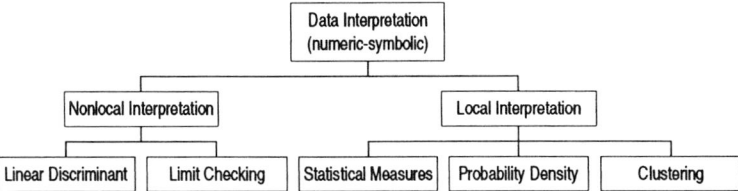

FIG. 5. Label class decision methods for data interpretation.

known process situations. Because of the complexity of chemical processes, most labels in this domain are context dependent.

Context-dependent situations often lead to a large-scale input dimension. Because the required number of training examples increases with the number of measured variables or features, reducing the input dimensionality may improve system performance. In addition, decision discriminants will be less complex (because of fewer dimensions in the data) and more easily determined. The reduction in dimensionality can be most readily achieved by eliminating redundancy in the data so that only the most relevant features are used for mapping to a given set of labels.

In addition to a large input dimension, complex processes also typically have a large output dimension (i.e., many possible plant behaviors). Although only a few events occur at any given time, identification of those few events requires consideration of a large subset of all possible events. Without careful management of both input and output dimensionality, interpretation performance deteriorates rapidly with scale and complexity.

D. CHALLENGES

Several significant challenges exist in applying data analysis and interpretation techniques to industrial situations. These challenges include (1) the scale (amount of input data) and scope (number of interpretations) of the problem, (2) the scarcity of abnormal situation exemplars, (3) uncertainty in process measurements, (4) uncertainty in process discriminants, and (5) the dynamic nature of process conditions.

1. Process Scale and Scope. Within the chemical process industry, a typical operator console will collect 2000 to 5000 measurements every time increment, typically 1 minute or less. Despite this large amount of process data being generated, relatively few events occur at any single time. The data must be used to determine which, out of a typical range of 300 to 600 possible descriptions (process labels), reflect current conditions. In addition to these "final" process labels, many useful, intermediate labels may also be developed. Thus, any approach must be able to manage the large dimensionality of the problem.

2. Lack of Abnormal Situation Exemplars. Because chemical processes are highly controlled, labeled sensor patterns that are available for inductive learning techniques are limited to primarily normal operating conditions. Although the abundance of process data provides for many process exemplars, very few are available for general abnormal situations—even fewer are available for a specific fault. Because most approaches to data analysis and interpretation depend heavily on using known patterns to define decision boundaries, the limited availability of these data severely affect performance and resolution. Ideally, the data would be well distributed (particularly in the vicinity of pattern class boundaries) so as to minimize uncertainty in the boundary regions.

3. Uncertainty in Process Measurements. Sensor measurements are always subject to noise, calibration error, and temporary signal loss, as well as various faults that may not be immediately detected. Therefore, data preprocessing will often be required to overcome the inherent limitations of the process measurements.

4. Uncertainty in Process Discriminants. Because processes operate over a continuum, data analysis generally produces distinguishing features that exist over a continuum. This is further compounded by noise and errors in the sensor measurements. Therefore, the discriminants developed to distinguish various process labels may overlap, resulting in uncertainty between data classes. As a result, it is impossible to define completely distinguishing criteria for the patterns. Thus, uncertainty must be addressed inherently.

5. Dynamic Changes to Process Conditions. Because of the dynamic nature of chemical processes, conditions change over time as a result of changes in production rates, quality targets, feed compositions, and equipment conditions. This in turn changes how certain process information may be interpreted. Thus, adaptive interpretation ability is often required.

E. Overview of Discussion

Data analysis and interpretation appears overwhelmingly complex when viewed as a single, unilateral approach (i.e., a single feature extraction step and a single interpretation step). This is because practical considerations demand (1) modularization to manage the input and output scale associated with a complex plant, (2) integration of a variety of data interpretation methods to address differing data analysis requirements throughout a plant, and (3) the use of multiple interpreters to map to specific labels.

Because the techniques for data analysis and interpretation are targeted to address different process characteristics, care must be taken in choosing the most appropriate set of techniques. For example, some techniques work best with abundant process data; others, with limited process data. Some can handle highly correlated data, while others cannot. In selecting appropriate methods, two practical considerations stand out:

1. Availability of good quality and adequate quantity of training data is essential for all empirical modeling methods.
2. Selecting the best method for a given task is not straightforward. It requires significant insight into the methods and the task.

In this chapter, we focus on recent and emerging technologies that either are or soon will be applied commercially. Older technologies are discussed to provide historic perspective. Brief discussions of potential future technologies are provided to indicate current development directions. The chapter substantially extends an earlier publication (Davis *et al.,* 1996a) and is divided into seven main sections beyond the introduction: Data Analysis, Input Analysis, Input–Output Analysis, Data Interpretation, Symbolic–Symbolic Interpretation, Managing Scale and Scope of Large-Scale Process Operations, and Comprehensive Examples.

Data Analysis presents a unifying framework for discussing and comparing input and input–output methods.

Input Analysis addresses input mapping approaches that transform input data without knowledge of or interest in output variables.

Input–Output Analysis considers feature extraction in the context of empirical modeling approaches.

Data Interpretation extends data analysis techniques to label assignment and considers both integrated approaches to feature extraction and feature mapping and approaches with explicit and separable extraction and mapping steps. The approaches in this section focus on those that form numeric–symbolic interpreters to map from numeric data to specific labels of interest.

Symbolic–Symbolic Intepretation addresses knowledge-based system (KBS) extensions to numeric–symbolic interpreters.

Managing Scale and Scope of Large-Scale Process Operations discusses the performance issues associated with data analysis and interpretation that occur as the scale of the problem increases (such as in complex process operations).

Finally, the chapter illustrates the integrated application of multiple analysis and interpretation technologies with several illustrative process examples.

This chapter provides a complementary perspective to that provided by Kramer and Mah (1994). Whereas they emphasize the statistical aspects of the three primary process monitoring tasks, data rectification, fault detection, and fault diagnosis, we focus on the theory, development, and performance of approaches that combine data analysis and data interpretation into an automated mechanism via feature extraction and label assignment.

II. Data Analysis: A Unifying Framework

Data analysis methods transform inputs into more meaningful and concentrated sources of information. As discussed in the introduction, measured variables can be analyzed by input or input–output transformation methods. These methods include a wide variety of linear and nonlinear modeling methods developed across a wide range of technical areas, including statistics, process simulation, control, and intelligent systems. Although the two categories of methods are quite different with respect to consideration of output variables, greater insight into similarities and differences can be achieved by comparing them in the context of a unified framework.

More specifically, input data analysis methods are similar to input–output methods, but rely on different strategies for extracting the relevant information. With reference to the general expression in Eq. (4), the resulting analyzed or latent variable for all input methods can be represented as

$$z_m = \phi_m(\alpha; x_1, \ldots, x_d), \quad (5)$$

where $x_1, \ldots, x_d$ are the input variables, α is the matrix of input transformation parameters, ϕ_m is the mth input transformation function, and z_m is a latent or transformed variable. The objective of input analysis methods is to determine α and ϕ to satisfy a selected optimization criterion. Univariate input analysis methods transform a single input variable at a time, whereas multivariate methods transform all inputs simultaneously. Equation (5) indicates that input analysis methods require decisions about the nature of the input transformation, ϕ, and the optimization criteria used for determining both α and ϕ. By definition, input analysis methods do not consider the outputs for determining the latent variables. Because the ultimate objective of data analysis and interpretation is to relate inputs to output labels, input analysis methods require an additional interpretation step to map the latent variables to the labels.

Sometimes more meaningful features may be obtained by considering the behavior of both input and output variables together in the analysis.

Typically, input-output analysis methods extract the most relevant signal features by relating the analyzed variables to process output variables, y_i,

$$y_i = \sum_{m=1}^{M} \beta_{mk}\theta_m(\phi_m(\alpha; x_1, x_2, \ldots, x_J)), \qquad (6)$$

where y_i is the ith predicted output or response variable, θ_m is the mth basis or activation function, and β_{mk} is the output weight or regression coefficient relating the mth basis function to the kth output. Equation (6) indicates that in addition to decisions about the nature of the input transformation and optimization criteria for the input transformation, input-output analysis methods require decisions about the type of basis functions, θ_m, and additional optimization criteria for the parameters, β_{mk}, and basis functions. With regard to interpretation, the y_i, β_{mk}, or the implicit latent variables, z_m, can be used as features.

Specific data analysis methods can be derived from Eqs. (5) and (6) depending on decisions about the input transformation, type of activation or basis functions, and optimization criteria. These decisions form the basis of a common framework for comparing all empirical modeling methods (Bakshi and Utojo, 1999).

Based on the nature of input transformation, both input and input-output analysis methods can be classified into the following three categories:

- *Methods based on linear projection* exploit the linear relationship among inputs by projecting them on a linear hyperplane before applying the basis function (see Fig. 6a). Thus, the inputs are transformed in combination as a linear weighted sum to form the latent variables. Univariate input analysis is a special case of this category where the single variable is projected on itself.
- *Methods based on nonlinear projection* exploit the nonlinear relationship between the inputs by projecting them on a nonlinear hypersurface resulting in latent variables that are nonlinear functions of the inputs, as shown in Figs. 6b and 6c. If the inputs are projected on a localized hypersurface such as a hypersphere or hyperellipse, then the basis functions are local, depicted in Fig. 6c. Otherwise, the basis functions are nonlocal, as shown in Fig. 6b.
- *Partition-based methods* address dimensionality by selecting input variables that are most relevant to efficient empirical modeling. The input space is partitioned by hyperplanes that are perpendicular to at least one of the input axes, as depicted in Fig. 6d.

With reference to Eq. (6), an input transformation is determined by the function, ϕ_m, and parameters, α, whereas the model relating the trans-

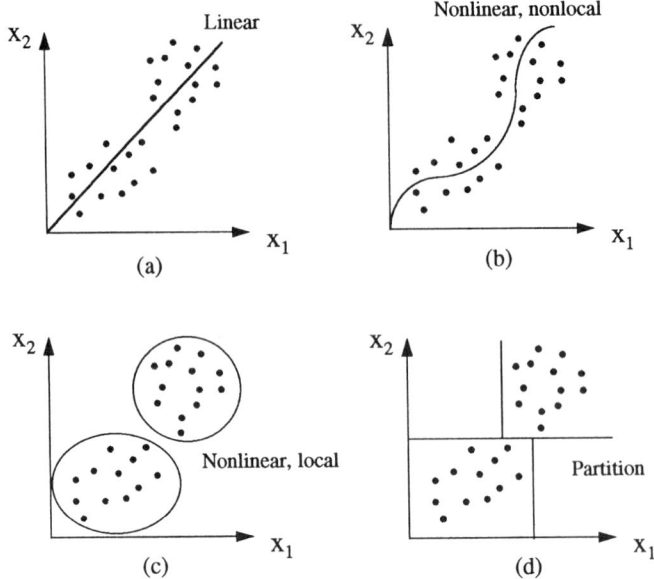

FIG. 6. Input transformation in (a) methods based on linear projection, (b) methods based on nonlinear projection, nonlocal transformation, (c) methods based on nonlinear projection, local transformation, and (d) partition-based methods. (From Bakshi and Utojo, 1998.)

formed inputs to the output is determined by the parameters, β_m, and basis functions, θ_m. Input–output methods often use different objective functions for determining the input transformation, and the transformed input–output relationship. This separation of optimization criteria provides explicit control over the dimensionality reduction by input transformation and often results in more accurate empirical models. Input–output methods therefore can be categorized depending on whether the optimization criterion for the input transformation contains information from

- *Only the inputs,* as in all input analysis methods
- *Both inputs and outputs,* as in input–output analysis and interpretation methods

An optimization criterion for determining the output parameters and basis functions is to minimize the output prediction error and is common to all input–output modeling methods. The activation or basis functions used in data analysis methods may be broadly divided into the following two categories:

- *Fixed-shape basis functions.* The basis functions are of a fixed shape, such as linear, sigmoid, Gaussian, wavelet, or sinusoid. Adjusting the

basis function parameters changes their location, size, and orientation, but their shape is decided a priori and remains fixed.
- *Adaptive-shape basis functions.* Some input–output modeling methods relax the fixed-shape requirement and allow the basis functions to adapt their shape, in addition to their location, size, and orientation, to the training and testing data. This additional degree of freedom provides greater flexibility in determining the unknown input–output surface and often results in a more compact model. Adaptive-shape basis functions are obtained through the application of smoothing techniques such as splines, variable span smoothers, and polynomials to approximate the transformed input–output space.

The nature of the input transformation, type of basis functions, and optimization criteria discussed in this section provide a common framework for comparing the wide variety of techniques for input transformation and input-output modeling. This comparison framework is useful for understanding the similarities and differences between various methods; it may be used to select the best method for a given task and to identify the challenges for combining the properties of various techniques (Bakshi and Utojo, 1999).

III. Input Analysis Methods

As stated, *input* analysis methods operate directly on the measurement input data set and map the numeric measurement input information into a modified numeric form. The objective is to emphasize the relevant process information that addresses or describes a particular set of behaviors while reducing or eliminating unwanted information. Input analysis methods belong to the categories of univariate and multivariate methods, depending on whether a single variable or multiple variables are analyzed. Univariate methods, also widely known as *filtering* methods, remove noise and errors based on their expected behavior in the time, frequency, or time–frequency domain. Multivariate methods reduce errors by exploiting redundancy in the variables. In either case, the principle of modeling the input data is the same: that is, to obtain a set of values that more clearly contain the relevant information present in the original set of measurements. In this section, we describe primary families of methods and illustrate their features with several specific examples. Consistent with the overall document, the objective of this section is to present key features of methods so it is clear how families of methods for input analysis relate to each other. We focus on two primary families: linear and nonlinear projection methods.

A. Methods Based on Linear Projection

Univariate input analysis or filtering is implemented as a mathematical algorithm where a data sequence corresponding to a single process variable, $x_i(n)$, is transformed by some mathematical computation into a different sequence, $z_m(n) = \phi_m x_i(n)$, where ϕ_m represents the mapping or transformation from the x_i into the new sequence z_m. A filtering operation usually generates a sequence z_m that is like x_i and maximizes the retention of relevant process information while minimizing the amount of interference that remains in the data. All filtering methods can be interpreted as analyzing data using a set of basis functions that are localized in the time and/or frequency space as

$$z_m = \sum h_i \Psi_i(s, u, \omega), \tag{7}$$

where $\psi_i(s, u, \omega)$ is the ith basis function at temporal location u, frequency location ω, and scale s. The scale parameter determines the extent of localization of the basis function in time and frequency. The input transformation function, Ψ_i, involves a linear or nonlinear transformation of the basis function coefficients, h_i. Key process considerations in determining filter characteristics are as follows:

- *Superposition* implies that whenever the filter input can be decomposed into a sum of two or more signals, the output can be formed as the sum of the responses of the filter to each of the components of the input acting one at a time.
- *Homogeneity* implies that a constant gain can be applied at either the input or the output of a system without changing the response.
- *Causality* requires that the filter response at time n be computed on the basis of present and past information and not require knowledge of either the future input or output. Thus, the computation of $z_m(n)$ involves only present and past (values of the) inputs and only past outputs.
- *Shift invariance* implies that the computational algorithm to determine the filter output does not change with time. Thus, a given sequence $x_i(n)$ produces the same output $z_m(n)$ regardless of the time at which it is applied.

Depending on how the previous measurements are combined in Eq. (7), univariate filtering methods can be categorized as linear or nonlinear. In terms of Eq. (7), linear filtering methods use a fixed scale parameter or are single-scale, whereas nonlinear filtering methods are multiscale. Figure 7 summarizes decompositions in terms of time and frequency.

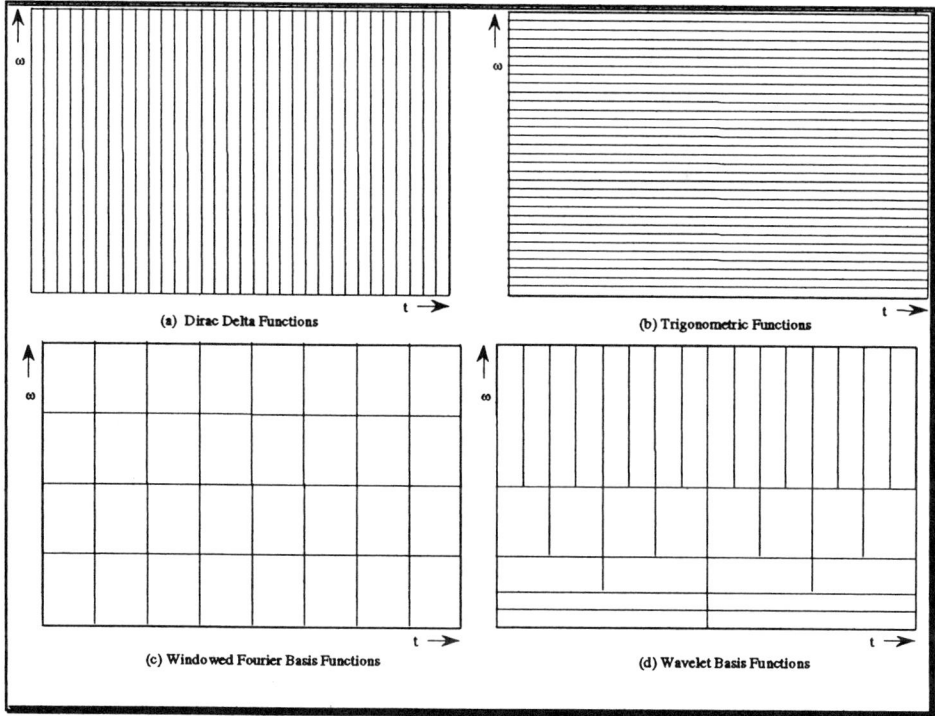

FIG. 7. Decomposition of the time (t)–frequenty (ω) by different basis functions.

1. Linear or Single-Scale Filtering

The general view just given offers a perspective on scale filtering methods that are in wide use. Newer methods expand on these.

In single-scale filtering, basis functions are of a fixed resolution and all basis functions have the same localization in the time–frequency domain. For example, frequency domain filtering relies on basis functions localized in frequency but global in time, as shown in Fig. 7b. Other popular filters, such as those based on a windowed Fourier transform, mean filtering, and exponential smoothing, are localized in both time and frequency, but their resolution is fixed, as shown in Fig. 7c. Single-scale filters are linear because the measured data or basis function coefficients are transformed as their linear sum over a time horizon. A finite time horizon results in *finite impulse response* (FIR) and an infinite time horizon creates infinite impulse response (IIR) filters. A linear filter can be represented as

$$z_m = \sum_{i=1}^{T} w_i x(n - i), \qquad (8)$$

where z_m is the transformed variable containing the information of interest for data interpretation. The w_i are the weights, T is the length of the time horizon. This equation is called the *convolution sum*. In practice, most FIR and IIR filters are linear, causal, and shift-invariant and are characterized by an *impulse response*, w_i, that is of finite duration and has zero value for $i < 0$ and $i > T$, the filter impulse response duration. An impulse response is defined as the output of a filter when the input is a sequence that is zero everywhere except for $x_i(0) = 1$ (the sequence element at $n = 0$). Knowledge of the impulse response is sufficient to specify the behavior of the filter. The impulse response can be interpreted as a weighting function that determines how past inputs affect the present filter output.

For a mean filter with a window width of T, the impulse response is given by

$$w_i = \begin{cases} \frac{1}{T}, & i = 1, \ldots T; \quad 0 \text{ otherwise.} \end{cases} \quad (9)$$

This filter produces a response at time n that is the average of the present plus $(n - 1)$ previous values of the input x_i.

For an IIR filter, the parameter T in Eq. (9) tends to infinity. IIR filters can be represented as a function of previous filter outputs and often can be computed with fewer multiplications and reduced data storage requirements compared to a FIR filter. A popular example of an IIR filter is the exponentially weighted moving average (EWMA) or exponential smoothing, which is represented as

$$z_m(n) = \alpha z_m(n - 1) + (1 - \alpha)x(n). \quad (10)$$

In general, the specifications that define filter behavior usually are based on their frequency domain behavior (Mitra and Kaiser, 1993). The frequency response of a system is therefore a useful tool for defining the desired behavior of a linear filter and for decomposing signals into features that might be beneficial for numeric–symbolic mapping. Frequency domain defines the effect of the filter on a sinusoid of a given frequency f_0. For any linear system, when a sinusoid is applied, the output is a sinusoid of the same frequency; only the amplitude and the relative location of the zero crossings change. The factor by which the amplitude changes is called the *filter gain* at f_0, and the shift in the zero locations as a fraction of a full period is the *phase shift* at f_0. Knowledge of the gain and phase at all possible frequencies completely defines the behavior of the filter. A linear system is said to pass a certain frequency if the amplitude of the sinusoid is substantially maintained in passing the signal through the filter. Con-

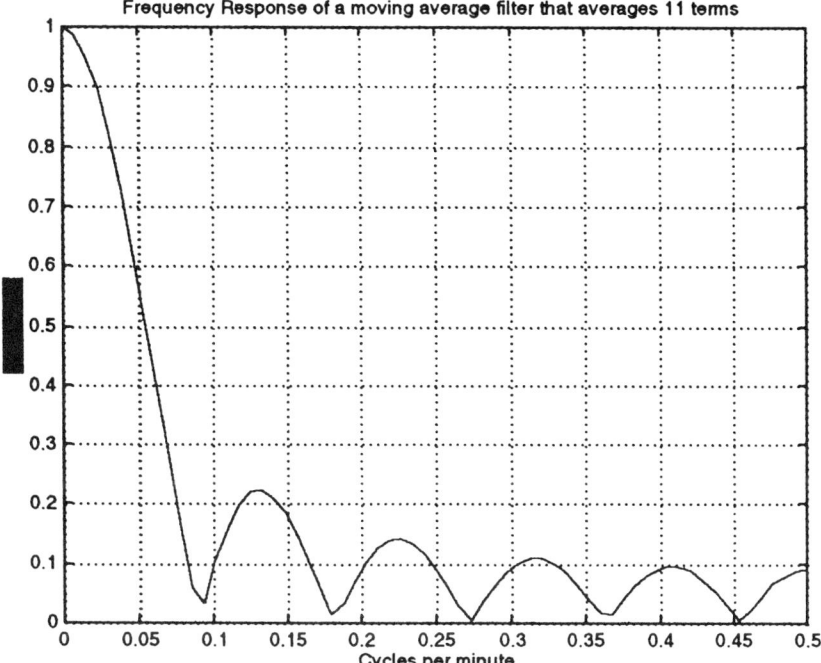

FIG. 8. Frequency response of a moving average filter.

versely, a frequency is rejected if the output amplitude drops below a specified threshold. The desired filter behavior is defined in terms of those frequencies passed and those rejected.

Filter specifications are matched to the response times of the process. For example, a given process has a time constant of 5 minutes. That means that it can respond over the frequency range of 0 to 1/20 of a cycle per minute. Higher frequencies are attenuated naturally by the process. Thus, if the data contain components beyond 0.05 cycles per minute, then those components are likely to be unwanted interferences. The linear filter would pass the frequencies between 0 and 1/20 and reject frequencies outside this range. The filter should attenuate frequencies higher than one decade above the break-point frequency. Process measurements processed by this filter are transformed to a new sequence with less interference than the original data. In this way, an input mapping has been defined.

To illustrate behaviors of different filters, consider a moving average filter that averages over 11 terms. Such a filter has the frequency response shown in Fig. 8. Note that this filter has a relatively low gain of 0.55 at the break-point frequency of 0.05 cycles per minute. So in the range of

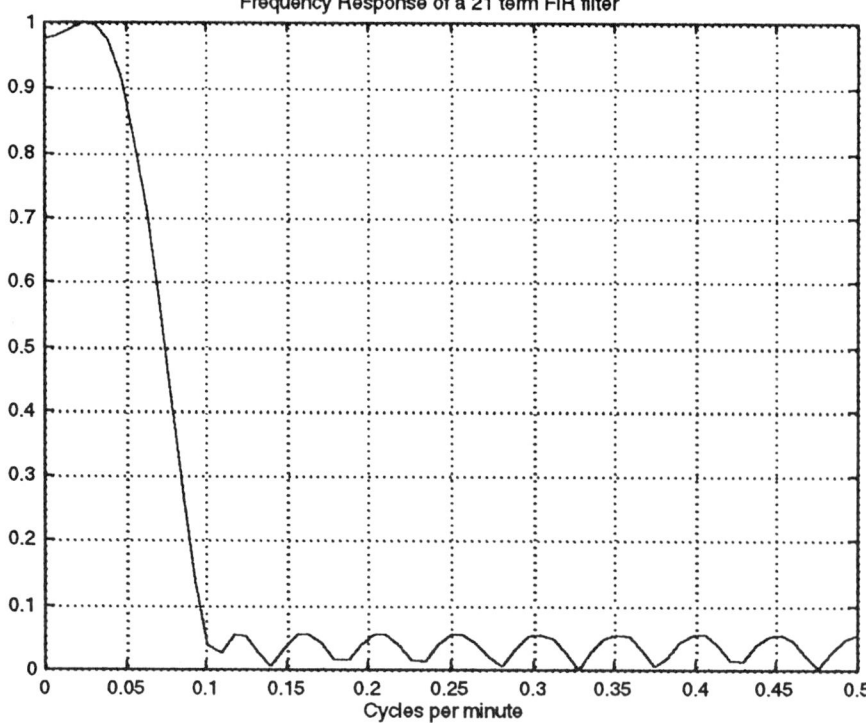

Fig. 9. Example of an FIR filter.

frequencies that it should pass, there is significant attenuation. In the frequency range beyond 0.05 where there is noise, the gain still can be quite large, thereby passing undesirable frequencies. Increasing the number of terms over which averaging occurs reduces the gain in the 0- to 0.05-cycle range and only makes the problem worse.

Compare this result with the FIR filter in Fig. 9, which is specifically designed to pass frequencies between 0 and 0.05 cycles per minute and to reject frequencies beginning at 0.1 cycles. Notice that the gain is much higher over the 0 to 0.05 range, allowing considerably more of the desired signal through the filter. Over the range from 0.1 to 0.5, the gain is much smaller than the moving average filter, removing the unwanted signal. This filter can be improved by increasing the number of impulse response coefficients. In terms of Eq. (8), FIR and IIR filtering is equivalent to expanding the signal on a set of time–frequency localized basis functions whose scale is determined by the window size T.

FIR and IIR filters both are able to meet a given filter specification and, in that sense, they are identical. However, FIR filters can be designed for *linear phase* where all input sinusoidals are delayed a fixed number of samples. Only the amplitude changes are determined by the filter gain. Thus, linear phase means that the filter output is the filter version of the input delayed by $(n - 1)/2$ samples, where $n - 1$ is the number of past values of x_i used in the computation. This is important if time correlations of the input measurements are critical. For example, assume there are two noisy signals corresponding to related process measurements. If an FIR filter is used, then the time correlation between the two is maintained because they are both delayed the same number of samples even though they are composed of different frequencies. IIR filters involve fewer multiples and achieve the same reductions of unwanted frequencies. However, they delay different frequency components by differing amounts. If a predictive model is needed, IIR filtering can change the underlying correlation structure of the input data and affect the final model form.

2. Nonlinear or Multiscale Filtering

The fixed resolution of single-scale filtering methods is inappropriate for signals containing contributions from events at multiple scales. Multiscale signals are very common in the chemical process industry because the underlying physical and chemical phenomena occur over different temporal and spatial horizons, making them inherently multiscale by nature. For slowly varying or steady portions of a signal, the temporal localization of the filter must be coarse. If this signal also contains any contribution from faster events, the coarse temporal localization will oversmooth the fine-scale events. In contrast, a fine temporal resolution will capture fast events accurately, but will retain too much noise when the changes are slow.

For example, consider a continuous process designed to operate at a specified steady state. The steady-state values may experience variation due to grade changes, throughput, or unmeasured disturbances. In this case, the data can be modeled effectively as a noisy measurement of a piecewise constant signal. The signal is constant until a new set point is defined, after which it takes on a new constant value. If single-scale filters are used, they will blur the points of transition because the high-frequency components of the transition are suppressed. In contrast, multiscale filters can be designed to remove this blurring at transition by adapting the filter resolution to the nature of the measured signal. Multiscale filters represent the next leap in input mapping technology by virtue of their ability to address the inherent multiscale nature of processes. Multiscale filters transform the measured data in either a nonlinear or a time-varying manner.

To illustrate, the FIR Median Hybrid (FMH) as described by Heinonen and Neuvo (1987) is particularly effective at removing noise from a piecewise constant signal corrupted by noise. Just as with FIR filters, its output depends on a finite number of present and past inputs. However, it produces an output that is the median if the data are rank ordered. The filter operates on $2n + 1$ data points to compute forward and backward predictions of the center or $(n + 1)$st value (as described next). n is the number of points in a selected window of observations. The filter output is then the median of these two predictions and the actual nth value. The main advantage of FMH filters over ordinary median filters is the reduced computational complexity. It involves the determination of the median of three elements rather than a sort of $2n + 1$ elements.

The FMH filter is configured by selecting a window half-width of size $n > 0$ but smaller than half the length of the shortest expected region of constant value in the data. Values of n that are too large result in a window width that is too large, leading to distortion of a signal whose constant region is shorter than n. However, there is a trade-off in that a large n contributes to improving the noise suppression capabilities because more data are used to compute the forward and backward predictions. Ideally, n should be chosen as large as possible without producing a window so large that the time length of constant values is smaller.

As stated, the computation involves a forward prediction and a backward prediction of the $(m + 1)$st data point. At a time instance m under consideration, the value of $x_i(m + 1)$ is the median of $x_i(m+1)$, $\hat{x}_i(m + 1)$, the average of n previous values, and $\tilde{x}_i(m + 1)$, the average of n future values. The backward prediction operates on a sequence, $\{x_i(m - n + 1), \ldots, x_i(n)\}$ to compute an average and uses that as an estimate of the next value. That is,

$$\hat{x}_i(m + 1) = \frac{1}{n} \sum_{k=m-n+1}^{n} x_i(k). \tag{11}$$

The forward prediction $\tilde{x}_i(m + 1)$ is computed by averaging the n sample values in the future of the $(m + 1)$st one. Since future data are required, this computation is not possible in real time and hence is noncausal.

$$\tilde{x}_i(m + 1) = \frac{1}{n} \sum_{k=1}^{n} x_i(m + 1 + k) \tag{12}$$

Mathematically, $(m + 1)$st FMH filter output is given as

$$x_i(m + 1) = \text{med}\{\hat{x}_i(m + 1), \tilde{x}_i(m + 1), x_i(m + 1)\}. \tag{13}$$

The filter can be made iterative by reapplying the same algorithm to the output of the filter. The advantage of doing this is to obtain greater reduction in process interference (Heinonen and Neuvo, 1987; Gallagher and Wise, 1981).

More recently, the development of wavelets has allowed the development of fast nonlinear or multiscale filtering methods that can adapt to the nature of the measured data. Multiscale methods are an active area of research and have provided a formal mathematical framework for interpreting existing methods. Additional details about wavelet methods can be found in Strang (1989) and Daubechies (1988).

In brief, wavelets are a family of functions of constant shape that are localized in both time and frequency. A family of discrete dyadic wavelets is represented as

$$\Psi_{mk}(t) = \frac{1}{\sqrt{d}} \Psi\left(\frac{t-u}{d}\right), \tag{14}$$

where d is the parameter that deletes or compresses the mother wavelet, ψ, and k is the translation parameter that translates to produce the multiscale character to the filter. Equation (14) can be discretized dyadically to

$$\psi_{dk}(t) = 2^{-d/2} \Psi(2^{-d}t - k), \qquad 0 \le d \le L, 0 < k \le N, \tag{15}$$

where L and N are the maximum number of scales and signal length, respectively. Wavelets decompose the time–frequency domain as shown in Fig. 7d, with the temporal localization being finer at higher frequencies. Figure 10 shows that any square integrable signal can be represented at multiple scales by decomposition on a family of wavelets and scaling functions. The signals in Figs. 10b through 10f represent information from the original signal in Fig. 10a, at increasingly coarse scales or lower frequency bands. Each point in these figures is the projection of the original signal on a wavelet. The last signal in Fig. 10f represents the coarsest scale or lowest frequencies.

Each wavelet is localized in time and frequency and covers a region of the time–frequency space depending on the value of its translation and dilation parameters. For orthonormal wavelets, fast filtering algorithms of wavelets proportional to the number of measurements have been developed for multiscale decomposition of a discrete signal of dyadic length (Mallat, 1989). Given these general properties, techniques for multiscale filtering rely on the observation that contributions from an underlying deterministic signal are concentrated in relatively few of the time–frequency localized basis functions, while the random errors are distributed over a large number of basis functions (Donoho et al., 1995). Thus, the problem of filtering a noisy signal is that of recovering coefficients that correspond to the underlying signal. The magnitudes of the wavelet coefficients corresponding to the

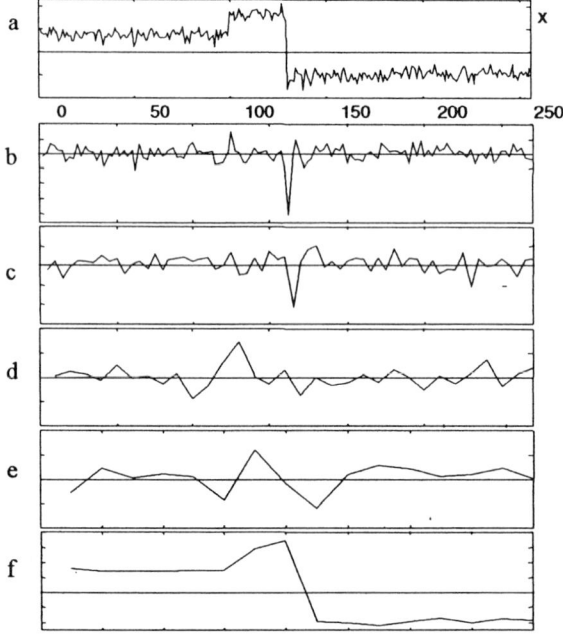

FIG. 10. Wavelet decomposition.

underlying noise-free signal are larger than the magnitude of the coefficients corresponding to the noise. The underlying signal can be recovered by eliminating wavelet coefficients smaller than a threshold. These properties are the basis of the wavelet thresholding approach developed by Donoho et al. (1995).

The methodology for multiscale filtering consists of the following three steps:

1. *Decompose* measured data on a selected family of basis functions
2. *Eliminate* coefficients smaller than a selected threshold to reduce contribution of the errors
3. *Reconstruct* the rectified or filtered signal

The multiscale basis functions capture the fast changes in coefficients corresponding to the fine-scale basis functions, while the slower changes are captured by the coarse-scale basis functions. Thus, the wavelet thresholding method adapts its resolution to the nature of the signal features and reduces the contribution of errors with minimum distortion of the features retained in the rectified signal.

The quality of the filtered signal depends on proper selection of the threshold, which can be determined based on knowledge about the variance of the errors, the stochastic nature of the errors, and the known properties of stochastic processes in the wavelet domain. The variance of the noise at each scale represents the energy of the stochastic process in the corresponding frequency band, which is equivalent to the power spectrum of the noise in the corresponding range of frequencies. If the noise is known to be uncorrelated, then the constant power spectrum at all frequencies indicates that the threshold should be identical at each scale and equal to the standard deviation of the noise in the original signal. If the noise is autocorrelated, then the threshold at each scale should change according to the variation of the power spectrum of the noise. Several methods have been suggested by statisticians for estimating the threshold for filtering from the measured data (Donoho et al., 1995; Nason, 1996). Several of these methods are practical for process data rectification. The *VisuShrink* method (Donoho et al., 1995) determines the threshold at each scale as

$$t_m = \sigma_m \sqrt{2 \log N}, \tag{16}$$

where σ_m is the standard deviation of the errors at scale m and N is the length of the signal. The error between the actual error-free signal and the rectified signal is guaranteed to be within $O(\log N)$ of the error between the error-free signal and the signal rectified with a priori knowledge about the smoothness of the underlying signal. Thus, the rectified signal by wavelet thresholding is nearly optimal for a wide variety of theoretical objectives such as various error norms and smoothness. These and other theoretical properties of wavelet thresholding are discussed in detail by Donoho et al. (1995).

The value of the standard deviation of the errors at each scale, σ_m, can be determined from knowledge of the power spectrum of the errors, if available. In most practical situations, information about the stochastic character and variance of the errors is not available, so σ_m must be estimated from the measured data. Because most of the small coefficients in the multiscale decomposition of a measured signal correspond to the errors, and those corresponding to the underlying signal are larger in magnitude, the variance of the errors at each scale can be estimated as a function of the wavelet coefficients at the selected scale. The accuracy of this estimation decreases for a noisy signal at coarser scales because the number of available data points decreases, and the contribution from the underlying signal may increase.

Several extensions of wavelets have been developed to improve their ability to solve practical problems. Wavelet packets (Coifman and Wickerhauser, 1992) are a library of basis functions that cover a wide variety of shapes. The library can be searched efficiently to select the best set of orthonormal basis functions for representing or filtering a signal. Wavelet packets overcome the fixed shape of wavelets and allow greater flexibility in decomposing the time–frequency space.

The multiscale filtering methods described in the preceding paragraphs can only be applied off-line to data sets of dyadic length because of the noncausal nature of the wavelet filters and the dyadic discretization of the translation and dilation parameters. Work by Nounou and Bakshi (1999) has focused on addressing these disadvantages with multiscale methods for online filtering of data of arbitrary length. This approach filters the data in a moving window of dyadic length.

3. Multivariate Input Analysis

Filters are designed to remove unwanted information, but do not address the fact that processes involve few events monitored by many measurements. Many chemical processes are well instrumented and are capable of producing many process measurements. However, there are far fewer independent physical phenomena occurring than there are measured variables. This means that many of the process variables must be highly correlated because they are reflections of a limited number of physical events. Eliminating this redundancy in the measured variables decreases the contribution of noise and reduces the dimensionality of the data. Model robustness and predictive performance also require that the dimensionality of the data be reduced.

Techniques for multivariate input analysis reduce the data dimensionality by projecting the variables on a linear or nonlinear hypersurface and then describe the input data with a smaller number of attributes of the hypersurface. Among the most popular methods based on linear projection is principal component analysis (PCA). Those based on nonlinear projection are nonlinear PCA (NLPCA) and clustering methods.

Principal component analysis encompasses the most important family of linear projection methods. In general, PCA models the correlation structure in a set of data assembled in a matrix X, where the rows are the process measurements at a fixed sampling time, and a column is a uniformly sampled variable. PCA produces a mapping of the data set X onto a reduced subspace defined by the span of a chosen subset of eigenvectors (*loadings*) of the variance–covariance matrix of the X data. The net effect is that a given vector can be represented approximately by another vector with fewer components. The components not included in the approximation capture the errors or noise in the data matrix, thereby providing PCA with the ability to perform multivariate filtering. In terms of the unifying framework described in Section II, PCA transforms the data by taking a linear combination of the measured variables that captures the variance in the data while being orthogonal to the previously determined transformed or latent variables. One advantage of this technique is that it permits the development of a linear model that produces an orthogonal set of pseudomeasurements (*scores*) that contain the significant variations of the X data. A thorough review of PCA can be found in the article by Wold *et al.* (1987a) and in books by Jolliffe (1986) and Jackson (1992).

PCA involves several steps. First, the data are mean-centered and often normalized by the standard deviation. Mean centering involves subtracting the average value for each variable from the corresponding measurement. Scaling is necessary to avoid problems associated with having some measurements with a larger range of values than others. This larger range may appear to be more meaningful than it really is. Scaling puts all the measurements on the same magnitude, which means multiplying the mean-centered data by an appropriate constant, usually the inverse of the standard deviation. If the variables are noisy, they must be normalized by the covariance matrix of the noise.

The variance–covariance matrix is generated from this normalized data by the relationship $X^T X$, where X is the normalized data matrix. The $X^T X$ matrix is positive semidefinite and its eigenvectors and their associated eigenvalues determine the variability in the X data. The eigenvectors define the directions in the X space where most of the variability occurs. The first eigenvector defines the direction along which the largest variations are located, and so forth. Overall, the eigenvectors define a subspace within the X space on which each data point is projected. These eigenvectors or directions form a matrix of the *principal component loadings* and are the input transformation parameters, or projection directions. The eigenvalue related to each eigenvector indicates the variance explained by the corresponding eigenvector.

The projection of the X data vectors onto the first eigenvector produces the first latent variable or pseudomeasurement set, Z_1. Of all possible directions, this eigenvector explains the greatest amount of variation in X. The second eigenvector explains the largest amount of variability after removal of the first effect, and so forth. The pseudomeasurements are called the *scores*, Z, and are computed as the inner products of the true measurements with the matrix of loadings, α:

$$Z = X\alpha. \tag{17}$$

Because the eigenvectors are ordered according to the maximum variability, a subset of the first $k < N$ scores define the most variability possible in a k-dimensional subspace. This subspace is referred to as the *score space*.

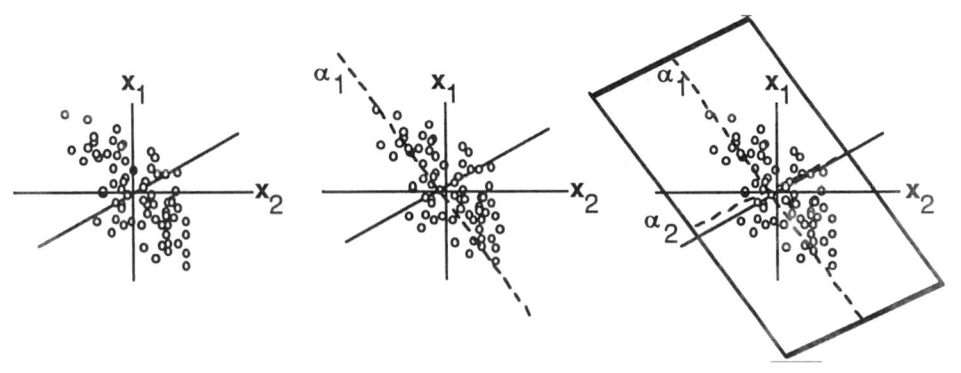

FIG. 11. Simple illustration of PCA. α_1 is the first principal component and α_2 is the second principal component.

The remaining $N - k$ variables capture the errors in the variables. The data matrix may then be represented as

$$X = Z_k \alpha_k^T + E_k, \tag{18}$$

where Z_k, α_k, and E_k are the matrices of the first k scores, loadings, and reconstruction errors, respectively.

PCA is useful not only from a dimension reduction standpoint, but also in that the scores themselves provide additional information important in data analysis (Kresta and MacGregor, 1991; Piovoso and Kosanovich, 1994). A simple example of PCA is illustrated in Fig. 11.

In this example, there are two measurements shown by the left panel. In general, the measurements define an N-dimensional hyperspace, where N is the rank of $X^T X$. Observe that the data are not scattered randomly in the variable space. Rather, they lie primarily along the dotted line drawn through the data shown in the middle panel. This line, α_1, is defined by the first eigenvector or principal component and represents those linear combinations of the data that capture the maximum variability of X. The projections of the data onto α_1, defined by their distances along it from the origin, constitute the scores. If this approximation is not sufficiently accurate as determined by the size of the residuals, a second eigenvector, α_2, can be found as a function of the residual data as shown in the right panel. The second eigenvector will be orthogonal to the first, in the direction of greatest variability of the residual $(X - z_1 \alpha_1^T)$.

The two eigenvectors define a plane in the original variable space. This process can be repeated systematically until the eigenvalue associated with each new eigenvector is of such a small magnitude that it represents the noise associated with the observations more than it does information. In the limit where the number of significant eigenvectors equals the number of variables, there is no reduction in the dimension of the variable space. If the eigenvalues reached exactly zero, it would suggest that the variables are linearly independent. This does not happen in practice because of noise in the data.

PCA is a data preprocessor in that it provides a mechanism to reduce data dimensionality and filter noise. Because only a few of the possible loadings are kept, data are projected into a lower dimensionality where most of the process information is located. Eliminating less significant loadings also eliminates the portions of the data space where the noise level is likely to be high. In this sense, PCA is a form of input mapping like filtering, but with extended capability. We note that PCA can be extended to handle data in three-dimensional arrays, making it also convenient for batch operations (Kosanovich *et al.,* 1995; Nomikos and Mac-Gregor, 1994; Wold *et al.,* 1987b). The three dimensions arise from batch trajectories that consist of batch runs, variables, and sample times. These data can be organized into an array X of dimension $(I \times J \times N)$, where I is the number of batches, J is the number of variables, and N is the number of time samples over the duration of the batch. This extended application of PCA, called *multidimensional PCA (MPCA)*, is equivalent to performing ordinary PCA on a two-dimensional matrix formed by unfolding X so that each of its vertical slices contains the observed variables for all batches at a given time. In this approach, MPCA explains the variation of variables about their mean trajectories.

This decomposition summarizes and compresses the data, with respect to both variables and time, into low-dimensional score spaces representing the major variability over the batches at all points in time. Each loading matrix summarizes the major time variations of the variables about their average trajectories over all the batches. In this way, MPCA uses not just the magnitude of the deviation of each variable from its mean trajectory, but also the correlations among them. To analyze the performance of a set of batch runs, an MPCA analysis can be performed on all the batches and the scores for each batch can be plotted in the space of the principal components. All batches exhibiting similar time histories will have scores that cluster in the same region of the principal component space. Batches that exhibit deviations from normal behavior will have scores falling outside the main cluster.

B. METHODS BASED ON NONLINEAR PROJECTION

Methods based on nonlinear projection are distinguished from the linear projection methods that they transform input data by projection on a nonlinear hypersurface. Although this general distinction is straightforward, the infinite number of possible nonlinear surfaces leads to a wide range of transformation behaviors. Broadly speaking, these methods can be further subdivided into nonlocal and local methods depending on the nature of the hypersurface. The hypersurface for nonlocal methods is unbounded along at least one direction, whereas that for local methods defines a closed region such as a hyperellipsoid.

While the basic objective of linear projection methods is to summarize the data by linear approximations, thereby minimizing the orthogonal deviations in the least-squares sense, the aim of nonlinear, nonlocal methods such as nonlinear principal component analysis is to fit a smooth curve through the data such that the data themselves are allowed to dictate the form of the nonlinear dependency. These methods should not be confused with techniques such as spline smoothers and kernel smoothers that produce a curve that minimizes the vertical deviations subject to some assumed smoothness constraint. Nonlinear PCA determines the principal curve that minimizes the shortest distance to the points with the assumption that the curves are self-consistent for a distribution or data set (Hastie and Tibshirani, 1990). A nonlinear principal component may not exist for every distribution and there is no guarantee that a unique set of nonlinear components exists for a given distribution. For example, the principal axes of an ellipsoidal distribution are the principal curves and, for spherically symmetric distributions, any line through the mean vector is a principal curve. The principal curve algorithm is not guaranteed to converge because of its nonlinear nature. Several algorithms have been developed for nonlinear PCA (Kramer, 1991; Hastie and Tibshirani, 1990; Dong and McAvoy, 1996; Tan and Mavrovouniotis, 1995).

Local methods, on the other hand, are characterized by input transformations that are approached using partition methods for cluster seeking. The overall thrust is to analyze input data and identify clusters of the data that have characteristics that are similar based on some criterion. The *objective* is to develop a description of these clusters so that plant behaviors can be compared and/or data can be interpreted.

Because of these characteristics, local methods are naturally used for interpretation. In this section, we present some of the fundamental elements of these methods as input analysis techniques. However, this discussion closely ties with the interpretation discussion in Section V.

The most commonly used family of methods for cluster seeking uses optimization of a squared-error performance criterion in the form

$$j = \sum_{k=1}^{N_c} \sum_{x \in S_k} \|x - t_k\|^2, \qquad (19)$$

where N_c is the possible number of data classes or clusters, S_k is the set of data corresponding to the kth class, and t_k is a prototype defined as the sample mean of the set S_k, which provides the cluster description. The performance index J corresponds to the overall sum of the squared error between the input data in all classes and the prototypes represented by the sample means for each class.

This index is employed by both the k-means (MacQueen, 1967) and the isodata algorithms (Ball and Hall, 1965), which partition a set of data into k clusters. With the k-means algorithm, the number of clusters are prespecified, while the isodata algorithm uses various heuristics to identify an unconstrained number of clusters.

In applying Equation (19), the objective is to find a set of cluster centers that minimize the Euclidean distance between each data point and the cluster center,

$$\min \sum_{k=1}^{N_c} \sum_{i=1}^{n} B_{ki} \|x_i - t_k\|^2, \qquad (20)$$

where B_{ki} is an $N_c \times n$ matrix of the cluster partition or membership function.

Of the several approaches that draw upon this general description, radial basis function networks (RBFNs) (Leonard and Kramer, 1991) are probably the best-known. RBFNs are similar in architecture to back propagation networks (BPNs) in that they consist of an input layer, a single hidden layer, and an output layer. The hidden layer makes use of Gaussian basis functions that result in inputs projected on a hypersphere instead of a hyperplane. RBFNs therefore generate spherical clusters in the input data space, as illustrated in Fig. 12. These clusters are generally referred to as receptive fields.

The algorithm originally proposed by Moody and Darken (1989) uses k-means clustering to determine the centers of the clusters. The hypersphere around each cluster center is then determined to ensure sufficient overlap between the clusters for a smooth fit by criteria such as the P-nearest neighbor heuristic,

$$\sigma_k = \left[\frac{1}{P} \sum_{i=1}^{P} \|x_i - t_k\|^2 \right]^{1/2}, \qquad (21)$$

where x_i are the P-nearest neighbors of the centers, t_k, resulting in hyperspherical receptive fields. Figure 12 illustrates this graphically. Results generated typically include numerous statistical measures such as distance and variance tables that enable the user to interpret the results and modify adjustable parameters.

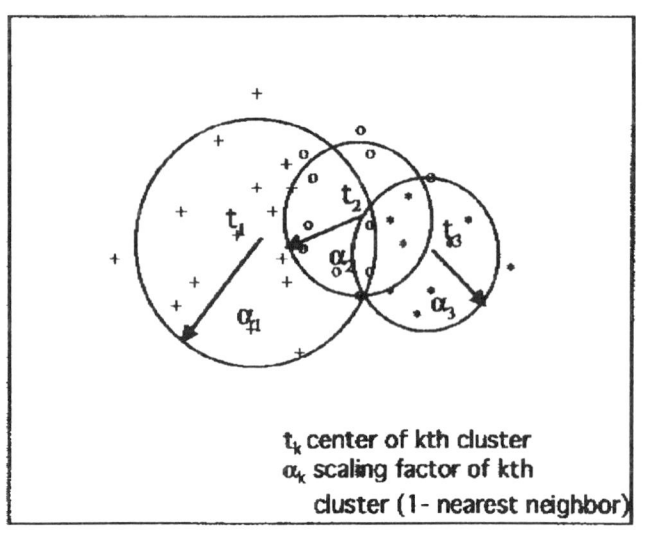

FIG. 12. Receptive fields in RBFNs.

t_k center of kth cluster
α_k scaling factor of kth cluster (1- nearest neighbor)

With respect to cluster seeking performance, RBFNs are influenced heavily by the occurrences of input data patterns because clustering effectively is averaged over individual pattern occurrences. It has been shown that this decreases the utility for interpretation (Kavuri and Venkatasubramanian, 1993). Current approaches also require a priori specification of the number of clusters, a drawback from an adaptation standpoint.

Ellipsoidal basis function networks (EBFNs) (Kavuri and Venkatasubramanian, 1993) can be regarded as modified RBFNs with an ellipsoidal basis function and a different learning algorithm. EBFNs also differ from RBFNs by constraining each input data set to one ellipsoid. Figure 13 illustrates EBFNs and compares them with RBFNs in representing the same region of data. The biggest advantage is there are more degrees of freedom to "shape" the clusters so they can better describe the input data. Both algorithms are best employed in an interactive manner because the algorithms are free to identify clusters without consideration of the analysis of interest.

The objective functions for both k-means clustering and the P-nearest neighbor heuristic given by Eqs. (20) and (21) use information only from the inputs. Because of this capacity to cluster data, local methods are particularly useful for data interpretation when the clusters can be assigned labels.

The Adaptive Resonance Theory (ART) family of approaches (described for process applications by Whiteley and Davis, 1994), although

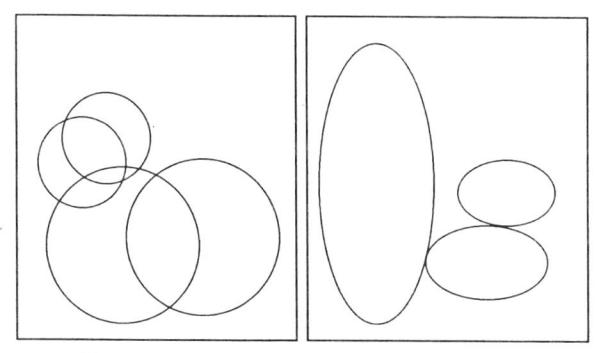

FIG. 13. A comparison of RBFNs and EBFNs.

functionally the same as RBFNs and EBFNs in a number of ways, represents an entirely different philosophy of cluster seeking and therefore embodies a much different algorithm. As shown in Fig. 14, ART forms clusters and regions of clusters in a manner different from EBFNs and RBFNs.

Input data mapping still corresponds to projection on a hypersphere; however, ART uses vector direction to assess similarity rather than using a distance measure as shown in Fig. 15. This translates into the use of hypercone clusters in a unit hypercube.

Unlike RBFNs and EBFNs, the ART approach is aimed at self-organizing properties that address the knowledge management issues associated with incorporating new input data without corrupting the existing information. This avoids the problem described in preceding paragraphs for RBFNs, in which the cluster performance is overly influenced by the number of occurrences of input data patterns. ART is the only clustering approach designed for evolutionary adaptation. This aspect of ART is discussed more fully in Section V, Data Interpretation.

Procedurally, an input data vector is presented to the ART network.

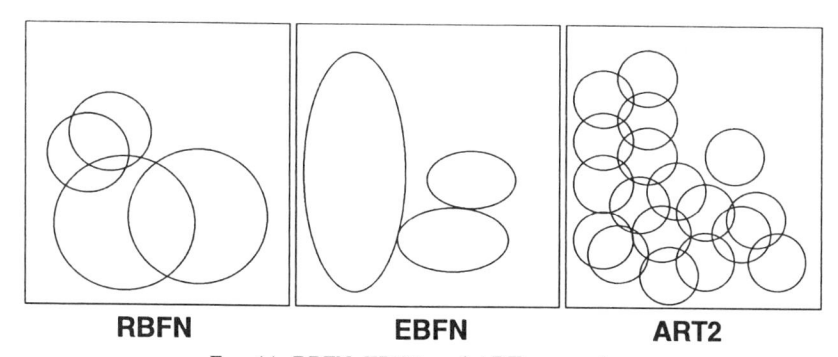

FIG. 14. RBFN, EBFN, and ART approaches.

Distance Measure   **Directional Similarity Measure**

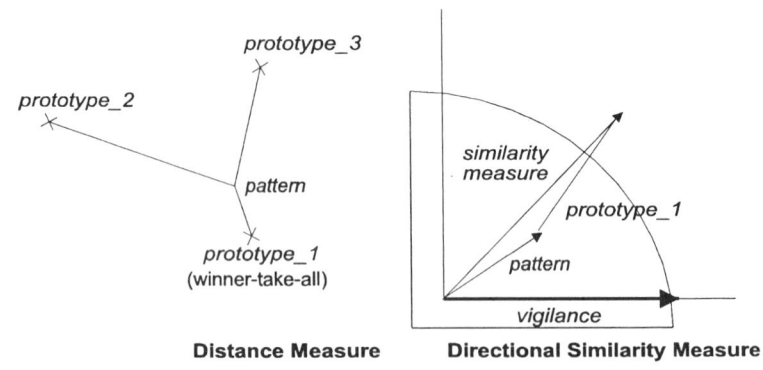

FIG. 15. Distance vs directional similarity measures.

After some specialized processing by the network, the input pattern is compared to each of the existing cluster prototypes in the top layer. The "winner" in the top layer is the prototype most similar to the input. If the similarity between the winner and the input exceeds a predetermined value (the vigilance parameter), learning is enabled and the winner is modified slightly to more closely reflect the input pattern. If the similarity between the winner and input is less than that required by the vigilance parameter, a new unit is initialized in the top layer and learning is enabled to create a prototype similar to the input. This ability to modify existing cluster definitions based on similarity criteria or to generate new cluster definitions provides ART with its adaptive capacity. This subject will be discussed in detail in Section V.

IV. Input–Output Analysis

In sharp contrast to input methods discussed in the last section, input–output analysis is aimed at the construction of models generally for the purpose of predicting output process behaviors given the input data. Although emphasis is on input–output modeling, several analysis methods account for the behavior of the outputs when extracting relevant input features. The analysis may therefore be used in two different ways: (1) the numeric output of the input–output model can be used as a predictive model, or (2) the transformed inputs or latent variables, z_m, which have been influenced by the output, can be used as data interpretation input. Although our focus is on interpretation, the close relationship between empirical modeling and data interpretation is recognized. In this section, we summarize those modeling methods that offer latent variables or transformations that could be used in interpretation. In Section V we focus specifically on the use of both input and input–output analysis for data interpretation.

Following the same generalized model given by Eq. (6), input–output methods may be broadly classified as either projection-based or partition-based methods, as listed in Fig. 2.

A. Methods Based on Linear Projection

Methods based on linear projection transform input data by projection on a linear hyperplane. Even though the projection is linear, these methods may result in either a linear or a nonlinear model depending on the nature of the basis functions. With reference to Eq. (6), the input–output model for this class of methods is represented as

$$y_j = \sum_{m=1}^{M} \beta_{mi} \theta_m \left\{ \sum_{i=1}^{d} \alpha_{mi} x_i \right\}. \tag{22}$$

The reference to *linear* stems from the linearity of the term in brackets even though the overall function can be nonlinear. Among the most popular linear methods for input–output analysis are ordinary least squares regression (OLS), principal component regression (PCR), and partial least squares regression (PLS). Popular nonlinear methods in this class are back propagation neural networks with one hidden layer (BPN), projection pursuit regression (PPR), nonlinear PCR (NLPCR), and nonlinear PLS (NLPLS).

The comparison of these methods based on the type of input transformation, the nature of the basis functions, and the optimization criteria is presented in Table I. These methods are all very similar, to such a degree that it is often difficult to select which would be most appropriate for a given application. In this section, we highlight the key differences that form the basis for choosing relative to the data in a particular application.

1. Ordinary Least Squares

OLS is also called *multiple linear regression* (MLR) and is a commonly used method to obtain a linear input–output model for a given data set. The model obtained by OLS for a single output is given by

$$y = \beta \alpha_1 x_1 + \beta \alpha_2 x_2 + \ldots + \beta \alpha_d x_d. \tag{23}$$

TABLE I
Comparison Matrix for Input–Output Methods

Method	Input transformation	Basis function	Optimization criteria
Ordinary least squares	Linear projection	Fixed shape, linear	α, maximum squared correlation between projected inputs and output β, minimum output prediction error
Principal component regression	Linear projection	Fixed shape, linear	α, maximum variance of projected inputs β, minimum output prediction error
Partial least squares	Linear projection	Fixed shape, linear	α, maximum covariance between projected inputs and output β, minimum output prediction error
Back propagation network (single)	Linear projection	Fixed shape, sigmoid	$[\alpha, \beta]$, minimum output prediction error
Projection pursuit regression	Linear projection	Adaptive shape, supersmoother	$[\alpha, \beta, \theta]$, minimum output prediction error
Back propagation network (multiple)	Nonlinear projection, nonlocal	Fixed shape, sigmoid	$[\alpha, \beta]$, minimum output prediction error
Nonlinear principal component analysis	Nonlinear projection, nonlocal	Adaptive shape	$[\alpha, \phi]$, minimum input prediction error
Radial basis function network	Nonlinear projection, local	Fixed shape, radial	$[\sigma, t]$, minimum distance between inputs and cluster center β, minimum output prediction error
Classification and regression trees	Input partition	Adaptive shape, piecewise constant	$[\beta, t]$, minimum output prediction error
Multivariate adaptive regression splines	Input partition	Adaptive shape, spline	$[\beta, t]$, minimum output prediction error

It can be considered a special case of Eq. (22) with only one linear basis function:

$$y_j = \beta_1 \sum_{i=1}^{d} \alpha_{i1} x_i. \qquad (24)$$

For multiple observations, Eq. (24) constitutes a system of linear equations,

$Y = X\beta\alpha$. If the number of input variables is greater than the number of observations, there is an infinite number of exact solutions for the least squares or linear regression coefficients, $\beta\alpha$. If the variables and observations are equal, there is a unique solution for $\beta\alpha$, provided that X has full rank. If the number of variables is less than the number of measurements, which is usually the case with process data, there is no exact solution for $\beta\alpha$ (Geladi and Kowalski, 1986), but α can be estimated by minimizing the least-squares error between the actual and predicted outputs. The solution to the least-squares approximation problem is given by the pseudoinverse as

$$\beta\alpha = (X^T X)^{-1} X^T Y. \qquad (25)$$

If the inputs are correlated, then the inverse of the covariance matrix does not exist and the OLS coefficients cannot be computed. Even with weakly correlated inputs and a low observations-to-inputs ratio, the covariance matrix can be nearly singular, making the OLS solution extremely sensitive to small changes in the measured data. In such cases, OLS is not appropriate for empirical modeling.

2. Principal Component Regression

Principal component regression (PCR) is an extension of PCA with the purpose of creating a predictive model of the Y-data using the X or measurement data. For example, if X is composed of temperatures and pressures, Y may be the set of compositions that results from thermodynamic considerations. Piovoso and Kosanovich (1994) used PCR and a priori process knowledge to correlate routine pressure and temperature measurements with laboratory composition measurements to develop a predictive model of the volatile bottoms composition on a vacuum tower.

PCR is based on a PCA input data transformation that by definition is independent of the Y-data set. The approach to defining the X–Y relationship is therefore accomplished in two steps. The first is to perform PCA on the X-data, yielding a set of scores for each measurement vector. That is, if x_k is the kth vector of d measurements at a time k, then z_k is the corresponding kth vector of scores. The score matrix Z is then regressed onto the Y data, generating the predictive model

$$Y = Z\beta + E_y, \qquad (26)$$

where β is the matrix of regression coefficients and E_y are the reconstruction errors. Using the orthogonality of the matrix of eigenvectors, α, and the relationship between X and α

$$Z = X\alpha, \qquad (27)$$

gives the relationship between X and Y as

$$Y = \alpha\beta X + E_y, \tag{28}$$

where $\alpha\beta$ are the principal component regression coefficients of X onto Y.

3. Partial Least Squares

PLS was originally proposed by Herman Wold (Wold, 1982; Wold *et al.*, 1984) to address situations involving a modest number of observations, highly collinear variables, and data with noise in both the X- and Y-data sets. It is therefore designed to analyze the variations between two data sets, $\{X, Y\}$. Although PLS is similar to PCA in that they both model the X-data variance, the resulting X space model in PLS is a rotated version of the PCA model. The rotation is defined so that the scores of X data maximize the covariance of X to predict the Y-data.

By relation, PLS is similar to PCR. Both decompose the X-data into a smaller set of variables, i.e., the scores. However, they differ in how they relate the scores to the Y-data. In PCR, the scores from the PCA decomposition of the X-data are regressed onto the Y-data. In contrast, PLS decomposes both the Y- and the X-data into individual score and loading matrices. The orthogonal sets of scores for the X- and Y-data, $\{T, U\}$, respectively, are generated in a way that maximizes their covariance. This is an attractive feature, particularly in situations where not all the major sources of variability in X are correlated to the variability in Y. PLS attempts to find a different set of orthogonal scores for the X-data to give better predictions of the Y-data. Thus, orthogonal vectors may yield a *poorer* representation of the X-data, while the scores may yield a *better* prediction of Y than would be possible with PCR.

Mathematically, X is decomposed into a model for the first principal component:

$$X = z_1\alpha_1 + E_{x,1}. \tag{29}$$

The Y-data are similarly decomposed,

$$Y = u_1 q_1 + E_{y,1}, \tag{30}$$

where Q is the matrix of eigenvectors such that

$$u = YQ. \tag{31}$$

The score vector, z_1, is found in a two-step operation to guarantee that the covariance of the scores is maximized. Once z_1, α_1, u_1, and q_1 have been found, the procedure is repeated for the residual matrices $E_{x,1}$ and $E_{y,1}$ to find $z_2, \alpha_2, u_2,$ and q_2. This continues until the residuals contain no useful information. The z's and u's are the scores, and the α's and q's are the loadings of the X- and Y-data, respectively. An excellent tutorial can be found in Geladi and Kowalski (1986).

Trying to preserve information in both the X- and Y-data puts constraints on the corresponding scores and loadings. To have orthogonal score matrices U and Z, two sets of loadings or basis matrices (generally termed Q and α) are needed to represent the X data. The Q matrix is orthogonal, but α is not. If α is forced to be orthogonal, then Z cannot be. Having Z orthogonal implies that the estimate of the inner relationship between the score matrix U for the Y-data and Z for the X-data is more reliable. That is, the variance of the estimates of the regression coefficients α are smaller than would be obtained otherwise.

In developing the relationship between the X- and Y-data, maintaining orthogonality of the scores is important and unlike PCR, is no longer guaranteed. Without orthogonality, simultaneous regression of all the X- and Y-data onto all the scores would have to be done. Not only is that computationally more expensive, but it can lead to a less robust model in that variances in the elements of the regression vector that relates the scores of the X-data to the scores of the Y-data are smaller. Again, having orthogonal Z's minimizes this variance. To generate orthogonality, an algorithm that is more complex than PCR is used. It guarantees that the scores of the X-data are orthogonal; however, to achieve this, orthogonality of the loadings is conceded. Examples of PLS are given by Piovoso *et al.* (1992a, b), where an estimate of the infrequently measured overhead composition in a distillation tower is obtained using many tray temperature measurements.

4. Nonlinear PCR and PLS

Linear PCR can be modified for nonlinear modeling by using nonlinear basis functions θ_m that can be polynomials or the supersmoother (Frank, 1990). The projection directions for both linear and nonlinear PCR are identical, since the choice of basis functions does not affect the projection directions indicated by the bracketed term in Eq. (22). Consequently, the nonlinear PCR algorithm is identical to that for the linear PCR algorithm, except for an additional step used to compute the nonlinear basis functions. Using adaptive-shape basis functions provides the flexibility to find the smoothed function that best captures the structure of the unknown function being approximated.

Similarly, extending linear PLS to nonlinear PLS involves using nonlinear basis functions. A variety of nonlinear basis functions have been used to model the inner relationship indicated in Eq. (22), including quadratic

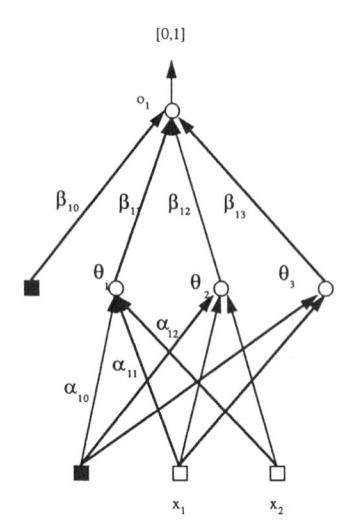

FIG. 16. Representation of BPN as a network of nodes.

functions (Wold et al., 1989), supersmoother (Frank, 1990), BPNs (Qin and McAvoy, 1992), and splines (Wold, 1992). Some nonlinear PLS algorithms prefer to determine the projection directions by linear PLS (Qin and McAvoy, 1992; Frank, 1990), which does not take the nonlinearity of the basis functions into account. This approach is computationally less expensive, but the model is usually less accurate than that obtained by the optimization criterion with the nonlinear basis function (see Table I).

5. Back Propagation Networks

Extensive study of BPNs shows they are very similar to nonparametric nonlinear statistical methods such as projection pursuit regression and nonlinear PLS (Ripley, 1994; Hwang et al., 1994; Bakshi and Utojo, 1998). From a statistical point of view, a BPN is a special case of Eq. (22) where the basis functions are of a fixed shape, usually sigmoidal. BPNs are used widely in the chemical industry, and they have already had a significant impact on solving problems in image and speech processing, inexact knowledge processing, natural language processing, sensor data processing, control, forecasting, and optimization (Maren et al., 1990; Miller et al., 1990).

The BPN usually is represented as a network of nodes as shown in Fig. 16. Although the figure illustrates a single output, the framework readily extends to multiple outputs. The input transformation parameters, β, are the weights of the edges connecting the input nodes to the first hidden layer, while the output parameters, α, are the weights on the edges connecting the hidden layer to the output nodes. The basis functions, θ, usually are sigmoidal, but can be of any other fixed shape. The training methodology for BPNs determines the model parameters for all the nodes simultaneously to minimize the mean squares error as shown in Table I. Different techniques exist for minimizing the error. A technique called *least mean squared error* (LMS), developed by Widrow and Hoff (1960) for the perception, was generalized for a multilayer BPN in Rumelhart *et al.* (1986) and is the most widely used neural network training method. It is based on a gradient descent approach to determine the changes in the weights needed to minimize the squared error. The LMS approach to training has the disadvantage that it can fail to converge or, conversely, converge too slowly if parameters are not set properly. There have been numerous empirical and heuristic approaches to overcome these difficulties, including the addition of a momentum term (Rumelhart *et al.*, 1986). Conjugate gradient techniques as an alternative have also been used (Leonard and Kramer, 1990).

Independent studies (Cybenko, 1988; Hornik *et al.*, 1989) have proven that a three-layered back propagation network will exist that can implement any arbitrarily complex real-valued mapping. The issue is determining the number of nodes in the three-layer network to produce a mapping with a specified accuracy. In practice, the number of nodes in the hidden layer are determined empirically by cross-validation with testing data.

6. Projection Pursuit Regression

PPR is a linear projection-based method with nonlinear basis functions and can be described with the same three-layer network representation as a BPN (see Fig. 16). Originally proposed by Friedman and Stuetzle (1981), it is a nonlinear multivariate statistical technique suitable for analyzing high-dimensional data, Again, the general input–output relationship is again given by Eq. (22). In PPR, the basis functions θ_m can adapt their shape to provide the best fit to the available data.

Optimization of the PPR model is based on minimizing the mean-squares error approximation, as in back propagation networks and as shown in Table I. The projection directions α, basis functions θ, and regression coefficients β are optimized, one at a time for each node, while keeping all other parameters constant. New nodes are added to approximate the residual output error. The parameters of previously added nodes are optimized further by backfitting, and the previously fitted parameters are adjusted by cyclically minimizing the overall mean-squares error of the residuals, so that the overall error is further minimized.

The PPR model usually has fewer hidden nodes than a BPN model and is easier and faster to train (Hwang *et al.*, 1994). The PPR basis functions are computed by smoothing the projected data vs the output by using a variable-span smoother such as the supersmoother (Friedman and Stuetzle, 1981). Determining the basis functions is an important, and often delicate, task, and researchers have suggested several alternatives to the supersmoother, including Hermite function regression (Hwang et al., 1994) and splines (Roosen and Hastie, 1994). The smoothness of the basis functions is usually determined via cross-validation.

B. Methods Based on Nonlinear Projection

This class of methods transforms the inputs in a nonlinear manner. The distinction is readily seen by referring once again to Eq. (22). This family of methods makes use of nonlinear functions both in the bracketed term and in the inner relation. Most of the popular methods project the inputs on a localized hypersurface such as a hypersphere or hyperrectangle.

Input–output analysis methods that project the inputs on a nonlocal hypersurface have also been developed, such as BPNs with multiple hidden layers and regression based on nonlinear principal components.

Among these, the most widely used is the radial basis function network (RBFN). The key distinctions among these methods are summarized in Table I and discussed in detail in Bakshi and Utojo (1998). An RBFN is an example of a method that can be used for both input analysis and input–output analysis. As discussed earlier, the basis functions in RBFNs are of the form $\theta(\|x_i - t_m\|^2)$, where t_m denotes the center of the basis function. One of the most popular basis functions is the Gaussian,

$$\theta_m = \exp\left(-\frac{(x_i - t_m)^2}{\sigma_m^2}\right), \tag{32}$$

where the parameter σ_m determines the extent of localization of the basis function. It should be noted that multidimensional Gaussians can be obtained by multiplying univariate Gaussians.

The most popular modeling method using RBFNs involves separate steps for determining the basis function parameters, σ_m and t_m, and the regression coefficients, β_{mi}. Without considering the behavior of the outputs, the basis function parameters are determined by k-means clustering and the p-nearest neighbor heuristic, as described earlier in the input analysis methods section. The regression parameters are then determined to minimize the output mean-squares error. The optimization criterion for determining the input transformation is given in Table I and considers the input space only. As Table I shows, training RBFNs by this method is analogous to PCR and PLS. Various approaches have been suggested for incorporating information about the output error in determining the basis function parameters (Chen *et al.*, 1991). Hierarchical methods also have been developed for modeling in a stepwise or node-by-node manner using Gaussian basis functions (Moody, 1989) and wavelets (Bakshi and Stephanopoulos, 1993).

Variations on RBFNs have been developed that project the inputs on hyperellipses instead of hyperspheres. These ellipsoidal basis function networks allow nonunity and unequal input weights, except zero and negative values, causing elongation and contraction of the spherical receptive fields into ellipsoidal receptive fields (Kavuri and Venkatasubramanian, 1993).

C. PARTITION-BASED METHODS

Partition-based modeling methods are also called *subset selection methods* because they select a smaller subset of the most relevant inputs. The resulting model is often physically interpretable because the model is developed by explicitly selecting the input variable that is most relevant to approximating the output. This approach works best when the variables are independent (De Veaux *et al.*, 1993). The variables selected by these methods can be used as the analyzed inputs for the interpretation step.

The general empirical model given by Eq. (22) can be specialized to that determined by partition-based methods as

$$Y = \sum_{m=1}^{M} \beta_m \theta_m(X \in R_m), \tag{33}$$

where R_m is the set of the selected inputs. The basis functions θ_m can be a fixed or adaptive shape. As in other input–output analysis methods, the regressions are usually determined to minimize the output prediction error.

1. Classification and Regression Trees

The basis functions in a CART or *inductive decision tree* model are given by

$$\theta_m(X) = \prod_{p=1}^{P_m} H[s_{pm}(x_{v(p,m)} - t_{pm})], \tag{34}$$

where H is the Heaviside or step function,

$$H[\eta] \begin{cases} 1 \text{ if } \eta \geq 0 \\ 0 \text{ otherwise} \end{cases}. \tag{35}$$

P_m is the number of partitions or splits; $s_{pm} = \pm 1$ and indicates the right or left of the associated step function; $v(p, m)$ indicates the selected input variables in each partition; and t_{pm} represents the location of the split in the corresponding input space. The indices p and m are used for the split and node or basis function, respectively. The basis functions given by Eq. (34) are of a fixed, piecewise constant shape.

The CART method selects the variable for partitioning the input space as the one that minimizes the output mean-square error of the approximation. Recursive partitioning of the input space is continued until a large number of subregions or basis functions are generated. After a region splits, it is removed from the model. Overfitting is avoided by penalizing the output prediction error for the addition of basis functions and by eliminating unnecessary splits by a backward elimination procedure (Breiman et al., 1984). The objective function used for determining all model parameters focuses entirely on minimizing the output error of approximation. Neither distribution of data in the input space nor relationships among the input variables are exploited. The CART model can be represented as a binary tree and is physically interpretable. However, the discontinuous approximation at the partitions prohibits its application to continuous input–output models. CART is often unable to identify interactive effects of multiple inputs.

2. Multivariate Adaptive Regression Splines

MARS overcomes the disadvantages of CART for multivariate regression (Friedman, 1991) by dividing the input space into overlapping partitions and keeping both the parent and daughter nodes after splitting a region. This prevents discontinuities in the approximation at the partition boundaries and produces continuous approximations with continuous derivatives. As in CART, the MARS model is represented by Eq. (34), but instead of using the fixed-shape and piecewise constant Heaviside or step functions as the basis functions, MARS uses multivariate spline basis functions obtained by multiplying univariate splines represented by a two-sided truncated power basis,

$$\theta_m(x) = \prod_{p=1}^{P_m} H[s_{pm}(x_{v(p,m)} - t_{pm})] + q, \qquad (36)$$

where q is the order of the splines. Comparison of Eqs. (34) and (36) shows that a value of $q = 1$ results in the CART basis functions. For $q > 0$, the approximation is continuous and has $q - 1$ derivatives.

Empirical comparison of MARS with BPN has shown that the performance of MARS on problems with correlated inputs is often inferior to that of BPN (DeVeaux *et al.*, 1993). This can be explained by the fact that the ability to approximate correlated inputs efficiently depends on the ability of the projection directions to assume any value. MARS restricts projection directions to assume 0 or 1 values to improve the physical interpretability of the model.

V. Data Interpretation

As discussed and illustrated in the introduction, data analysis can be conveniently viewed in terms of two categories of numeric–numeric manipulation, *input* and *input–output*, both of which transform numeric data into more valuable forms of numeric data. Input manipulations map from input data without knowledge of the output variables, generally to transform the input data to a more convenient representation that has unnecessary information removed while retaining the essential information. As presented in Section IV, input–output manipulations relate input variables to numeric output variables for the purpose of predictive modeling and may include an implicit or explicit input transformation step for reducing input dimensionality. When applied to data interpretation, the primary emphasis of input and input–output manipulation is on feature extraction, driving extracted features from the process data toward useful numeric information on plant behaviors.

Figure 4 in the introduction is a defining view showing *data interpretation* as an extension of data analysis where labels are assigned to extracted features by establishing decision criteria or discriminants based on labeled or annotated data (known and observed) that exhibit sufficiently similar characteristics. Input and input–output manipulations therefore provide the critical feature extraction capability. This publication refers to input and input–output mapping extended to label assignment as *numeric–symbolic mapping; feature mapping* is used to describe the extension of feature extraction to label assignment.

Numeric–symbolic approaches are particularly important in process applications because the time series of data is by far the dominant form of input data, and they are the methods of choice if annotated data exist to develop the interpretation system. With complete dependence on the annotated data to develop the feature mapping step, numeric–symbolic mappers can be used to assign labels directly. However, as the amount and coverage of available annotated data diminishes for the given label of interest, there is a need to integrate numeric–symbolic approaches with knowledge-based system (KBS) approaches where the numeric–symbolic approaches are used to produce intermediate labels as input to the KBS approaches.

Figure 3 illustrates a comprehensive data interpretation system that maps directly from an input time series of data, X, to the desired vector of labels, Ω. After data are preprocessed to produce a more valuable form of the input data, the numeric–symbolic interpreter is shown to consist only of a feature extractor and feature mapper. This implies the availability of a sufficient amount of annotated data, i.e., labeled training exemplars, $\Omega \rightarrow X$, from existing operating experiences, such that the labels can be assigned with certainty to arbitrary data from the current operation of the plant. This implies that if there is sufficient operating history to provide the labeled data, then it is desirable to rely on this information for data interpretation because it reflects the actual operation. With sufficient operating history, the burden of label assignment is on feature extraction.

In practice, there may not be sufficient operating experience and resultant data to develop a numeric–symbolic interpreter that can map with certainty to the labels of interest, Ω. Under these circumstances, if sufficient knowledge of process behaviors exists, it is possible to construct a KBS in place of available operating data. But the KBS maps symbolic forms of input data into the symbolic labels of interest and is therefore not sufficient in itself. A KBS depends on intermediate interpretations, Ω', that can be generated with certainty from a numeric–symbolic mapper. This is shown in Fig. 4. In these cases, the burden of interpretation becomes distributed between the numeric–symbolic and symbolic–symbolic interpreters. Figure 4 retains the value of input mapping to preprocess data for the numeric–symbolic interpreter.

General considerations of data availability lead immediately to the recognition that detection systems are more likely to be designed as comprehensive numeric–symbolic interpreters as illustrated in Fig. 3. State description systems may be configured as shown in either Fig. 3 or Fig. 4. Fault classification systems are most likely to require the symbolic–symbolic mapping to compensate for limited data as shown in Fig. 4. Many practical data interpretation problems involve all three kinds of interpreters. In all situations, there is a clear need for interpretation systems to adapt to and evolve with changing process conditions and ever-increasing experience.

The front-end and back-end boundaries on each analysis and interpretation component can be defined so that a particular approach or methodology can be determined based on the specific mapping requirements. Practical data interpretation applications often involve the integration of multiple technologies as required by these three distinct forms of data mapping. This section focuses on several families of numeric–symbolic pattern-recognition methods. The input and input–output mapping approaches discussed previously apply to data interpretation and data analysis in the same way.

Numeric–symbolic pattern recognition employs deterministic and statistical methods when patterns can be represented as vectors or probability density functions (PDFs). These methods are also called *geometric* approaches because they identify regions in a mathematical or statistical representation space that are used to distinguish classes of data (Duda and Hart, 1973; Tou and Gonzales, 1974). Numeric–symbolic approaches often can be characterized by closely linked feature extraction and feature mapping steps. Unlike KBS approaches, geometric approaches are characterized by a heavy emphasis on feature extraction followed by a critical definition for a discriminant or test to assign the labels. Often the feature extraction and mapping steps are embedded and not easily separated.

The number of numeric–symbolic approaches is too vast to address individually in this section. Discussion requires categorization and analysis by key characteristics. As a result, a decomposition by categories brings out the practical advantages and limitations and provides a useful map for selecting approaches for particular problems. Note that, in general, no one family of methods stands out uniformly as the best. Rather, different approaches offer different capabilities that can be used to advantage when addressing particular problem characteristics. Problem characteristics change dramatically from process to process and from label to label. It is therefore critical to approach a complex problem expecting to configure, integrate, and distribute multiple methods and technologies to construct a particular interpretation system.

Data interpretation methods can be categorized in terms of whether the input space is separated into different classes by local or nonlocal boundaries. Nonlocal methods include those based on linear and nonlinear projection, such as PLS and BPN. The class boundary determined by these methods is unbounded in at least one direction. Local methods include probabilistic methods based on the probability distribution of the data and various clustering methods when the distribution is not known a priori.

As introduced earlier, inputs can be transformed to reduce their dimensionality and extract more meaningful features by a variety of methods. These methods perform a numeric–numeric transformation of the measured input variables. Interpretation of the transformed inputs requires determination of their mapping to the symbolic outputs. The inputs can be transformed with or without taking the behavior of the outputs into account by univariate and multivariate methods. The transformed features or latent variables extracted by input or input–output analysis methods are given by Eq. (5) and can be used as input to the interpretation step.

Alternatively, some input–output methods may be used to analyze and interpret the data by directly determining a model between the numeric inputs and symbolic class labels.

The features extracted by univariate and multivariate methods based on linear projection are global, while those extracted by methods based on localized nonlinear projection are restricted to a closed region in the input space, as shown in Fig. 6. With a focus on data interpretation or label assignment, the extracted features are classified based on various distance measures such as simple limit checking and Mahalanobis or Euclidean distances (discussed in detail later). Local methods, including pattern clustering approaches, are distinguished by their direct emphasis on feature extraction for nearest neighbor similarity or pattern feature similarity, a distinction that brings out spatial similarities rather than, direct feature similarity. When extended for interpretation, all data analysis methods require explicit or implicit methods for establishing decision surfaces for assigning labels to data patterns with similar characteristics in the context of the extracted features: that is, interpreting the θ_m and the corresponding weights β_{mi} and parameters α_i. Methods based on nonlocal projection can exhibit desirable properties for useful extrapolation, but also can lead to dangerous misclassifications without any warning about the unreliable nature of the predicted outputs. It is possible to establish decision surfaces that exist as hyperplanes and that are infinite in one or more dimensions. Problems with extrapolation are present when these methods are used to directly map the inputs to the output classes or labels. The potential for extrapolation error necessarily must be managed by estimating the reliability of the predicted values. This is particularly important for process applications where data associated with specific operational classes can be limited or process units can transition into novel operating states.

Alternatively, methods based on nonlocal projection may be used for extracting meaningful latent variables and applying various statistical tests to identify *kernels* in the latent variable space. Figure 17 shows how projections of data on two hyperplanes can be used as features for interpretations based on kernel-based or local methods. Local methods do not permit arbitrary extrapolation owing to the localized nature of their activation functions.

In general, a given numeric–symbolic interpreter will be used in the context of a locally constrained set of labels defining and limiting both the input variables and possible output interpretations. In this context, feature extraction is intended to produce the features that are resolved into the labels. The numeric–symbolic problem boundary is defined backward from the labels of interest so that feature extraction is associated only with the input requirements for a given approach to produce the relevant features needed to generate the label. The distinctions of *relevant features* and

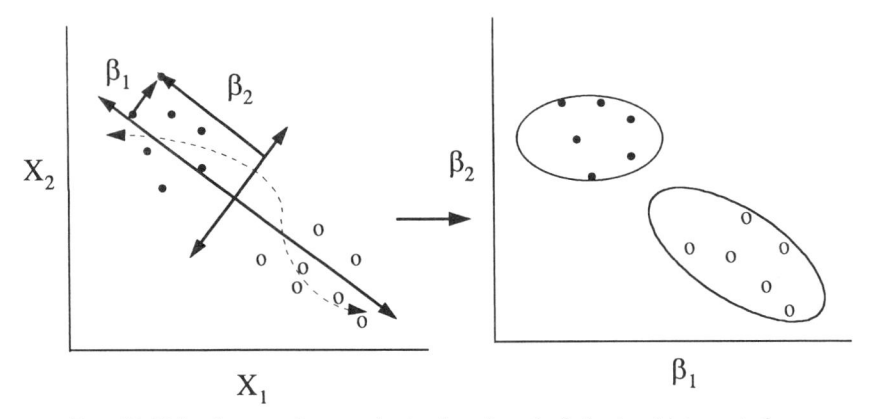

FIG. 17. Using features from projection-based methods for local interpretations.

required input data are made because a separate data preprocessing component may be necessary to condition the input so that interpretation performance is more effective, i.e., filtering. This preprocessing is a more generalized application of feature extraction.

Work on dimension reduction methods for both input and input–output modeling and for interpretation has produced considerable practical interest, development, and application, so that this family of nonlocal methods is becoming a mainstream set of technologies. This section focuses on dimension reduction as a family of interpretation methods by relating to the descriptions in the input and input–output sections and then showing how these methods are extended to interpretation.

A. NONLOCAL INTERPRETATION METHODS

Among nonlocal methods, those based on linear projection are the most widely used for data interpretation. Owing to their limited modeling ability, linear univariate and multivariate methods are used mainly to extract the most relevant features and reduce data dimensionality. Nonlinear methods often are used to directly map the numerical inputs to the symbolic outputs, but require careful attention to avoid arbitrary extrapolation because of their global nature.

1. Univariate Methods

Univariate methods are among the simplest and most commonly used methods that compose a broad family of statistical approaches. Based on anecdotal observation of their widespread use, these methods represent the state of the art in practical data interpretation.

The most commonly implemented method of limit checking is the *absolute value check* as referred to by Iserman (1984),

$$\phi(x_i)_{\min} < \phi(x_i(t)) < \phi(x_i)_{\max}, \qquad (37)$$

where $\phi(x_i(t))$ is the value of the extracted feature at any time t, and $\phi(x_i)_{\min}$ and $\phi(x_i)_{\max}$ are the lower and upper mapping limits, respectively. As shown in Fig. 18, limit checking can be obtained from Eq. (5) by specializing it to the form $z = \phi(x_i)$. Interpretation is achieved when ω_1 is the classification if $\phi(x_i)_{\min} < z_m < \phi(x_i)_{\max}$ and ω_2 otherwise. Note that each input x_i is considered to be a feature. No matter what the value of x_i, the class z is determined by relation to the specified range on x_i. This form of limit checking is easy to implement and places the interpretation emphasis on the specification of the limits.

In the extended family of limit-checking approaches, there is a variety of definitions for $\phi(x_i)$. All definitions are designed to render a time series of data points into a feature that can be appropriately labeled with static limits. These include an individual point, statistically averaged data points, integral of the absolute value of the error (IAE), moving average filtering, cumulative sum (CUSUM), Shewart Charts, and exponentially weighted moving average (EWMA) models. A survey of the use of these methods for statistical process control (SPC) is given by MacGregor and Kourti (1995) and for fault detection by Basseville and Benveniste (1986). The interpretation capability of these approaches is concentrated in establishing appropriate values for $\phi(x_i)_{\min}$ and $\phi(x_i)_{\max}$ that account for the fluctuations in the operation relative to the class labels of interest.

Although usable in many situations, limit-checking methods have limited applicability. First, these methods are univariate and depend on reducing the time series of data points into a single feature. In cases where the

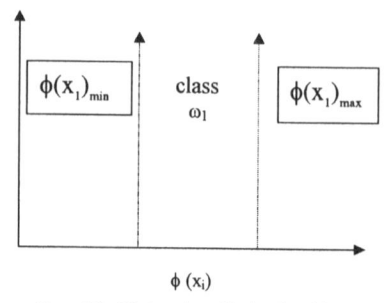

FIG. 18. Univariate limit checking.

process operation is not in a pseudo-steady-state, this has the effect of eliminating much of the useful information contained in the data. These approaches work very well during periods of pseudo-steady-state behavior where various averaging methods represent the true behavior exhibited by the data. However, without the ability to make use of information during transient periods, it is difficult to interpret information during critical periods such as startup, shutdown, and process transitions. Secondly, these methods work best for data interpretations that require static limits. However, many interpretations require changing limits. For example, "normality" interpretation requires definition of a fixed range of normality. When process conditions change and therefore the normality range changes, either the limits must be set at a maximum width to accommodate an entire range of conditions, but with a corresponding degradation in interpretation capability, or the limits must change dynamically. The latter case has been implemented with reasonable success in the form of context-dependent interpretations. This method requires considerable definition of the process operation.

2. Multivariate Linear Discriminant Methods

Multivariate linear discriminant methods use the structure of patterns in the representation space to perform pattern recognition. Patterns in a common class are assumed to exhibit similarities with the analytic result that different pattern classes occupy different regions in the representation space. Such a pattern structure enables the representation space to be partitioned into different pattern classes using linear decision surfaces (hyperplanes). There is therefore an important up-front assumption that the pattern classes must be linearly separable. For an arbitrary pattern, a decision on class membership is made by comparing the projection of the pattern on the hyperplane with some threshold value. The global nature of this family of approaches is associated with the infinite length of the decision surfaces, regardless of the data patterns that were used to establish their positions. Figure 19 illustrates a hyperplane (line) in a two-dimensional space.

Because a hyperplane corresponds to a boundary between pattern classes, such a discriminant function naturally forms a decision rule. The global nature of this approach is apparent in Fig. 19. An infinitely long decision line is drawn based on the given data. Regardless of how closely or distantly related an arbitrary pattern is to the data used to generate the discriminant, the pattern will be classified as either ω_1 or ω_2. When the arbitrary pattern is far removed from the data used to generate the discriminant, the approach is extremely prone to extrapolation errors.

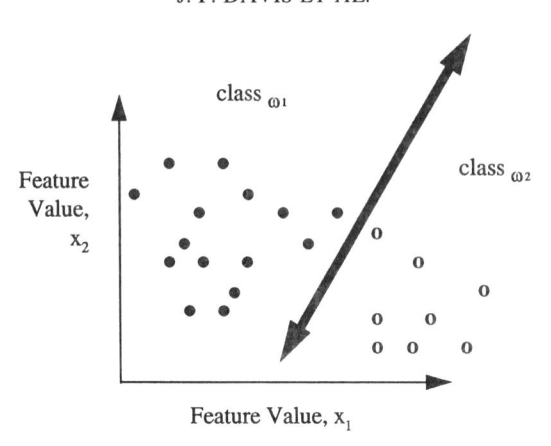

FIG. 19. Partitioning of representation space into regions of discrete pattern class using linear discriminants.

Figure 20 shows more definitively how the location and orientation of a hyperplane is determined by the projection directions, α_i, and the bias, α_0. Given a pattern vector **x**, its projection on the linear discriminant is in the α direction and the distance is calculated as $|d(\mathbf{x})|/\|\alpha\|$. The problem is the determination of the weight parameters for the hyperplane(s) that separate different pattern classes. These parameters are typically learned using labeled exemplar patterns for each of the pattern classes.

Once determined, these parameters can be used to create linear discriminant functions of the form

Linear discriminant described by

$$d(\mathbf{x}) = \alpha^T \mathbf{x} + \alpha_0$$

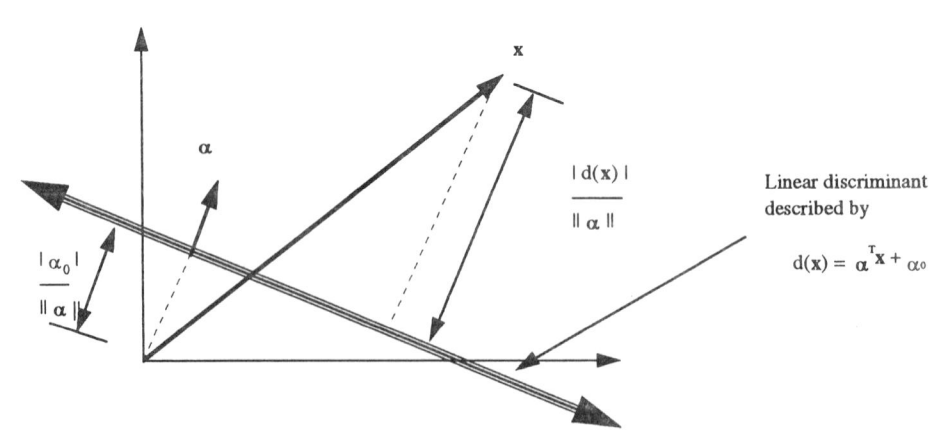

FIG. 20. Orientation and location of linear decision surface.

$$z_m = \sum_{i=1}^{n} \alpha_j x_j + \alpha_0. \tag{38}$$

This function provides a convenient means of determining the location of an arbitrary pattern **x** in the representation space. As shown, patterns above the hyperplane result in $z(\mathbf{x}) > 0$, while patterns below generate values of $z(\mathbf{x}) < 0$. The simplest form of the decision rule then is

$$\begin{array}{l} \omega_1 \text{ if } z_m(\mathbf{x}) > 0 \\ \text{else} \\ \omega_2 \text{ if } z_m(\mathbf{x}) < 0 \end{array} \tag{39}$$

With this approach, all that matters is on which "side" of the hyperplane the pattern lies. Because the relative position between the pattern and a discriminant is arbitrary in that the discriminant is determined by available data and is scaled arbitrarily, it is easy to misclassify any pattern outside the range of training data. Successful application of this method is dependent on (1) linear separability of the pattern classes and (2) determination of the proper weight parameters to describe each hyperplane. Assuming the pattern classes are linearly separable, several iterative procedures have been developed to adaptively learn the hyperplane parameters from representative pattern class exemplars, including the perceptron algorithm (Rosenblatt, 1962), the Widrow–Hoff algorithm (1960), and the Ho–Kashyap algorithm (Simon, 1986).

By definition, the exemplar patterns used by these algorithms must be representative of the various pattern classes. Performance is tied directly to the choice and distribution of these exemplar patterns. In light of the high dimensionality of the process data interpretation problem, these approaches leave in question how reasonable it is to accurately partition a space such as R^{6+} (six-dimensional representation space) using a finite set of pattern exemplars. This degradation of interpretation performance as the number of possible labels (classes) increases is an issue of output dimensionality.

Furthermore, the pattern structures in a representation space formed from raw input data are not necessarily linearly separable. A central issue, then, is feature extraction to transform the representation of observable features into some new representation in which the pattern classes are linearly separable. Since many practical problems are not linearly separable (Minsky and Papert, 1969), use of linear discriminant methods is especially dependent on feature extraction.

With regard to linear projection based methods, the latent variables or scores determined by linear multivariate statistical methods such as

PCA and PLS can be used to interpret data by defining distance metrics for classification. The scores are the extracted features, and significant changes in them with respect to a set of reference scores enables assignment of labels. Reference scores are developed from a careful selection of process data, i.e., normal operating conditions. From a training perspective, these methods require that the relationship between variables be sufficiently linear and/or the calibration sample span a narrow enough portion of space for a linear approximation. The methods are therefore sensitive to variations in the data and outliers and work best for steady-state behaviors. It is particularly important to remove variables that have global effects because they appear in the first principal component can mask other information about the state of the process (Kosanovich *et al.*, 1995).

The most serious problem with input analysis methods such as PCA that are designed for dimension reduction is the fact that they focus only on pattern representation rather than on discrimination. Good generalization from a pattern recognition standpoint requires the ability to identify characteristics that both define and discriminate between pattern classes. Methods that do one or the other are insufficient. Consequently, methods such as PLS that simultaneously attempt to reduce the input and output dimensionality while finding the best input–output model may perform better than methods such as PCA that ignore the input–output relationship, or OLS that does not emphasize input dimensionality reduction.

The global nature of these nonlocal modeling methods manifests itself in the determination of the principal components, which are vectors of infinite length. A new data vector that is far removed from the reference data still will be scored against the principal component eigenvectors. As a result, it is important to manage extrapolation that could lead to incorrect classifications. For example, it is possible that, for a reasonable value of the proximity measure, the sampled variance could be very large, implying significant unexplained variability. By calculating the sampled residual variance, it is possible to estimate the probability that the sample residuals are similar to the data from which the reference score set was developed. This gives a direct indication that the data vector is an outlier or is a member of the reference, or that the process has been modified such that the scores are outside the range defined by the calibration set.

Nonlinear methods based on linear projection also can be used for data interpretation. Since these methods require numeric inputs and outputs, the symbolic class label can be converted into a numeric value for their training. Proposed applications involving numeric to symbolic transformations have a reasonably long history (e.g., Hoskins and Himmel-

PROCESS DATA ANALYSIS AND INTERPRETATION 53

Pattern Recognition Using a
BP Network
[0,1]

Output Layer

Hidden Layer

Input Layer

Pattern Recognition Using
Linear Discriminants

If $d(\theta) > 0$ then
else class 2

Linear Discriminant Function

$d(\theta) = \beta_{11}\theta_1 + \beta_{12}\theta_2 + \beta_{13}\theta_3 + \beta_{10}$

$d(\theta) = \beta^T\theta$

Transformed Feature Vector

$\theta = (f_1(x), f_2(x), f_3(x), 1)^T$

$\theta_1 = f_1(x)$

Feature Vector

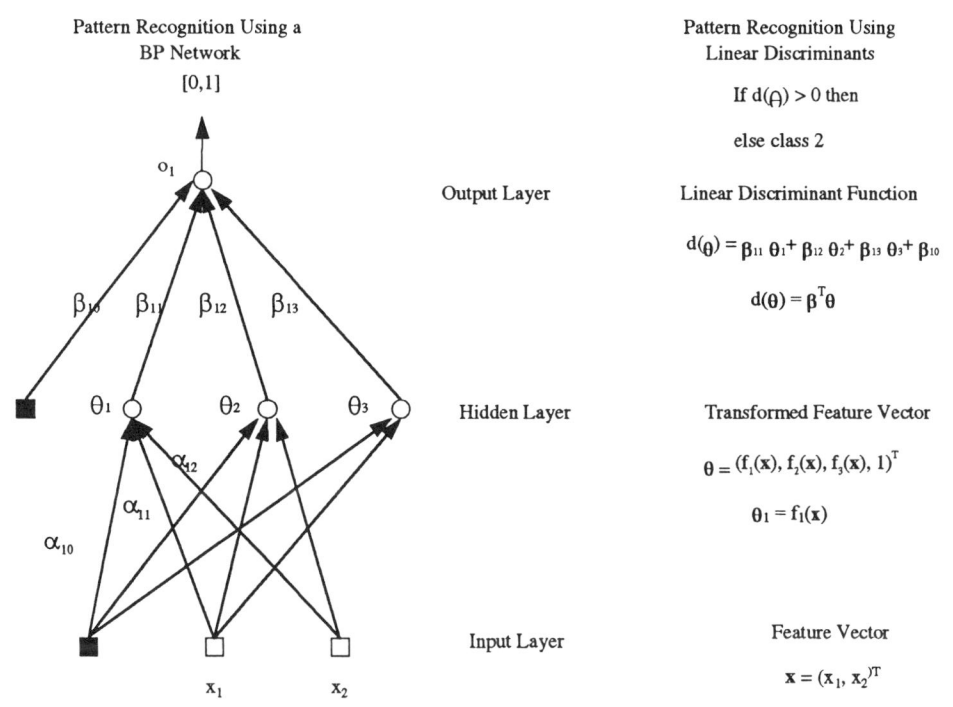

$x = (x_1, x_2)^T$

FIG. 21. Correspondence between BP networks and traditional pattern recognition using linear discriminants.

blau, 1988; Hoskins *et al.*, 1991; Naidu *et al.*, 1989; Ungar *et al.*, 1990; Whiteley and Davis, 1992a and b). Figure 21 illustrates the correspondence between the neural network representation of methods based on linear projection and pattern recognition using linear discriminants. Each step of the linear discriminant approach to pattern recognition has a mathematical equivalent in its neural network representation. Both methods utilize a numeric feature vector **x** as input. Patterns are represented by two observable features so that the network contains two input units plus a bias unit for the hidden layer. The presence of a hidden layer indicates that the problem is linearly nonseparable based solely on the observable features. Consequently, it is necessary to perform feature extraction with the objective of identifying higher order features that yield a distribution of patterns in the transformed feature space that are linearly separable. With respect to Fig. 21, three higher order features are required. These scalar-valued features are generated by transformation functions $\theta = f(\mathbf{x})$ in the traditional pattern recognition approach. In the corresponding network, each higher order feature is represented by a single hidden unit. Consequently, the hidden layer contains three processing units and one bias unit.

The feature transformations $\theta_m(\mathbf{x})$ can take any form depending on the selected method. In Fig. 21, the higher order features are generated by applying the sigmoid function to a linear combination of the observable pattern features. The weights, α_{ij}, associated with the input layer of a BP network define the nonlinear feature extraction performed by the network. The next step is identification of a set of parameters that defines a hyperplane or decision surface that separates the pattern classes in the transformed feature space R^3. These parameters are represented by the vector β in traditional pattern recognition on the right side of Fig. 21. In the BP network, these parameters correspond to the connection weights, β_{mi}, between the units in the hidden and output layers. Pattern classification using the traditional approach involves application of a discriminant function that is similar to a hard limit. The BP network approximates this decision function using a sigmoid. Consequently, the sigmoid function plays two different roles in BP networks with respect to pattern recognition: one associated with transforming the feature space into linearly separable features, and another associated with formation of the linear discriminant as a decision surface.

The BP network is ideally suited to implementation of the linear discriminant approach to pattern recognition. However, the great potential for using BP networks does not reside in the architecture, but rather in the generalized delta rule. The most difficult problem associated with the linear discriminant method is feature extraction, and this is where BP networks make their most significant contribution. By using the generalized delta rule to learn the weights between the input and hidden layers, BP networks effectively automate the feature extraction process. The generalized delta rule also is used to learn the weights associated with the linear discriminant. However, this represents less of a contribution, because there are numerous procedures available when pattern classes are linearly separable.

Given the power of BP networks as methods for automating feature extraction and for combining feature extraction with the linear discriminant, we note that BP networks are ultimately in the family of linear discriminant methods. Therefore, they suffer from the same problems with extrapolation errors associated with infinitely long decision surfaces, as described previously. In general, feedforward networks offer effective feature extraction and interpretation capabilities. Their primary drawback is that a linear discriminant decision surface grossly overstates the size of the true class regions as they are described by the existing data. Data patterns outside the data distributions used to define the decision surfaces are subject to misclassification. As with projection methods, it is critical to employ safeguards to manage the potential for extrapolation errors. Great care must be taken to use BP networks to interpret only those data that are within the coverage of the training exemplars.

B. LOCAL INTERPRETATION METHODS

Local interpretation methods encompass a wide variety of approaches that resolve decisions about input data relative to annotated data or known features that cluster. By characterizing the cluster or grouping, it is possible to use various measures to determine whether an arbitary pattern of data can be assigned the same label as the annotated grouping. All approaches are statistical, but they vary in terms of measures, which include statistical distance, variance, probability of occurrence, and pattern similarity.

1. Statistical Measures for Interpretation

If the probability distribution of the data is or assumed Gaussian, several statistical measures are available for interpreting the data. These measures can be used to interpret the latent variables determined by a selected data analysis method. Those described here are a combination of statistical measures and graphical analysis. Taken together they provide an assessment of the statistical significance of the analysis.

The Q-statistic or square of predicted errors (SPE) is the sum of squares of the errors between the data and the estimates, a direct calculation of variability:

$$Q_i = \sum (x_{ik} - \hat{x}_{ik})^2. \tag{40}$$

Here $\hat{x}_{ik}$ is an estimated value of a variable at a given point in time. Given that the estimate is calculated based on a model of variability, i.e., PCA, then Q_i can reflect error relative to principal components for known data. A given pattern of data, x, can be classified based on a threshold value of Q_i determined from analyzing the variability of the known data patterns. In this way, the Q-statistic will detect changes that violate the model used to estimate $\hat{x}$. The Q-statistic threshold for methods based on linear projection such as PCA and PLS for Gaussian distributed data can be determined from the eigenvalues of the components not included in the model (Jackson, 1992).

The Mahalanobis distance measures the degree to which data fit the calibration model. It is defined as

$$h_i = \{(x_i - \bar{x}_i)^T S^{-1}(x_i - \bar{x}_i)\}^{1/2}, \qquad (41)$$

where S is the estimate of the covariance matrix and $\bar{x}_i$ is the estimate of the mean. If the calibration model data represent process operation at one set of operating conditions, and the process has shifted to a different set, then the statistic will show that data at this new operating condition cannot be classified with the calibration data. The use of this statistic in monitoring and analysis is illustrated in Examples 1 and 2 in Section VIII.

2. Probability Density Function Methods

PDF approaches represent a statistically formal way of accomplishing local kernel definition. Although intent and overall results are analogous to defining kernels of PCA features, considerable work currently is required for PDF approaches to be viable in practice. It is presently unrealistic to expect them to adequately recreate the underlying densities. Nevertheless, there are advantages to performing data interpretation based on direct PDF estimation and, as a result, work continues.

From both a theoretical and practical view, it is ideal to use Bayesian Decision Theory because it represents an optimal classifier. From a theoretical perspective, Bayesian Decision Theory offers a general definition of the pattern recognition problem and, with appropriate assumptions, it can be shown to be the basis of many of the so-called non-PDF approaches. In practice, however, it is typically treated as a separate method because it places strong data availability requirements for direct use compared to other approaches.

In brief, the Bayesian approach uses PDFs of pattern classes to establish class membership. As shown in Fig. 22, feature extraction corresponds to calculation of the a posteriori conditional probability or joint probability using the Bayes formula that expresses the probability that a particular pattern label can be associated with a particular pattern.

Class membership is assigned using some decision rule that is typically some inequality test performed on $P(\omega_i|x_k)$.

$$P(\omega_i|x_k) = \frac{P(x_k|\omega_i) P(\omega_i)}{P(x_k)}$$

Posteriori probability that given data set, x_k its class label is ω_i

Conditional probability of x_k given ω_i

Prior probability of ω_i

Prior probability of x_k

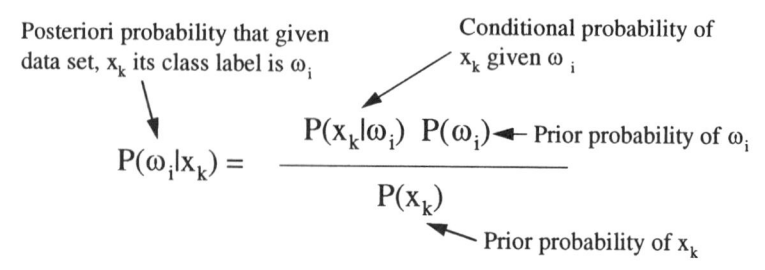

FIG. 22. Bayesian decision theory for pattern recognition.

The knowledge required to implement Bayes' formula is daunting in that a priori as well as class conditional probabilities must be known. Some reduction in requirements can be accomplished by using joint probability distributions in place of the a priori and class conditional probabilities. Even with this simplification, few interpretation problems are so well posed that the information needed is available. It is possible to employ the Bayesian approach by estimating the unknown probabilities and probability density functions from exemplar patterns that are believed to be representative of the problem under investigation. This approach, however, implies supervised learning where the correct class label for each exemplar is known. The ability to perform data interpretation is determined by the quality of the estimates of the underlying probability distributions.

Estimation of class-conditional densities is difficult. If the form of the probability function is known, then the problem becomes one of parameter estimation. Unfortunately, for process data interpretation, there is no basis for assuming the form of the function and one is forced to use nonparametric methods such as Parzen windows or the k-nearest neighbor method. More recently, radial basis function neural networks have been used to estimate probability density functions (Leonard and Kramer, 1991). These methods construct an approximation of the class-conditional densities by sampling the representation space containing the available training exemplars. The quality of these estimations is a strong function of the number of available exemplar patterns. For problems characterized by a large number of available exemplars and relatively small dimensionality, use of nonparametric techniques may be feasible. However, data interpretation is generally characterized by a large dimensionality and limited numbers of exemplars.

We include Bayesian Belief Networks (Kramer and Mah, 1994) in this category because they combine the Bayesian classification approach with a knowledge-based approach in the form of a semantic network. Ultimately, the approach attempts to calculate probabilities of events and subsequently assign labels to those events with the highest probabilities. Each node in the network represents the probability of some event. Events are linked based on some defining relationship, i.e., causality. Data interpretation is done by calculating and propagating probabilities exhaustively through the network. Although an exact computation may be intractable for realistic problems, several approximate inferencing methods have been proposed (Henrion et al., 1991). Although a number of desirable properties of the belief network have been demonstrated and discussed, in practice the major consideration is availability of probability information. It requires considerable probability information that simply does not exist. Prototype demonstration results have indicated that rough and subjective estimates of proba-

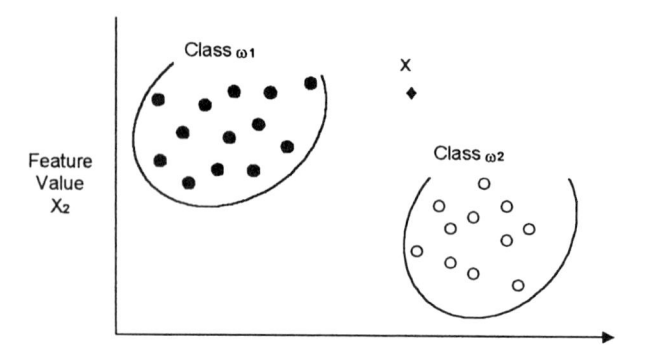

FIG. 23. Clustering characteristics of a two-class problem involving two observable features.

bilities may suffice, and therefore data requirements may be somewhat reduced.

3. Clustering Methods

As described by Oja (1995), kernel-based or clustering approaches are a general category distinguished by an ability to map highly nonlinear input data into prototype vector descriptions that form a multidimensional lattice. Unlike the nonlocal methods that draw upon latent variables generated from the application of dimension reduction and modeling techniques, clustering approaches are designed directly for pattern recognition. The underlying assumption is that patterns in a common pattern class exhibit similar features and that this similarity can be measured using an appropriate proximity index. Using this concept, patterns are assigned to the class of the pattern(s) to which they are most similar. This is illustrated in Fig. 23, where the concept is illustrated for two pattern classes in a two-dimensional representation space, R^2.

The training objective is to define an appropriate prototype vector description for a given set of sufficiently similar input data patterns and to establish a topological mapping of the prototypes such that each has a set of neighbors. Feature extraction is associated with forming the prototype descriptions and then placing them in an abstract feature space. Mapping is based on the similarity or nearness of a given pattern with labeled prototype descriptions, as shown in Fig. 23.

Clustering lacks the strict Bayesian requirement that input patterns be identifiable with a known prototype for which underlying joint distributions or other statistical information is known. Rather, the clustering approach leverages similarities between pattern features so that patterns that have never been seen before and for which no statistical data are available still can be processed. This is illustrated in Fig. 23, by the diamond representing an arbitrary pattern, x.

The technology of proximity indices has been available and in use for some time. There are two general types of proximity indices (Jain and Dubes, 1988) that can be distinguished based on how changes in similarity are reflected. The more closely two patterns resemble each other, the larger their *similarity* index (e.g., correlation coefficient) and the smaller their *dissimilarity* index (e.g., Euclidean distance). A proximity index between the *i*th and *j*th patterns is denoted by $D(i, j)$ and obeys the following three relations:

1. $D(i, i) = 0$ for all *i*, using dissimilarity index
   $D(i, i) \geq \max_k D(i, j)$ for all *i*, using a similarity index
2. $D(i, j) = D(j, i)$ for all *i, j*
3. $D(i, j) \geq 0$ for all *i, j*

The Euclidean distance is the dissimilarity index most frequently used. It is characterized by invariance to translation and rotation,

$$D(z, x) = \|z - x\| = [(z - x)^T(z - x)]^{0.5}, \tag{42}$$

where z and x are two pattern vectors. Other proximity indices can be used, including the Mahalanobis distance (see section on statistical measures), the Tanimoto measure, and the cosine of the angle between x and z (Tou and Gonzalez, 1974).

The most straightforward decision rule that can be employed is what is referred to as the *1-nearest neighbor* (1-NN) rule. This rule assigns a pattern x of unknown classification to the class of its nearest neighbor z_i from the set of known patterns and pattern classes;

$$x \in \omega_i \text{ iff } D(z_{i, x}) = \min\{D(z_j, x)\} \text{ for } j = 1, N, \tag{43}$$

where *N* is the number of possible pattern classes. This was illustrated previously in Fig. 15.

This rule can be easily extended so that the classification of the majority of *k* nearest neighbors is used to assign the pattern class. This extension is called the *k-nearest neighbor* rule and represents the generalization of the 1-NN rule. Implementation of the *k*-NN rule is unwieldy at best because it requires calculation of the proximity indices between the input pattern x and all patterns z_i of known classification.

For process systems, one key practical advantage of clustering is the ability to work with limited and poorly distributed pattern data, typical of many labels of interest. Rather than attempting to partition unknown regions of the representation space with projection-based methods, clustering approaches simply identify the structure in the existing representation space based on available data. These kernal-based methods are more robust to unreliable extrapolation, but they give up the potential for extrapolation. The focus on local data structures produces bounded decision regions and identifies regions of indecision (Kavuri and Venkatasubramanian, 1993; Davis et al., 1996a). The notion of proximity or distance can be used to make additional decisions about the data patterns, such as novel pattern identification, or to establish gradations of uncertainty (Davis and Wang, 1995).

Algorithms fundamental to kernel-based approaches involve cluster seeking, the task of identifying class prototypes. In cluster seeking, as described in the Data Analysis section, pattern classes are defined based on discovery of similarities between patterns. For interpretation, the situation is reversed; that is, the pattern classes are defined a priori and the task is to identify the attributes of the exemplar patterns that cause an expert to assign them to specified classes. Supervised training is necessary to build the interpretation system based on an expert's collected knowledge in the form of annotated training sets. In practice, generating the annotated training set is probably the largest barrier to developing the system.

The applicability of a clustering algorithm to pattern recognition is entirely dependent upon the clustering characteristics of the patterns in the representation space. This structural dependence emphasizes the importance of representation. An optimal representation uses pattern features that result in easily identified clustering of the different pattern classes in the representation space. At the other extreme, a poor choice of representation can result in patterns from all classes being uniformly distributed with no discernible class structure.

Even when the patterns are known to cluster, there remain difficult issues that must be addressed before a kernel-based approach can be used effectively. Two of the more fundamental conceptual issues are the number and size of clusters that should be used to characterize the pattern classes. These are issues for which there are no hard and fast answers. Despite the application of well-developed statistical methods, including squared-error indices and variance analysis, determining the number and size of clusters remains extremely formidable.

Performance requires a careful balance between *generalization* and *memorization* (Davis and Wang, 1995). On one hand, a cluster can be defined too broadly, leading to misclassification of data. On the other hand, clusters can be so tightly defined that similar patterns in the same pattern classes cannot be classified with confidence. The appropriate balance is dependent on the desired functionality.

Differences in approaches rest with how the prototype vectors are generated, how the topological map is formed, and how the proximity indices are calculated. Most network performance comparisons have been made against back propagation neural networks (Kavuri and Venkatasubramanian, 1993; Leonard and Kramer, 1990; Whiteley and Davis, 1992a, b). Clustering approaches clearly outperform BPN primarily because of extrapolation errors. There have been some comparisons between clustering networks (e.g., Kavuri and Venkatasubramanian, 1993). These comparisons do not lead to any general recommendations based on classification performance. All are highly dependent on available data, and it appears that relatively close classification performances can be obtained when the various approaches are optimized for a particular static set of data patterns. From a practical point of view, differences manifest themselves more dramatically in the ability to initially establish and train the network and to accommodate new information as adaptive units.

It is possible to group approaches along several possible dimensions that include distance vs directional similarity measures, variations of winner-take-all strategies as illustrated earlier in Fig. 15, and fixed vs variable clustering capability. Fixed vs variable clusters appears to have the greatest practical significance because these categories reflect perspectives on initial set up of a cluster approach and adaptation, so they are considered here as the two primary categories.

(i) *Fixed Cluster Approaches.* This category represents those approaches that require a fixed number of clusters to be specified a priori. Probably the most commonly discussed approach is the radial basis function network (Moody and Darken, 1989; Leonard and Kramer, 1991). Referring to the earlier discussion of cluster seeking approaches in Section III, RBFNs using the exact same techniques can be readily extended for feature mapping. As discussed, an RBFN represents a neural network mapping procedure performing a nonlinear transformation at the hidden layer followed by a linear combination at the output layer. In terms of general structure, the RBFN is the same as the BPN. The extension to interpretation is accomplished in the same way as for BPN, namely, by declaring the numeric output to be 1 or 0, corresponding to the presence or absence of a particular interpretation. Distinctions are associated only with the basis functions identified with the hidden layer of nodes. The use of a Gaussian basis function rather than a sigmoid gives RBFNs local interpretation characteristics.

As described in detail in Section III, Input Analysis, training an RBFN for interpretation generally can be regarded as a two-step process consisting of clustering and supervised learning. A topological structure is formed by specifying the number of clusters that will be used and then using training data to converge on an optimal set of cluster centers via k-means clustering.

The selection of cluster number, which is generally not known beforehand, represents the primary performance criterion. Optimization of performance therefore requires trial-and-error adjustment of the number of clusters. Once the cluster number is established, the neural network structure is used as a way to determine the linear discriminant for interpretation. In effect, the RBFN makes use of known transformed features space defined in terms of prototypes of similar patterns as a result of applying k-means clustering.

As new data patterns become available, RBFNs can be updated incrementally, but the number of clusters cannot be changed without retraining the new network with all the data. This is a serious limitation from an adaptation standpoint. The second performance consideration stems from the formation of the cluster structure itself. Once the cluster centers have been established using k-means clustering, the cluster structure is formed using some nearest neighbor heuristic, i.e., 2-NN (Duda and Hart, 1973). Clustering based on k-NN approaches is effectively forming clusters by proximity and not class. As a result it is possible and likely to have mixed clusters that diminish the utility of the radial unit.

Ellipsoidal basis functions networks (EBFNs) represent a rather complex approach to addressing some of the shortcomings of RBFNs for interpretation (Kavuri and Venkatasubramanian, 1993). An EBFN is structured with a version of k-means clustering. Rather than establishing a cluster center based only on the winning cluster, the EBFN approach uses a fuzzy membership cluster neighborhood concept that is an extension of the Kohonen algorithm (Kohonen, 1989). The result is that there is no single winning cluster; rather, there is a degree of member for each cluster. The claim is that cluster centers are moved into different positions that enhance the classification performance. The problem of mixed clusters that result from the application of the k-NN algorithm is avoided by explicitly clustering by class. Increased performance using ellipsoidal rather than radial basis functions is not established. RBFN and EBFN approaches are compared

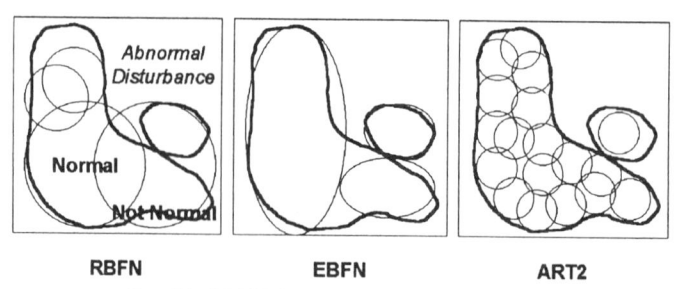

FIG. 24. RBFN, EBFN, and ART approaches.

graphically in Fig. 24 in terms of how they represent regions of normal and not normal plant operation.

We mention qualitative trend analysis (QTA) (Vedam *et al.*, 1998) in this section on clustering. QTA is a syntactic approach to numeric–symbolic mapping in that input data is mapped into a grammar of letters and words that capture a sequence of component shapes to form a pattern. Interpretation is accomplished by comparing the sequence of letters describing the input pattern signatures for known situations. A separate set of decision rules is invoked to monitor changes in the magnitudes of patterns.

(ii) *Variable Cluster Approaches.* The distribution of patterns available in chemical processes predestines the incompleteness of available data. Evolutionary adaptation is therefore necessary for online use because a practical data interpretation system must adapt smoothly to an ever-increasing amount of sensor data. The most effective measure of the maximum coverage of the interpretation capacity of a system is its ability to be updated on an ongoing basis.

To be effective, the kernel-based approach must be adaptive to the situations that have not occurred previously while maintaining information already residing in the system. This requires that the system be plastic so that it can incorporate new knowledge while maintaining a level of stability relative to the existing knowledge. An appropriate stability–plasticity balance is critical to adaptation. This implies that the number of clusters should be variable and that exiting clusters should be adjustable. Variable cluster approaches address this with the capacity for adding new clusters or removing unused or unimportant clusters.

Adaptive Resonance Theory (ART) (and, in particular, ART2), developed for both binary and analog input patterns, is the only kernel-based architecture that addresses variable clustering capacity and supports the information management components required for determining whether a new cluster is needed or whether an existing cluster should be modified. ART represents a philosophy as much as a model and continues to be refined with time (Grossberg, 1982, 1987a, 1987b, 1988). The theory has been substantiated in a series of neural network models called ART1, ART2, and ART3 (Carpenter and Grossberg, 1987a, 1987b, 1988, 1990).

ART2 forms clusters from training patterns by first computing a measure of similarity (directional rather than distance) of each pattern vector to a cluster prototype vector, and then comparing this measure to an arbitrarily specified proximity criterion called the *vigilance*. If the pattern's similarity measure exceeds the vigilance, the cluster prototype or "center" is updated to incorporate the effect of the pattern, as shown in Fig. 25 for pattern 3. If the pattern fails the similarity test, competition resumes without the node

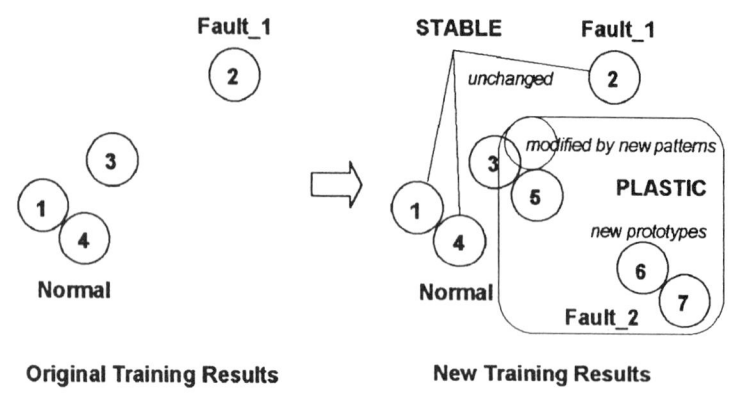

FIG. 25. Clustering in ART.

corresponding to the cluster. This match/reset operation produces a crisp decision boundary for each cluster. When no match is found between an input pattern and an existing cluster, a new, uncommitted node is allocated for the input, which then becomes the prototype for the new cluster. This is shown in Fig. 25 in terms of the formation of new clusters 5, 6, and 7. No existing node is altered by the pattern. The new cluster then participates in future competitions with previous nodes. The size of the hyperspherical cluster is determined by the vigilance value and characteristics of the training data. Stability–plasticity issues are addressed because the algorithm comprises the information management mechanisms to decide when to form a new cluster and how to adjust an existing cluster. Unlike RBFNs and EBFNs, the algorithm asks which cluster is most similar and then asks and answers the question about the pattern being close enough.

No single ART2 equation or set of equations provides the desired adaptive properties for the system. Rather, it is the synergistic effect of this architecture as a whole that gives rise to these properties. A description of the ART network applied to process situations can be found in Whiteley and Davis (1996); a complete description of an ART network can be found in Carpenter and Grossberg (1987a).

VI. Symbolic–Symbolic Interpretation: Knowledge-Based System Interpreters

If there is sufficient operating history to provide the labeled data and sufficient resources to label it, then it is desirable to rely on this information for data interpretation because it reflects the actual operation. With respect to Fig. 3, this is the case of using a numeric–symbolic interpreter for assigning the labels of interest. However, when operating histories (or trustworthy simulations) do not provide sufficient exemplars to develop an adequate mapping using training based methods alone, knowledge-based systems (KBSs) can be used for interpretations, and they are especially useful for extending intermediate interpretations to the labels of interest (McVey *et al.,* 1997).

There are degrees of capacity, however, that need to be taken into account. If operating history is less than fully sufficient, but is nevertheless plentiful, training based methods may be able to carry *most* of the burden of interpretation by providing effective intermediate labeling. Then the requirements for the KBS become very modest. At run time, the KBS will perform the final mapping to appropriate labels, typically through some form of *table lookup*: e.g., simple rule sets or a decision tree. On the other hand, if the intermediate labeling requires considerably more analysis to reach the desired classifications, then the KBS portion might be very extensive. This sequential integration of a numeric–symbolic interpreter and the KBS interpreter is illustrated in Fig. 4, where the measurement data is mapped into intermediate interpretations that are then input to a KBS interpreter. The KBS interpreter ultimately produces the final interpretations of interest.

In lieu of data exemplars, KBS approaches use expert descriptions to accomplish the mapping. Explicit items of knowledge provide pathways between combinations of symbolic input values and possible output labels. KBS approaches provide both means for capturing this knowledge, and control constructs for exploiting efficient strategies used by experts. There have been many reports on knowledge-based system constructs (e.g., Quantrille and Liu, 1991; Rich and Knight, 1991; Shapiro and Eckroth, 1987; Stephanopoulos and Davis, 1990, 1993; McVey et al., 1997; Stephanopoulos and Han, 1994). A review of intelligent systems in general is given in Davis *et al.* (1996b).

Consider the example in Fig. 26 of determining if there is a problem with the feed injection system of a fluidized catalytic cracking unit (Ramesh *et al.,* 1992). In this example, there is a set of rules that relate combinations of process observations to establish or reject this possibility.

The label of interest is *Feed Injection System Problem*. The *if, and,* and *or* statements relate specific process observations that can establish that there is a likelihood of an injection system problem. *Injector header pressure* is a process measurement and *abnormal* is an intermediate label of interest. The label *abnormal* can be determined by developing a numeric–symbolic interpreter that maps injector header pressure data as either normal or abnormal.

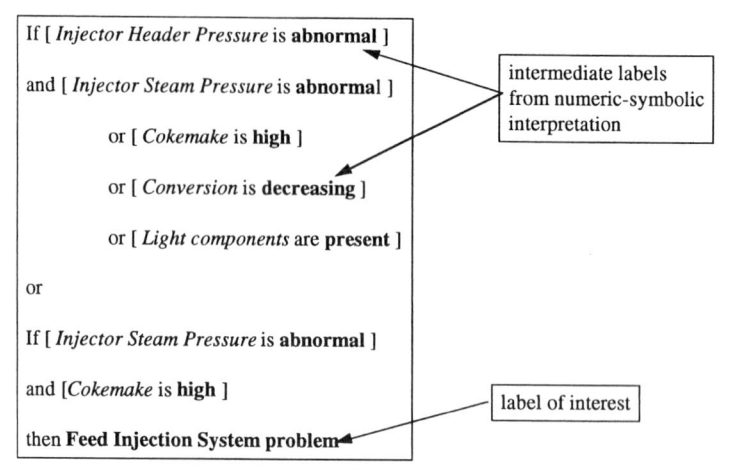

```
If [ Injector Header Pressure is abnormal ]

and [ Injector Steam Pressure is abnormal ]          intermediate labels
                                                     from numeric-symbolic
    or [ Cokemake is high ]                          interpretation

    or [ Conversion is decreasing ]

    or [ Light components are present ]

or

If [ Injector Steam Pressure is abnormal ]

and [Cokemake is high ]                              label of interest

then Feed Injection System problem
```

FIG. 26. A fluidized catalytic cracking unit example of combining numeric–symbolic and symbolic–symbolic interpretation (Ramesh *et al.*, 1992).

It is common to view KBSs from a mechanistic perspective only and, as a result, to oversimplify the construction of a system and miss out on potential capabilities. At a mechanistic level, the constructs for representing knowledge are fundamentally lists, rules, objects, and so on, and constructs for manipulating knowledge are matching, chaining, attribute inheritance, and so forth. This view is in danger of missing the importance of higher-level analysis that considers types of knowledge, types of problem-solving activity, and ways of organizing knowledge. Diagnostic reasoning, for example, relies on forms of knowledge and has processing strategies distinct from the knowledge forms and processing strategies most useful for, for example, response planning. Constructing a KBS by choosing a mechanism, prior to higher-level analysis of the task to be performed or of the reasoning strategies and types of knowledge required, is likely to lead to disappointment, especially when constructing large systems (McVey *et al.*, 1997).

A useful characteristic of KBS interpreters is that of generality of coverage: That is, they use representations and manipulations to reach correct conclusions without having to explicitly enumerate all possible relationships between input combinations and conclusions. Situations can be covered that are theoretically or realistically possible, but that have not been observed or recorded. Moreover, a well-designed KBS degrades gracefully when presented with novel or unanticipated situations. The mapping performed by a KBS might be direct, essentially table lookup, or it might employ problem-solving using semantic representations and thereby be capable of providing broad generality of coverage.

Many applications of knowledge based systems have been described (e.g., Antsaklis and Passino, 1993; Badiru, 1992; Mavrovouniotis, 1990; Myers *et al.*, 1990; Reklaitis and Spriggs, 1987; Rippin *et al.*, 1994). The absence of extensive training data, together with the value of the anticipated conclusions, explains the wide interest in the use of KBS systems for fault classification and corrective action planning.

A. FAMILIES OF KBS APPROACHES

Representing knowledge requires selecting a set of conventions for describing devices and relations (i.e., causal relations, processes, and so on). This selection strongly affects the ability of the KBS to make appropriate connections between relations. The advantages and limitations of various knowledge representation schemes vary considerably with the characteristics of the problem, the types of knowledge, and the requirements for manipulation. In general, no one representation emerges as uniformly best.

Two broad categories of approaches have emerged and are in wide use as symbolic–symbolic interpreters: *semantic networks* and *tables*. Semantic networks refer to a very large family of representations comprising nodes to represent objects or concepts and nodal links representing the relations between them. Semantic networks are particularly useful for representing taxonomies, spatial relations, causal relations, and temporal relations. In general, semantic networks take on many forms, including hierarchies, decision trees, relationship structures, directed graphs, and others (Stephanopoulos and Davis, 1990). The other very popular representation is the table. In tables, input and output labels (variables) are structured as one-to-one relations. In the simplest case, a given input label maps directly into an output. The strategy involves simply searching the input table for the relevant conditions and then noting the output. Most applications are more complicated in that input–output conditions exhibit more complex relations. For example, an input can be associated with more than one output, an input can only partially explain an output, or multiple inputs can explain an output. With these kinds of complex relations, the procedure for mapping is more complicated and involves constructing a best explanation for a given set of inputs by accounting for implications and incompatibilities (Josephson and Josephson, 1994).

With the view that a KBS interpreter is a method for mapping from input data in the form of intermediate symbolic state descriptions to labels of interest, four families of approaches are described here, each offering inference mechanisms and related knowledge representations that can be used to solve interpretation problems: namely, model-based approaches, digraphs, fault trees, and tables. These methods have been heavily used for diagnostic applications in process operations. Of these, at one extreme are model-based methods, using quantitative and qualitative simulation that does very little to constrain the allowable possibilities. By using behavioral models for the individual components in a process operation, model-based methods enable consideration of many kinds of possible behaviors. The advantage is that by manipulating models it is possible to discover input–output relationships that have not been pre-enumerated. The disadvantage is that descriptions of large numbers of behaviors can be generated, many of which will not be realistic or relevant under the circumstances and, therefore, the number of output considerations must be carefully managed. At the other extreme are table methods that fully constrain the possible input–output situations. Such systems do not allow for relation-path considerations at run time because they compile relation pathways into direct input–output relations. The advantage here is that table lookup is very direct and performance levels can be very high. The disadvantage is that correct performance is completely dependent on the ability to preenumerate all the relations. Digraphs and fault trees take intermediate positions in the trade-off between flexibility and efficiency.

1. Model-Based

The term *model-based* can be a source of confusion because descriptions of any aspects of reality can be considered to be models. Any KBS is model based in this sense. For some time, researchers in KBS approaches (Venkatasubramanian and Rich, 1988; Finch and Kramer, 1988; Kramer and Mah, 1994; McDowell and Davis, 1991, 1992) have been using model-based to refer to systems that rely on models of the processes that are the objects of the intent of the system. This section will avoid confusion by using the term *model* to refer to the type of model in which the device under consideration is described largely in terms of components, relations between components, and some sort of behavioral descriptions of components (Chandrasekaran, 1991). In other words, model-based is synonymous with *device-centered*. Figure 27 shows a diagram displaying relationships among components. The bubble shows a local model associated with one of the components that relates input–output relationships for flow, temperature, and composition.

A large number of model-based systems use either qualitative or quantitative simulation, such as FAULTFINDER (Kelly and Lees, 1986) or EAGOL (Roth, Woods, and Pople, 1992). These systems simulate normal behavior and compare the simulation results with observations, they simulate faults and compare simulation results with detected symptoms, or they interleave simulation with observation comparing the two to dynamically track normal and abnormal states. It is computationally very expensive to

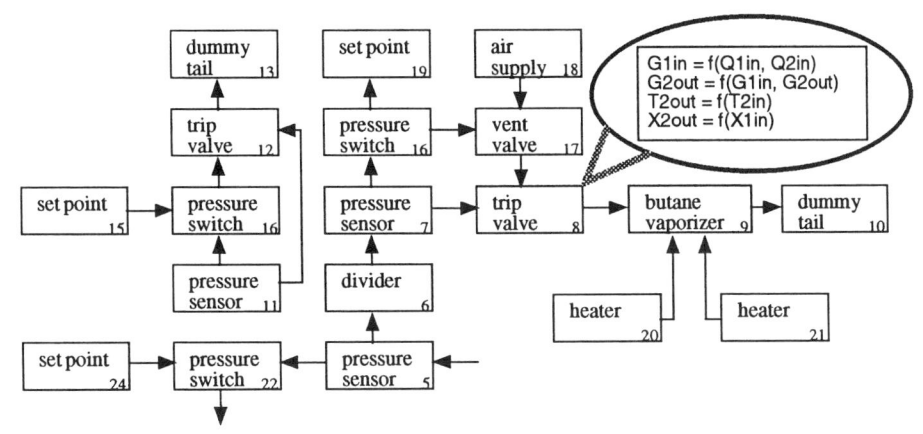

FIG. 27. Components and connections. [Drawn from J. S. Mullhi, M. L. Ang, and F. P. Lees, *Reliability Engineering and System Safety*, **23**, 31 (1988).

apply high-detail quantitative simulation to large and complex systems because of the large number of causal events that must be computed. Furthermore, quantitative simulations often do not capture malfunction situations adequately. Qualitative simulation, while capturing an appropriate level of fault description and saving computation compared with quantitative simulation, still typically results in generating too many malfunction candidates for large-scale systems, but has the advantage of flexibility, allowing causal relations to propagate forward and backward readily (Kuipers, 1994; Catino and Ungar, 1995; McDowell and Davis, 1991; Venkatasubramanian and Vaidhyanathan, 1994). One main source of computational expense in qualitative simulation arises because lack of precision leads to ambiguity in projecting the future, which leads to considering combinations of alternative future possibilities. Of course, local use of simulation is practical for large-scale systems when it is used sparingly in the focused pursuit of specific subgoals.

2. Fault Trees and Digraphs

Fault trees and digraphs were the earliest structures used for capturing causal relations in diagnostic KBS. A fault tree is a graph representing the causal dependencies of symptoms and faults. Confidence values are used to control and optimize the inference mechanism as well as to provide additional information for decision making. Early examples of the use of fault trees in diagnostic expert systems includes the work of Ulerich and Powers (1987) and Venkatasubramanian and Rich (1988). Because the relationship between faults and symptoms is a directed graph, not a tree (Kramer and Mah, 1994), a basic event in a fault tree may result from more than one event and typically can contribute to more than one event. This creates a tangled graph, which may contain loops that make diagnostic inference complicated. Figure 28 provides an illustrative segment of a fault tree.

A digraph contains equivalent information to fault trees, but is written in terms of system variables; a variable is given as a node and the relationship between two variables is given as an edge. The digraph also can contain nodes that represent fault conditions rather than system variables—identical to fault trees. Positive and negative influences and their magnitudes, or conditional probabilities, are represented by annotations associated with the edges. Examples of digraphs include belief networks (Pearl, 1988; Kramer and Mah, 1994), possible cause–effect graphs (Wilcox and Himmelblau, 1994), and signed directed graphs (SDGs) (Mohindra and Clark, 1993). Diagnosis is carried out essentially by determining the maximal strongly connected subgraph. Special operators are used to handle loops. The inference mechanism involves enumerating combinations of the unmeasured nodes. For each unmeasured node, the system generates a set of patterns that correspond to all possible states. Execution time is exponential in the number of unmeasured nodes. However, the unmeasured nodes are necessary to maintain the diagnostic resolution. For a large and complex system with many unmeasured variables, diagnosis using only this approach is too time consuming. Figure 29 illustrates a digraph where the arcs represent directed causal connections and the nodes represent possible events.

3. Tables

Widely used in rule-based systems, input–output matching approaches are used in conventional expert systems going back to the earliest systems.

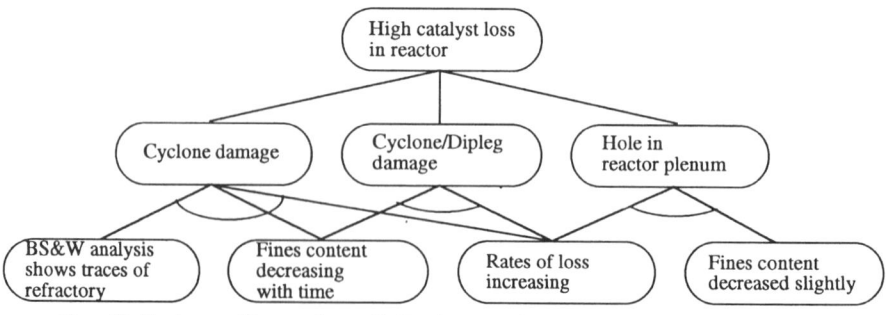

FIG. 28. Faul tree. [Drawn from G. Stephanopoulos, Ed., *Knowledge-Based Systems in Process Engineering,* Vol. 1, Cache Case Studies Series, CATDEX, Cache Corporation, Austin, TX (1988).

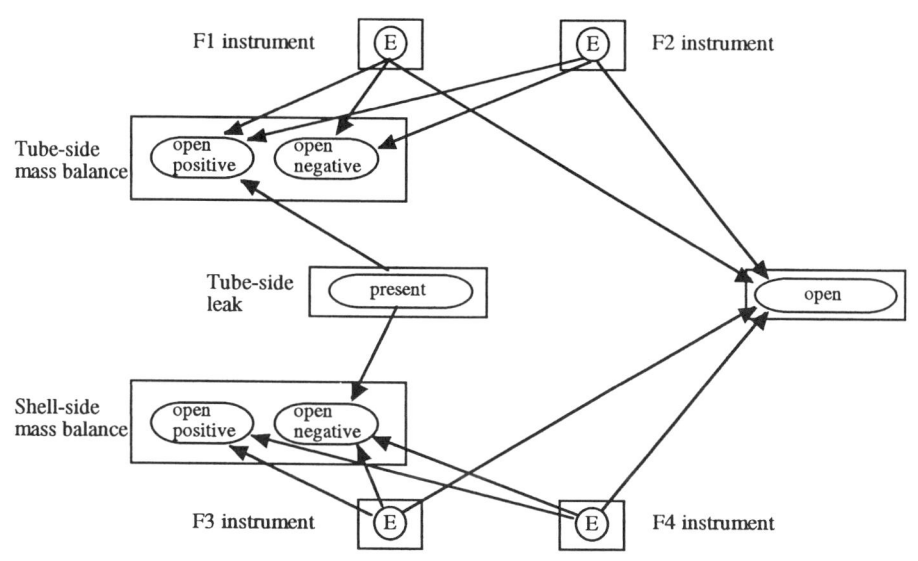

FIG. 29. Digraph. [Drawn from N. A. Wilcox and D. M. Himmelblau, *Comput. CHem. Eng.* **18**, 103 (1994).

There has been some interest in compiling digraphs and fault trees into rules for use in solving diagnostic problems. Kramer and Palowitch (1987) use a method that analyzes the digraph structure offline to derive simple diagnostic tests to be used online. Ragheb and Gvillo (1993) have developed a methodology for using fault-tree analysis techniques to represent knowledge, which is in turn translated to goal trees and then encoded as production rules. Computational load during run time is reduced considerably. For a large and complex system, however, compilation itself is computationally very expensive.

Figure 26, shown earlier, is a simple form of input mapping called table lookup. A more complicated inference mechanism is illustrated in Fig. 30. Here we see a simple example from a fluidized catalytic cracking unit in which multiple product quality attributes can be explained by multiple operating parameters (Ramesh *et al.*, 1992).

Overall, each of these KBS approaches has its place, depending on the needs of the problem and the characteristics of the process. However, the advantages, limitations, and roles of the various methods must be clarified so that problem-solving efficiency can be achieved along with the ability to reach correct conclusions despite challenging circumstances: knowledge that is incomplete; data that are incomplete, unreliable, or extremely plentiful; or situations that are untested or completely unanticipated.

Product Quality to be Explained

[Conversion: Low]
[Cokemake: High]
[H/C Ration: High]

Explanatory Relationships

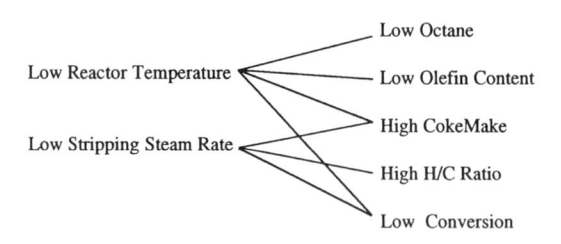

Low Reactor Temperature
- Low Octane
- Low Olefin Content

Low Stripping Steam Rate
- High CokeMake
- High H/C Ratio
- Low Conversion

Low Stripping Steam Rate is the better explanation given the current data

FIG. 30. A form of table lookup requiring assembly of best hypothesis.

VII. Managing Scale and Scope of Large-Scale Process Operations

KBSs can be viewed with increasing levels of commitment to problem solving. At the level described in the previous section, a KBS accomplishes symbolic–symbolic mappings between input and output variables analogous to the numeric–symbolic mappings of approaches such as neural networks and multivariate statistical interpreters. For each problem-solving task, the particular numeric–symbolic or symbolic–symbolic approach is based on the task and the knowledge and data available.

At another level, certain KBS approaches provide the mechanisms for decomposing complex interpretation problems into a set of smaller, distributed and localized interpretations. Decomposition into smaller, more constrained interpretation problems is necessary to maintain the performance of any one interpreter and it makes it possible to apply different interpretation approaches to subparts of the problem. It is well recognized that scale-up is a problem for all of the interpretation approaches described. With increases in the number of input variables, potential output conclusions, complexity of subprocess interactions, and the spatial and temporal distribution of effects, the rapidity, accuracy, and resolution of interpretations can deteriorate dramatically. Furthermore, difficulties in construction, verification, and maintenance can prohibit successful implementation.

PROCESS DATA ANALYSIS AND INTERPRETATION 73

A. DEGRADATION OF INTERPRETATION PERFORMANCE

Consider the following simulation-based study of the performance degradation of an Adaptive Resonance Theory numeric–symbolic interpreter for identifying malfunctions from input process data (Ali *et al.*, 1997). The data for this study were generated from a training simulator of a fluidized catalyst cracking unit (FCCU). The model simulates a stacked regenerator/disengager design. The regenerator, at the bottom, operates at a higher pressure. An external riser standpipe, a single-stage steam stripper standpipe, and a single-stage steam stripper complete the converter. A steam turbine powered centrifugal air blower supplies the regeneration air. The waste heat boiler develops steam from the surplus heat in the flue gas. There is also one main fractionator with a light cycle oil (LCO) stripper as part of the model.

The simulator models the FCCU, generating output from 110 sensors every 20 seconds. In all, 13 different malfunction situations were simulated and are available for analysis. There are two scenarios for each malfunction, slow and fast ramp. Table II provides a list and brief description of each malfunction. A typical training scenario for any fast ramp malfunction simulation had the landmarks listed in Table III. Similarly, a typical training scenario for any slow ramp malfunction simulation is shown in Table IV. For both the fast and slow ramp scenarios, there was data corresponding to 10 min of steady-state behavior prior to onset of the faulty situations.

With reference to Figs. 24 and 25, Fig. 31 illustrates a pattern representa-

TABLE II
LIST OF MALFUNCTIONS AVAILABLE FOR ANALYSIS

I.D.	Malfunction description
kldft	KO drum level transmitter drifts.
mfdft	Main fractionator transmitter drifts.
absd	Air blower speed drop.
ftdft	Air blower flow transmitter drifts.
cinc	Carbon in the gas oil feed increases.
spar	Slurry pump performance degrades.
lpar	LCO pump performance degrades.
npar	Naphtha pump performance degrades.
hpar	HCO pump performance degrades.
sdft	Slurry flow sensor drifts.
ldft	LCO flow sensor drifts.
ndft	Naphtha flow sensor drifts.
hdft	HCO flow sensor drifts.

TABLE III
Fast Ramp Training Scenario

Time	Landmarks
0:00	Simulation begins.
1:00	The faulty ramp starts.
3:00	The ramp is over and controllers take over.
10:00	The simulation ends.

tion space in ART for an evolving data set. The numbers represent patterns of data that occur over time and are compared to known pattern descriptions represented by the clusters, labeled *Normal, fault₁*, and *fault₂*.

As shown, the data patterns 1 and 2 are classified as Normal with high certainty, as they lie within the boundaries of the normal class. However, 3 and 4 are classified as normal with medium certainty, as they lie outside the normal region, but their similarities are closest to the normal cluster. Similarly, the data pattern represented by 5 is classified as fault$_2$ with medium certainty and the data patterns represented by 6, 7, and 8 are classified as fault$_2$, and so on. If the data are collected every 20 seconds as in the case study, the dynamic interpretation is tabulated as shown in Table V, with the labels in italics representing the correct class and appropriate certainty. An x means there was not an interpretation with this certainty.

To illustrate that increasing output dimensionality can lead to performance problems for a numeric–symbolic interpreter, the ART interpreter was trained with various subgroupings of the 13 malfunction scenarios. With 13 malfunction scenarios and one normal scenario, the maximum output dimension is 14.

The subgroups were determined based on a hierarchical decomposition strategy for a FCCU (Ramesh *et al.*, 1992). The FCCU was decomposed into four separate units: feed.system, catalyst.system, reactor/regenerator.system, and separation.system as shown in Fig. 32. Each of these units was further divided into more detailed functional, structural, or behavioral

TABLE IV
Slow Ramp Scenario

Time	Landmarks
0:00	Simulation begins.
1:00	The faulty ramp starts.
46:00	The ramp is over and controllers take over.
1:00:00	The simulation ends.

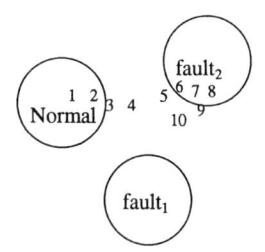

FIG. 31. ART pattern representation space for an evolving data set.

categories, going into three or, sometimes, four levels of detail. Each tip node corresponds to a specific malfunction such as feed temperature controller problem or stripper level problem. The particular segments of the decomposition that involve the 13 malfunctions are also shown in Fig. 32. As illustrated, the 13 malfunctions reside in three distinct groups.

Group 1: Loss of wet gas flow. This grouping of malfunctions is concerned with events leading to the loss of flow through the compressor which draws off the fractionator overhead vapors.

Group 2: Coke imbalance. This grouping considers malfunctions leading to a difference between the rate at which coke accumulates on the catalyst and the rate at which it is burned off. A coke imbalance is associated with a reduction of oxygen, which can be caused by a loss of combustion air or through an increase in the conradson carbon in the gas oil feed to the unit.

Group 3: Loss of pump around flow. This grouping comprises malfunctions associated with the circulation of fluids through and around the main fractionator. The loss of one or more pump flows (slurry, light cycle oil, naphtha, and heavy cycle oil) leads to the loss of some of the fractionator heat sink.

Table VI shows an example for detecting and isolating problems due to an air blower speed drop (fast ramp) *absd*. The behavior illustrated by this

TABLE V
DATA INTERPRETATION TABLE ONLINE

Time (min)	High certainty	Medium certainty
0.00–0.67	Normal	
1.00–1.33	x	Normal
1.67–1.67	x	$fault_2$
2.00–2.67	$fault_2$	
...	...	...

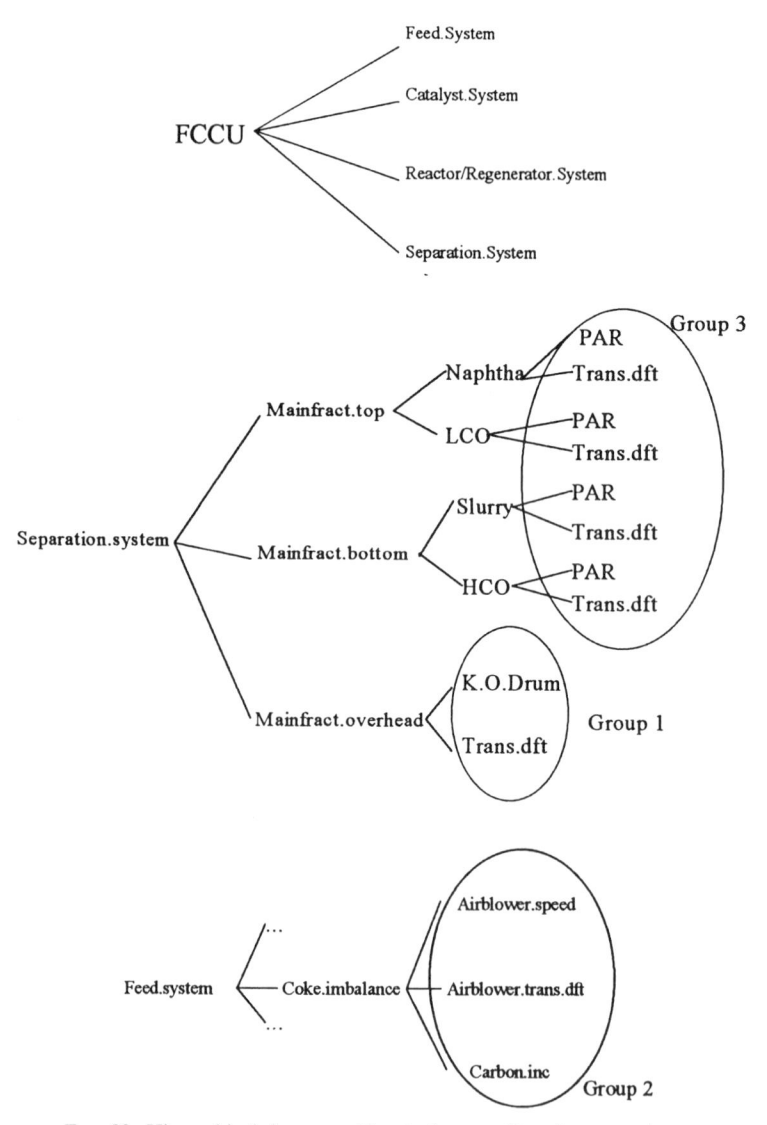

FIG. 32. Hierarchical decomposition to form malfunction groupings.

example was consistently observed for the other malfunction scenarios. When trained only with group 2 malfunction scenarios, an output dimension of 4, the air blower malfunction is both detected with high certainty and isolated with medium certainty within 4.67 min. When the interpreter is asked to provide extended coverage for both group 1 and group 2 malfunc-

TABLE VI
Performance of Art for Air Speed Drop
(Fast Ramp) *absd*

Time	High certainty	Med certainty
\multicolumn{3}{c}{Grouped training (Group 2) output dimensionality is 4.}		
0.00–4.33	*Normal*	mixed
4.67–26.00	x	**absd**
\multicolumn{3}{c}{Training with group 2 and 1 output dimensionality is 6.}		
0.00–1.00	*Normal*	mixed
1.33–4.67	*mixed*	Normal
5.00–26.00	x	**absd**
\multicolumn{3}{c}{Training with group 2 and 3 output dimensionality is 11.}		
0.00–1.00	*Normal*	mixed
1.33–4.67	*mixed*	Normal/cinc
5.00–16.33	*x*	**absd**
16.67–26.00	*mixed*	ndft/ldft
\multicolumn{3}{c}{Complete training output dimensionality is 14.}		
0.00–2.33	*Normal*	mixed/cinc
2.67–4.33	*mixed*	ldft
4.67–16.00	x	**absd**
16.33–21.33	x	*mixed*
21.67–26.00	*mixed*	hdft

tions, raising the output dimension to 6, some mixed clusters are formed (clusters constructed from patterns from both normal and malfunction classes), leading to a mixed high certainty normal classification from 1.33 to 4.67 min and a deterioration in detection and isolation times to 5.00 min. However, when the dimension is extended to 11 with group 2 and group 3 malfunctions, performance significantly deteriorates. Over a period of 26 minutes, detection of changes from normal remains mixed; the correct interpretation with medium certainty is made between 5.00 and 16.33 minutes, but changes to an incorrect mixed conclusion about naphtha and LCO flow sensor drifts. When the dimension is raised to 14, further deterioration with mixed and incorrect medium-certainty conclusions are drawn.

The example clearly brings out the performance degradation as output scale increases. This particular example uses ART as the numeric–symbolic interpreter. Similar degradation in performance can be shown with other interpreters such as PCA and PLS (Ali *et al.,* 1997; Wold *et al.,* 1996).

The example brings out performance degradation associated with a numeric–symbolic interpreter. The principles apply to symbolic–symbolic interpreters as well. Both digraph and fault tree methods have many problems when applied to physically complex systems (Wilcox and Himmelblau, 1994; Chen and Modarres, 1992). Kramer (1987) discussed scale-up problems with symptom pattern matching methods using quantitative models with non-Boolean reasoning. Catino and Ungar (1995) described the increase in time and memory required for qualitative simulations as the device complexity increases. In general, it is not possible to accurately predict when scale-up problems will become significant because performance is heavily dependent on the interactions occurring in the particular plant operation.

The complexity of large-scale process operations sometimes calls for applying, within a single application, a variety of methods to address different information-processing requirements for interpreting diverse behaviors of diverse components, to realize a range of different types of interpretations, or to provide redundancy to raise confidence in conclusions. The inclusion of components reflecting multiple methods and the demands of developing large-scale software essentially require some form of decomposition to manage and distribute decision making into a set of focused, but coordinated, problem-solving modules. While most numeric–symbolic interpretation methods are relatively effective at reducing input complexity, the scale of complex processes still requires constraining input variables into appropriate groupings to reduce the number of variables considered by any one interpreter. Similarly, a large number of possible output interpretations virtually requires that considerations be localized so that interpretations from any one interpreter are not confounded by too many possible conclusions. Thus, it is advantageous if the overall problem, and the knowledge needed to do it, can be decomposed into manageable modules, where each module is constructed by adopting the best available method for its local subproblem. A global perspective must also be maintained so that interpretations from localized modules can be coordinated and a global evaluation produced (Prasad *et al.*, 1998).

One key approach is to design a decision-support system made up of modules that mirror the modular structure of the process system. Interpretations then can be isolated on the basis of local behaviors, goals, and locally relevant input without generating long chains of inferences or hypothesizing complex combinations of behaviors. An important benefit of this organizing principle is that it facilitates knowledge acquisition for symbolic–symbolic interpreters or the training and configuration effort associated with numeric–symbolic interpreters. With knowledge about the process partitioned into manageable chunks and associated with named and identifiable process-system elements, the development of interpretation systems then becomes modular itself. Each modular component directs a small, manageable effort focused on the behaviors and potential malfunctions of its corresponding system element. This same modular property facilitates debugging and maintenance of a plantwide decision-support system; debugging and maintenance are especially important design considerations for constructing large systems.

B. HIERARCHICAL MODULARIZATION

Engineered processes typically reflect hierarchical whole–part designs (system–subsystem, devices–components). Any behavior in a given system always can be associated with one or more of its devices or device components. So it is convenient to associate process behaviors, various normal and abnormal situations, and so on with corresponding systems and/or devices at various levels, and it is convenient and natural to use the whole–part decomposition as a basis for modularizing a large-scale interpretation system. For example, typically what is wanted in diagnosis is a fault description that is as specific as possible. So if diagnostic reasoning establishes that a device is malfunctioning, it is usually desirable to know if the malfunction can be further localized to a part of the device, or even to a specific malfunction mode of a specific part. A reasonable way to modularize large-scale processes, therefore, is a hierarchy of categories, ordered by interpretation specificity, where most levels of the hierarchy mirror the whole–part organization of the process system and bottom levels represent specific components or specific component behaviors. Such a hierarchy of malfunction categories need not exhibit a strict tree structure, because a component may belong to more than one subsystem.

With care, these categories can be designed to respect strict set-inclusion semantics so that category–subcategory will correspond faithfully with set–subset. Besides providing clarity and discipline for knowledge acquisition and debugging, respecting set-inclusion semantics has distinct advantages for interpretation. What is required is that if a malfunction belongs to a class in the hierarchy, then it belongs to any superclass in the hierarchy. That is, for malfunction classes X and Y, and for the partial-order relation whereby elements of the malfunction hierarchy are ordered child-to-parent, we require that

$$\forall\ X\ \forall\ Y \text{ such that } X \propto Y, x \in X \to x \in Y. \qquad (44)$$

This requirement is logically equivalent to the requirement that if a malfunction does not belong to a certain category, then it does not belong to any subcategory of it.

For example, if a valve is not malfunctioning, then it is not stuck closed, and if the cooling system is not malfunctioning, then no component is malfunctioning. This may seem somewhat counterintuitive, because in general a device might continue to function properly, even though a component has malfunctioned; e.g., because of redundancy or compensation. Yet the requirement becomes reasonable if the convention is adopted that a component is not considered to be malfunctioning unless it impairs the functioning of the device of which it is a component. Alternatively, the convention may be that a device is considered to be functioning properly only if every component is functioning properly. In either case, if a malfunction does not belong to a category, then it does not belong to any subcategory. This property permits a diagnostic strategy where ruling out a high-level malfunction category rules out all of its subcategories. If it can be determined that there is no apparent problem with the cooling system, then problems with cooling-system components need not be considered.

A variety of hierarchical structures going back many years have been used to manage scale and achieve modularization. Finch and Kramer (1988) used an approach for diagnosis in which each hypothesis in the hierarchy corresponds to a control-loop system. Interdependencies among the systems are identified by postulating system malfunctions and determining which other systems are directly affected by a malfunction. The inference mechanism is similar to that of a sign-directed graph, in which the dependencies between systems are arcs. Even though a degree of modularization is achieved, the overall system suffers the same scaling problems as digraph systems because of the inference mechanism. Chen and Modarres (1992) developed a diagnostic and correction-planning system, FAX, which uses a goal-tree/success-tree hierarchical model. Malfunction categories are defined using the top plant goal or objective, and then decomposing the goal vertically downward to progressively more detailed subgoals. Table lookup and Bayes' theorem are used in the inference mechanism. Davis (1994); McDowell and Davis (1992); Ramesh *et al.* (1988); Prasad and Davis (1992); and Prasad *et al.* (1998) have extensively used an approach called *hierarchical classification* (HC) where the event or situation is classified as belonging to one or more categories, the categories being hierarchically organized according to levels of specificity. The hierarchical categories make use of combinations of system–subsystem, component–subcomponent, component–mode, and event–subevent decompositions customized for the problem of interest.

Together, these decomposition approaches argue for combining functional, goal, and structural perspectives of the plant, as needed, in constructing interpretation systems. With any of these hierarchical approaches, each node in the hierarchy can be construed to define a localized interpreter.

Interpretation requirements can be clearly defined and methods can be customized to meet the needs of the local decision. That is, a localized interpreter can be constructed to use a method that is most effective, given available knowledge and information, for inferring conclusions about the problem defined by the node—malfunction category rejected, goal achieved, subsystem functioning properly, and so on. Each such module will then provide information that influences the overall problem solving in a well-understood way. Figure 33 shows a portion of a hierarchical classification system for diagnostic decision support with a bubble showing a local module associated with one of the nodes (see Fig. 26).

In practice, high-level malfunction categories typically are monitored and evaluated continuously. Lower-level categories are evaluated only if high-level categories are established. The evaluation of malfunction situations may then proceed through the hierarchy from general to specific. For support of decision making, notification may be given to a higher level situation, but advice will be given based on the lowest level, which relates most directly to specific corrective actions that can be taken.

With a similar intent of decomposing problems to reduce dimensionality and to make interpretations more manageable, Wold, Kettaneh, and Tjessem (1996) and MacGregor *et al.* (1994) have reported on an approach called *blocking*, used in conjunction with multivariate PLS and PCA. The motivation for developing the technique of blocking stems from the problem that PLS and PCA models are difficult to interpret with many variables, loadings, and scores. Referred to as *hierarchical multiblock* PLS or PCA models, these hierarchical models group variables into meaningful groups or blocks. Blocking then leads to two levels where the upper level models

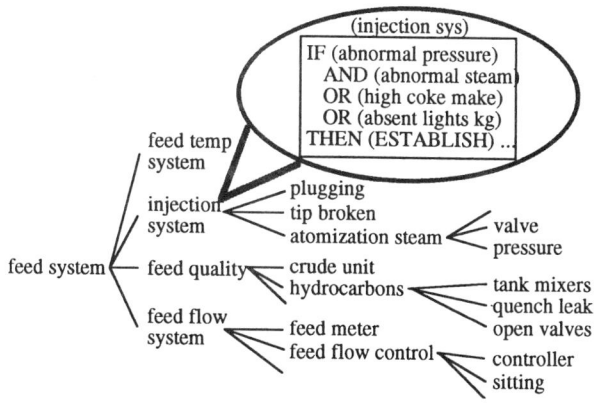

FIG. 33. Hierarchical classification. [Drawn from T. S. Ramesh, J. F. Davis, and G. J. Schwenzer, *Comput. Chem. Eng.* **16**, 109 (1992).]

the relation between blocks and the lower level models the details of each block.

As shown in Fig. 34, the approach involves breaking variables into blocks or groupings X_k and mapping the variables in each block into a set of block scores, r_k, and loadings, p_k. The block scores themselves are then aggregated, R, and become a variable block themselves from which a set of super scores, t, and loadings, w, can then be calculated. Although the algorithms for extracting the various features for hierarchical blocks have been developed and demonstrated, a systematic method of blocking variables is not yet available. It is clear that blocking based on knowledge of the process is necessary because it leads to logical groupings of variables. This observation suggests a future avenue of investigation of knowledge-based decomposition methods as described in preceding paragraphs. The strength of the KBS decompositions is that they consider the organization and structure of the process knowledge explicitly.

In general, there is no plug-in technology available for determining the best hierarchical decomposition of a plant. Prasad *et al.* (1998) have put forth some guidelines to assist in the development of taxonomies based on hierarchical classification. Given the vast variety of process plants, the design of an effective decomposition is not a trivial task. Moreover, the decomposition must be designed in concert with designing the localized interpreters. Yet achieving an effective modular design is crucial to the overall performance, ease of construction, and effective maintenance of a data interpretation system.

VIII. Comprehensive Examples

With reference to Fig. 4, this section presents several examples that demonstrate how various technologies can be assembled into data analysis and interpretation systems in practical application.

A. Comprehensive Example 1: Detection of Abnormal Situations

This example illustrates a system for monitoring and detection in a continuous polymer process (Piovoso *et al.*, 1992b). In the first stage of the process, several chemical reactions occur that produce a viscous polymer

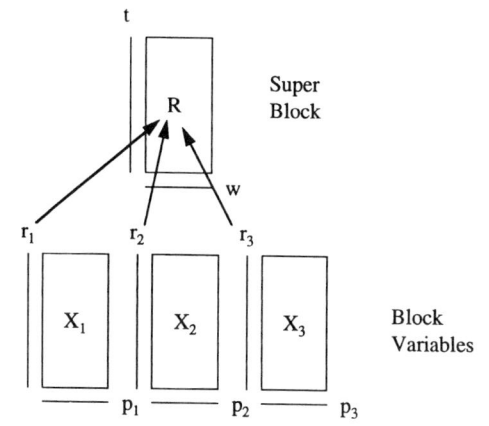

FIG. 34. A hierarchical multiblock principal component model.

product; in the second stage, the polymer is treated mechanically to prepare it for the third and final stage. Critical properties such as viscosity and density, if altered significantly, affect the final product, resulting in a loss of revenue and operability problems for the customer. Moreover, it is difficult to determine which of the three stages is responsible when the quality is not within acceptable limits. The situation is confounded by a lack of online sensors to measure continuously the critical properties. This makes it impossible to relate any specific changes to a particular stage. Indeed, property changes are detected by laboratory measurements that may have delays of 8 hours or more. The results of the analysis represent history; consequently, the current state of operations and the laboratory analysis of the process state are not aligned.

Such infrequent measurements make the control of the product quality difficult. At best, the operators have learned a set of heuristics that, if adhered to, usually produces a good product. However, unforeseen disturbances and undetected equipment degradations not accounted for by the heuristics still occur and affect the product. In addition, there are periods of operation when the final process step produces a degraded product in spite of near-perfect upstream operations.

Since stage 2 operates in support of stage 3 as well as to compensate for errors in the first stage, this example focuses on process monitoring and detection in stage 2. A simplified description of stage 2 is as follows. The product from the first stage is combined with a solvent in a series of mixers operated at carefully controlled speeds and temperatures. A sample of the mixture is taken at the exit of the final mixer for lab analysis. The material leaving the mixer enters a blender whose level and speed are maintained to impart certain properties to the final product. After exiting the blender the material is then filtered to remove undissolved particulates before it is sent to the final stage of the process.

A significant amount of mechanical energy is necessary to move the highly viscous fluid through a complex transport network of pipes, pumps, and filters. Consequently, the time in service of the equipment is unpredictable. Even identical types of equipment placed in service at the same time may need replacement at widely differing times. Problems of incipient failure, pluggage, and unscheduled downtimes are an accepted, albeit undesirable, part of operation. To prolong continuous operations, pumps and filters are installed in pairs so that the load on one can be temporarily increased while the other is being serviced. Tight control of the process fluid is desirable because the equipment settings in the final stage of the process are preset to receive a uniform process fluid. As such, small deviations in the fluid properties may result in machine failure and nonsalable product.

Frequent maintenance on the equipment is not the only source of operating problems. Abrupt changes also occur due to the throughput demands in the final stage of the process. For example, if there is a decrease in demand due to downstream equipment failure, or a sudden increase due to the addition of new or reserviced equipment, the second stage must reduce or increase production as quickly as possible. It is more dramatic when throughput must be turned down because the process material properties change if not treated immediately, especially in the blender. These situations are frequent and unscheduled; thus, the second stage of this process moves around significantly and never quite reaches an equilibrium. Clearly, throughput is a dominant effect on the variability in the sensor values and process performance.

To address this situation, a data interpretation system was constructed to monitor and detect changes in the second stage that will significantly affect the product quality. It is here that critical properties are imparted to the process material. Intuitively, if the second stage can be monitored to anticipate shifts in normal process operation or to detect equipment failure, then corrective action can be taken to minimize these effects on the final product. One of the limitations of this approach is that disturbances that may affect the final product may not manifest themselves in the variables used to develop the reference model. The converse is also true—that disturbances in the monitored variables may not affect the final product. However, faced with few choices, the use of a reference model using the process data is a rational approach to monitor and to detect unusual process behavior, to improve process understanding, and to maintain continuous operation.

We can expect throughput to have a major effect on all the measure-

Scores of Principal Component 1

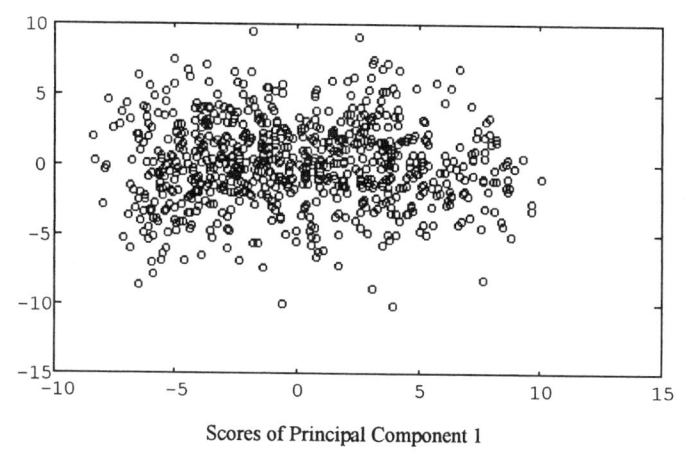

FIG. 35. Scores of Principal Component 1 vs scores of Principal Component 2; PCA model.

ments. A PCA analysis would, as a result, not only be overwhelmed by this single effect, but would obscure other information about the state of the process. An effective strategy then is to eliminate the throughput effect using a PLS analysis. Analysis of the residuals using PCA can then reveal other sources of variations critical to process operations.

To construct the reference model, the interpretation system required routine process data collected over a period of several months. Cross-validation was applied to detect and remove outliers. Only data corresponding to normal process operations (that is, when top-grade product is made) were used in the model development. As stated earlier, the system ultimately involved two analysis approaches, both reduced-order models that capture dominant directions of variability in the data. A PLS analysis using two loadings explained about 60% of the variance in the measurements. A subsequent PCA analysis on the residuals showed that five principal components explain 90% of the *residual* variability.

In this example, we focus on the PCA of the residuals. Figure 35 shows the observations of the residual data plotted in the score space of principal components 1 and 2. Although the observations are spread over a wide region of operating conditions and include a wide range of rate settings, Fig. 35 clearly shows how the data cluster when presented as principal components. The detection system exploits this cluster by identifying operations that fall within or outside of the cluster region based on a discriminant using the Mahalanobis distance as a measure of the goodness of fit (see Section V).

Figure 36 shows the performance of the interpretation system during a period of 30 hours. To detect an abnormal operation, the Mahalanobis

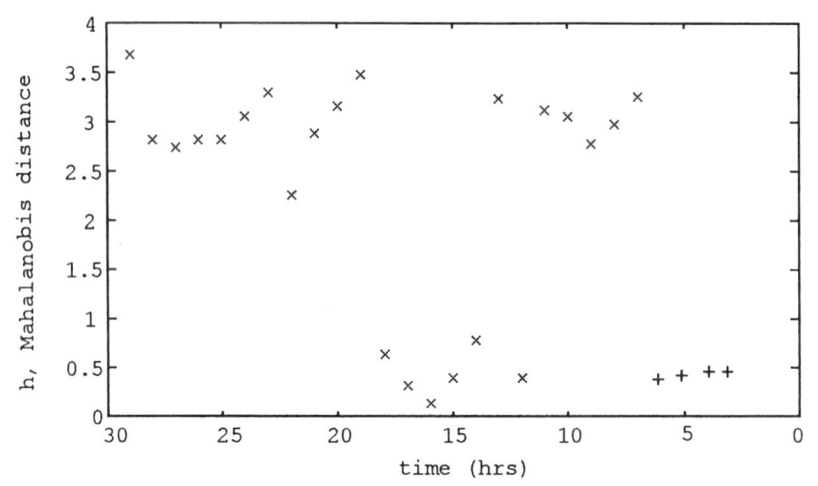

FIG. 36. Thirty hours of online operation. +, most recent 4 hours; ×, most distant history.

distance, h, expresses how close a given data pattern matches data used to form the principal component cluster in Fig. 35. The abscissa represents the most recent data to the left and the oldest data to the right. It was determined that a value greater than 1 implies that a given data pattern is not similar to the reference cluster. It is easy for operators to observe from figures such as this when the process is not performing well ($h > 1$) and to identify and correct the cause of the problem. In the first case (2–6 hours), a filter plugged and, in the second (12–18 hours), a pump failed. More details about this example can be found in Piovoso *et al.* (1992a).

B. Comprehensive Example 2: Data Analysis of Batch Operation Variability

This example (Kosanovich *et al.*, 1995) builds on the previous example and illustrates how multivariate statistical techniques can be used in a variety of ways to understand and compare process behavior. The charge to the reactor is an aqueous solution that is first boiled in an evaporator until the water content is reduced to approximately 20% by weight. The evaporator's contents are then discharged into a reactor in which 10 to 20 pounds of polymer residue can be present from the processing of the previous batch.

This batch reactor is operated according to a combination of prespecified reactor and heat source pressure profiles and timed stages. The time to complete a batch is approximately 120 minutes. Key process checkpoints (e.g., attaining a specific temperature within a given time) determine when one processing stage ends and the next one begins.

Heat is applied to the reactor to further concentrate the reactants and to supply the energy to activate the polymerization reactions. At the outset, the reactor temperature and pressure rise rapidly. Sensor measurements indicate the existence of a temperature gradient having as much as a 40°C difference between material at the top and at the bottom of the reactor. Shortly after the pressure reaches its setpoint, the entire mixture boils and the temperature gradient disappears. The solution is postulated to be well mixed at this time. The cumulative amount of water removed is one indication of the extent of polymerization.

The reactor pressure is reduced to 0 psig to flash off any remaining water after a desired temperature is reached. Simultaneous ramp up of the heat source to a new setpoint is also carried out. The duration spent at this second setpoint is monitored using CUSUM plots to ensure the batch reaches a desired final reactor temperature within the prescribed batch time. The heat source subsequently is removed and the material is allowed to continue reacting until the final desired temperature is reached. The last stage involves the removal of the finished polymer as evidenced by the rise in the reactor pressure. Each reactor is equipped with sensors that measure the relevant temperature, pressure, and the heat source variable values. These sensors are interfaced to a distributed control system that monitors and controls the processing steps.

In this example, data interpretations are based on Q-statistic limits. These are computed by assuming the data are normally distributed in the multivariate sense. The diagnostic limits are used to establish when a statistically significant shift has occurred. Charts based on these statistics and used in this manner are analogous to conventional SPC charts.

As in example 1, the explained variance (the total variance minus the residual variance) is calculated by comparing the true process data with estimates computed from a reference model. This explained variance can be computed as a function of the batch number, time, or variable number. A large explained variance indicates that the variability in the data is captured by the reference model and that correlations exist among the variables. The explained variance as a function of time can be very useful in differentiating among phenomena that occur in different stages of the process operations.

Process data for the same polymer recipe were analyzed for 50 nonconsecutive, sequential batches. As before, the data were preprocessed to remove outliers and sorted to reflect normal operation. The data are sampled at 1-minute intervals during production of each batch. Fillering and normalization of the data is done prior to analysis. The final polymer quality

88   J. F. DAVIS *ET AL.*

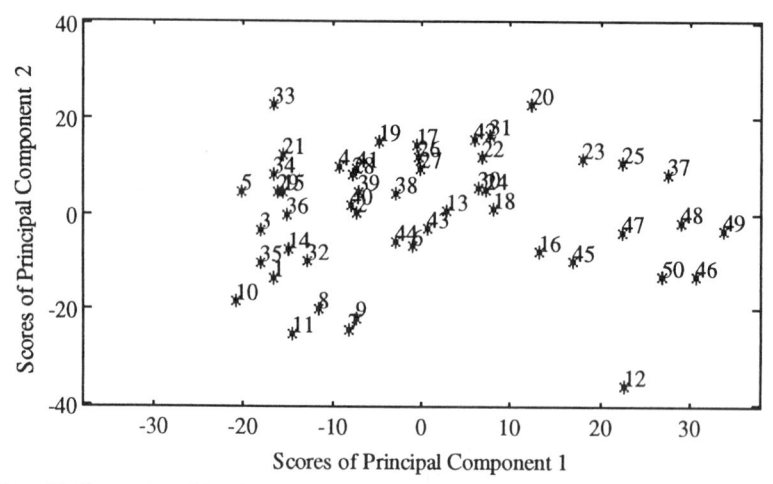

FIG. 37. Score plot of batch polymer reactor data (numbers indicate batch numbers).

data are obtained from laboratory measurements of polymer molecular weight and endgroups—one reading for each batch. As shown in Fig. 37, the PCA score plots for the reactor provide a summary of process performance from one batch to the next. All batches exhibiting similar time histories have scores that cluster in the same region of this particular principal component space.

Figure 38 shows the variance explained by the two principal component (PC) model as a percentage of each of the two indices: batch number and time. The lower set of bars in Fig. 38a are the explained variances for the first PC, while the upper set of bars reflects the additional contribution of the second PC. The lower line in Fig. 38b is the explained variance over time for the first PC and the upper line is the combination of PC 1 and 2. Figure 38a indicates, for example, that batch numbers 13 and 30 have very small explained variances, while batch numbers 12 and 33 have variances that are captured very well by the reference model after two PCs. It is impossible to conclude from this plot alone, however, that batches 13 and 30 are poorly represented by the reference model.

Additional insight is possible from Fig. 38b. Here we see that the magnitude of the explained variance accounted for by the second PC has noticeably increased after minute 70. This is consistent because, from process knowledge, it is known that removal of water is the primary event in the first part of the batch cycle, while polymerization dominates in the later part, explaining why the variance profile changes around the 70-minute point.

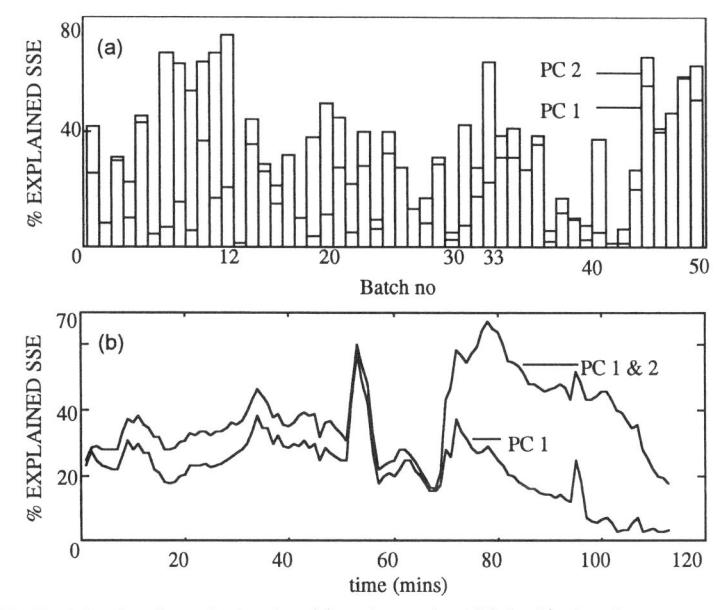

FIG. 38. Explained variance by batches (a) and over time (b) for batch polymer reactor data.

This interpretation of Fig. 38b leads to the conclusion that better insight into the variations can be obtained by separating the data into two time histories—the first covers time $k = 1$ to 53, and the second covers time $k = 54$ to 113 (all time units are in minutes). Incorporating this extended perspective, Fig. 39 provides score plots for two PC models where the score plots are now divided into these two main processing phases. Observe the

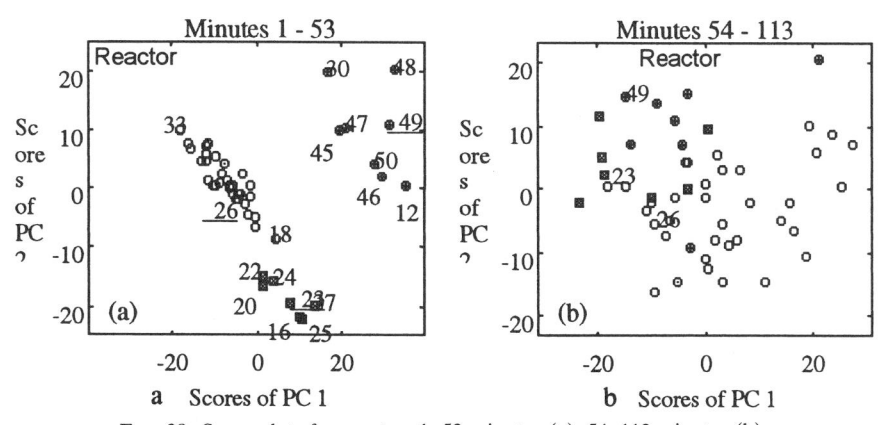

FIG. 39. Score plots for reactor: 1–53 minutes (a); 54–113 minutes (b).

presence of clusters in Fig. 39a for data from time $k = 1$ to 53, and the scatter in Fig. 39b for $k = 54$ to 113. A closer inspection of the operating conditions for these later batches indicates they were processed at a different heat transfer rate than the others. One possible conclusion is that the processing steps in the later stages are not influenced by the same factors that led to formation of clusters in the earlier stages. Conversely, it is possible that the processing steps in the later stage remove whatever led to the differentiation in the earlier stage.

Q-statistics can provide a multivariate way to view the batch processes. Computation of the Q-statistics as a function of batch number are shown in Fig. 40 for data taken from time $k = 1$ to 53. The Q-statistic for batches 30 and 37, for example, exceeds the 95% limit for *both* PCs, indicating that the reference model does not capture the variations in these batches. That is, the correlation structure of batches 30 and 37 are different than the majority of the batches. The Q-statistic as a function of time in Fig. 41 shows that deviations from the model subspace occur primarily in the first 35 minutes. Note that although the batches start out with a large Q-statistic, they end up within the 95% confidence limit.

The overall interpretation comes together when we recall that batches 13 and 30 had small explained variances. We note that the Q-statistic for batch 13 indicates that it is within the 95% limit for both PCs while the Q-statistic of batch 30 is not. The conclusion is that the variations in batch 13 are small random deviations about the average batch. In the case of batch 30, larger variations occur that are not well explained by the reference model. These variations are either large random fluctuations or variations that are orthogonal to the model subspace. Hence, the quality of batch 30, with a high probability, will not be within the specified limits.

C. COMPREHENSIVE EXAMPLE 3: DIAGNOSIS OF OPERATING PROBLEMS IN A BATCH POLYMER REACTOR

This example illustrates how numeric–symbolic interpreters are combined with symbolic–symbolic interpretation for root cause diagnosis. It

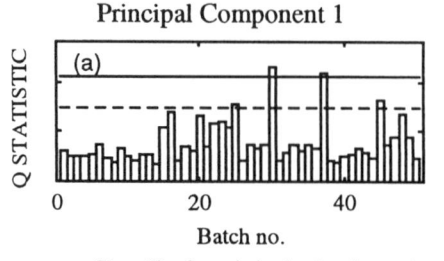

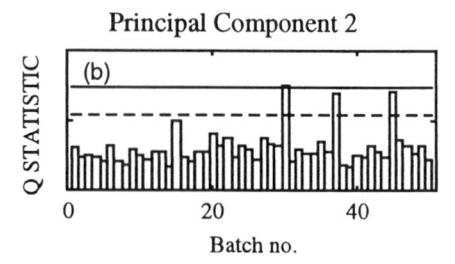

FIG. 40. Q-statistics by batch number for $k = 1$ to 53 minutes for reactor data.

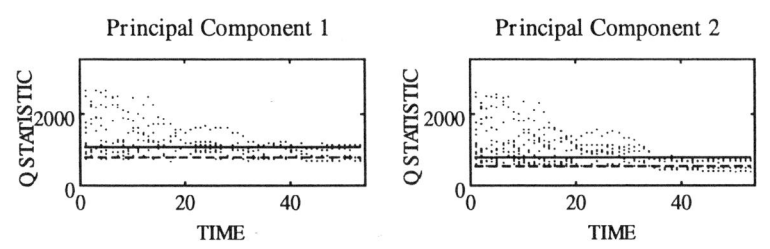

FIG. 41. Q-statistics over time for $k = 1$ to 53 minutes for reactor data.

also shows how hierarchical classification manages output dimensionality by decomposing the process to form a distributed set of data interpretation systems.

Consider again a batch polymerization process where the process is characterized by the sequential execution of a number of "steps" that take place in the two reactors. These are steps such as initial reactor charge, titration, reaction initiation, polymerization, and transfer. Because much of the critical product quality information is available only at the end of a batch cycle, the data interpretation system has been designed for diagnosis at the end of a cycle. At the end of a particular run, the data are analyzed and the identification of any problems is translated into corrective actions that are implemented for the next cycle. The interpretations of interest include root causes having to do with process problems (e.g., contamination or transfer problems), equipment malfunctions (e.g., valve problems or instrument failures), and step execution problems (e.g., titration too fast or too much catalyst added). The output dimension of the process is large with more than 300 possible root causes. Additional detail on the diagnostic system can be found in Sravana (1994).

The overall objective of the system is to map from three types of numeric input process data into, generally, one to three root causes out of the possible 300. The data available include numeric information from sensors, product-specific numeric information such as molecular weight and area under peak from gel permeation chromatography (GPC) analysis of the product, and additional information from the GPC in the form of variances in expected shapes of traces. The plant also uses univariate statistical methods for data analysis of numeric product information.

As is typical of process systems, the plant runs with a very high availability and product quality is normally maintained. For diagnostic data interpretation, there is very little data for developing any kind of numeric–symbolic interpreter that maps directly from the input data to the output diagnostic conclusions. It is possible, however, to map with a great deal of confidence from the input numeric data to a set of useful intermediate interpretations. With respect to the sensor data, there is considerable information for mapping between the input data and labels such as *high, low, increasing,* and *decreasing*. In this particular case study, all these interpretations are accomplished by limit checking.

In addition, the GPC trace, an example of which is shown in Fig. 42, reflects the composition signature of a given product and reflects the spectrum of molecular chains that are present. Analysis of the area, height, and location of each peak provides valuable quantitative information that is used as input to a CUSUM analysis. Numeric input data from the GPC is mapped into *high, normal,* and *low,* based on variance from established normal operating experience. Both the sensor and GPC interpretations are accomplished by individual numeric–symbolic interpreters using limit checking for each individual measurement.

In addition, operators can observe significant shape deviations in the GPC data such as *double peak* and *shoulder on low molecular weight side*. As shown in Fig. 42, the GPC trace can be subdivided into several regions. Each region is associated with an expected shape, e.g., Gaussian or flat. The features to be identified are variations in these expected shapes. These variations are complex in that they are characterized by skew to the right or left and extra distinct peaks. Figure 43 shows the expected shape in the Gaussian region and several examples of shape variation.

Development of a GPC shape interpreter is characterized by a large number of training examples of normal shapes and a few training examples of each of the abnormal shapes. Because of the existence, but limited availability, of abnormal examples, Adaptive Resonance Theory (see Sec-

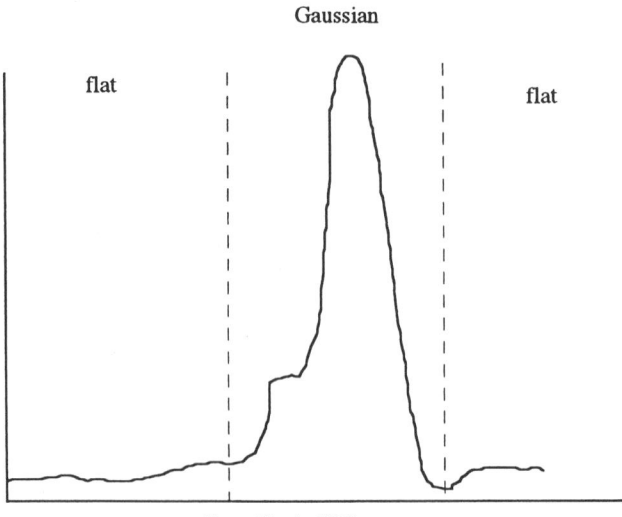

FIG. 42. A GPC trace.

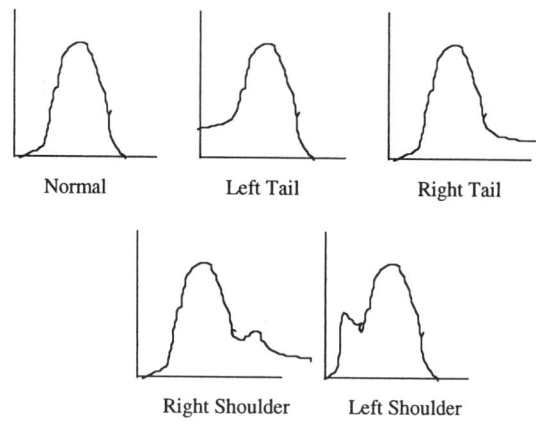

FIG. 43. Examples of Gaussian region shape deviations.

tion V) was selected for constructing a numeric–symbolic interpreter. Inputs to the interpreter were the numeric points that formed the GPC trace. The output of the interpreter were the labels *normal* and the various abnormal shapes. Figure 44 illustrates GPC pattern recognition using ART. Label assignment is determined by whether x falls or does not fall within the boundaries of defined fault classes that group patterns of data with similar characteristics. If an arbitrary pattern falls within a class boundary, then it is labeled that fault class with high certainty. If it falls outside, but near, a class boundary, then it is labeled the nearest fault class, but with medium certainty. The resulting interpretations in the form of labels such as *Gaussian Region—Left Tail—medium certainty* are used along with sensor interpretations as intermediate label inputs to a symbolic–symbolic interpreter.

Given the intermediate labels from the GPC pattern and sensor interpreters, a KBS is used to map them into the diagnostic labels of interest. However, because the output dimension of the problem is large (at over 300 possible malfunctions), a hierarchical classification system is used to decompose the process into a hierarchical set of malfunction categories that

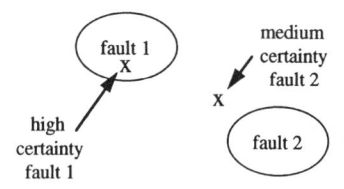

FIG. 44. Data interpretation in ART.

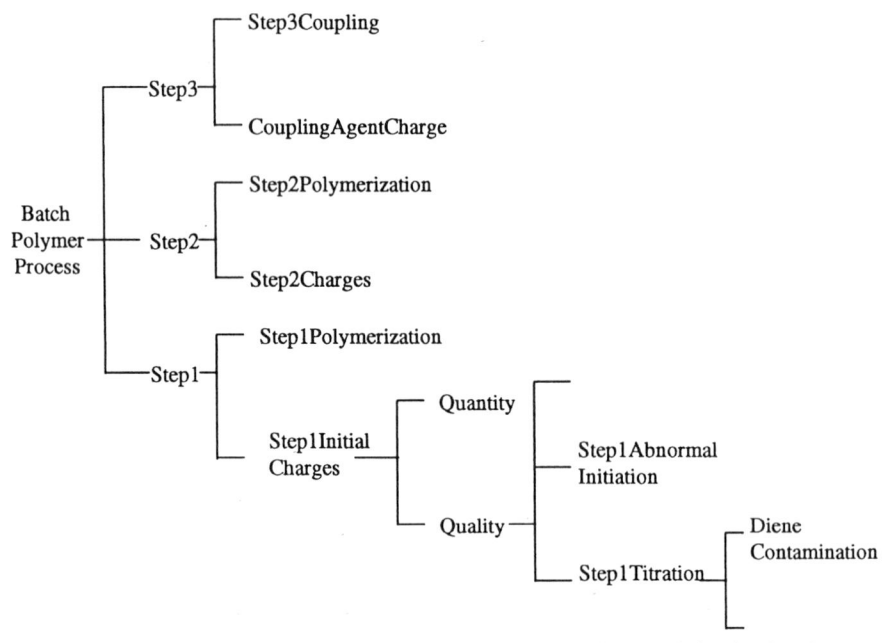

FIG. 45. Top-level organization of malfunction hypotheses for an existing batch polymer reactor with a focus on Step 1: Initial Charge problems.

provide a mechanism for generating and testing malfunction hypotheses in a distributed and localized manner.

As shown in Figure 45, the polymerization process can be organized at the top level according to major procedural steps, which in this case are called *Step1, Step2,* and *Step3* rections. At this level, problems with the process are associated with completion of the intended functions. Considering completion in more detail leads to consideration of problems with either the initial charges or the polymerization reaction itself. For example, in the case of Step1: Initial Charges, the next level of consideration is either the quantity or the quality of charge, and so forth. This decomposition continues as illustrated for diene contamination in Fig. 46 to tip nodes that deal with problems in specific equipment items, operator errors, and so forth.

Details of the hierarchical decomposition used to manage the scope of the problem are given in Sravana (1994). There are many considerations in developing a particular decomposition, but the objective is to decompose the process into smaller, more localized malfunction considerations.

The decomposition and localization of the problem effectively produces a distributed set of data interpretation problems, each with a particular set

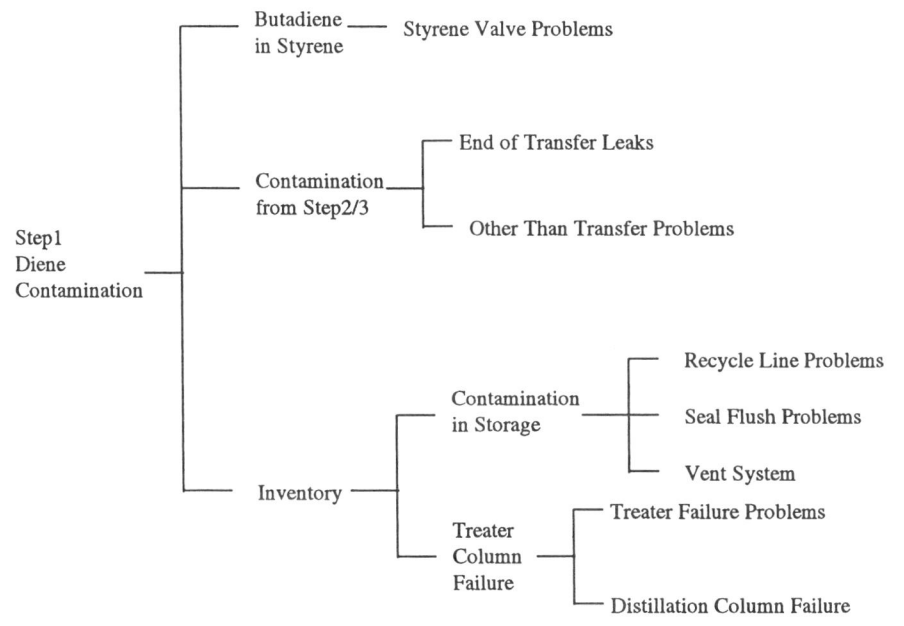

```
                          ┌── Butadiene ──── Styrene Valve Problems
                          │    in Styrene
                          │
                          │                      ┌── End of Transfer Leaks
                          │    Contamination ────┤
                          │    from Step2/3      │
Step1                     │                      └── Other Than Transfer Problems
Diene ────────────────────┤
Contamination             │
                          │                                           ┌── Recycle Line Problems
                          │                     Contamination         │
                          │                  ┌── in Storage ──────────┤── Seal Flush Problems
                          │                  │                        │
                          │                  │                        └── Vent System
                          └── Inventory ─────┤
                                             │                        ┌── Treater Failure Problems
                                             │   Treater              │
                                             └── Column ──────────────┤
                                                 Failure              │
                                                                      └── Distillation Column Failure
```

FIG. 46. Diene contamination branch.

of possible output conclusions and a set of input variables that provides information for the interpretation.

Each of the nodes represents an individual symbolic–symbolic interpretation problem of determining whether there is a malfunction associated with that particular element of the process. As illustrated in Fig. 47, the node Step1: Diene Contamination represents an interpretation problem in which there are two possible output conclusions—*yes* or *no*. There are three input variables, the values of which (intermediate interpretations) are determined by individual numeric–symbolic interpreters using process data, GPC data, and a GPC shape interpretation. At each node, mapping between the input variables and the output conclusions is accomplished by simple table lookup. To reach a detailed diagnostic conclusion, the objective is to pursue the diagnostic branches in greater and greater detail as diagnostic hypotheses indicated by the nodes are accepted or rejected using at each a local set of numeric–symbolic and symbolic–symbolic interpreters.

In this way, data interpretation is accomplished by a set of nested numeric–symbolic and symbolic–symbolic interpreters. Note that the hierarchical decomposition results in a distributed set of symbolic–symbolic interpretation problems represented by nodes. Each problem requires intermediate interpretations of numeric data as input to the symbolic–

```
                  If catalyst charge normal
                  and
                  If molecular weight
                  indicates slow 1xn
                  and
                  If Gaussian low molecular
                  weight shoulder
                  then Step 1:
                  Diene contamination
```

                                              Butadiene in styrene
       Step 1:   Contamination
       Diene contamination ─────────── Steps 2/3

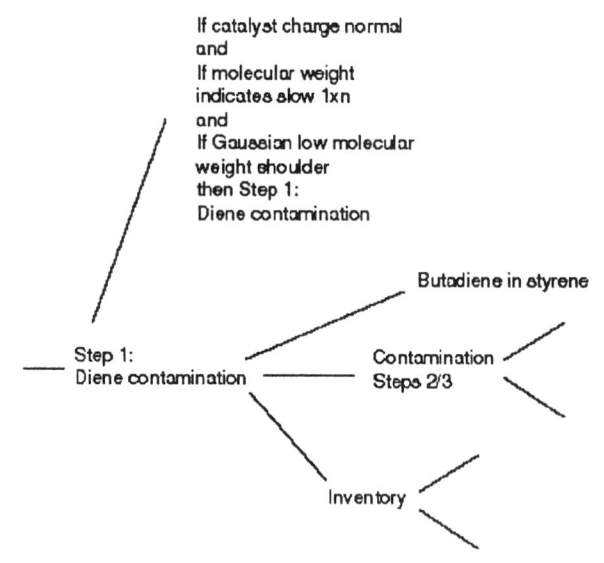

                                              Inventory

FIG. 47. Local diene contamination subproblem.

symbolic problem. The intermediate interpretations use individual numeric–symbolic interpreters tailored for the specific interpretation of interest.

IX. Summary

The widespread interest and success in Statistical Process Control over the past 15 years has introduced the value and importance of data analysis and interpretation in process operations. In response to much more stringent demands on production, quality, and product flexibility, multivariate statistical and intelligent system technologies and a multitude of associated integrated technologies have produced a new wave of more powerful approaches that are proving their economic worth. These approaches are not only providing considerably improved ways of analyzing data, but are also providing the mechanisms for interpretation that enhance how the essential human element can interact better with the process operation.

Figure 4 is a defining view of data analysis and interpretation illustrating four major families of technologies that must be considered for any given interpretation of numeric process data. With this view, the construction of any single data interpretation system is a complicated integration of technologies in itself. This is especially brought out in Fig. 2 and 5, in which the scope of empirical analysis and interpretation techniques, respectively, become clear. However, when we further overlay this view of a single interpretation system with the fact that the number of interpretations possible is highly constrained by performance and that a complex process operation is likely to require many of these interpretation systems used locally, the problem is daunting. Nevertheless, progress on individual technologies and the relatively recent emphasis on integration have brought these complex analysis and interpretation systems into reality. Focused applications of analysis and interpretation systems for specifically defined process situations are coming into wide use at the present time. Comprehensive systems for entire process units are beginning to be considered.

The success of this chapter lies not in detailed listings, descriptions, or comparisons of individual approaches, nor in an exhaustive review of the literature, but in a perspective for integrating a very wide-ranging array of technologies and managing complexity in comprehensive analysis and interpretation systems. This integrated perspective is, in fact, the product of a considerable struggle by the four authors, who are representative of these widely varying technology perspectives. To this end, the chapter has taken a unique look at alternatives to quantitative behavioral model approaches from the point of view of interpretation. The emphasis on how to manipulate knowledge and data to clarify process behaviors and to assign interpretations based on experiences with the operation offers a practical view for linking the diverse technology areas.

Acknowledgments

The chapter is a compilation of a great deal of work in this area by the four authors and a large number of graduate students who have contributed over a period of more than 15 years, as well as the many contributions of a large technical community. We recognize and gratefully acknowledge these contributions, without which this chapter would not have been possible. We also specifically thank David Miller, a postdoctoral student at The Ohio State University, for his detailed review, and Lynda Mackey for her editorial assistance.

REFERENCES

Ali, Z., Ramachandran, S., Davis, J. F., and Bakshi, B., On-line state estimation in intelligent diagnostic decision support systems for large scale process operations, *AIChE Mtg.*, Los Angeles, CA, November (1997).

Antsaklis, P. J., and Passino, K. M., Eds., "An Introduction to Intelligent and Autonomous Control." Kluwer Academic Publishers, 1993.

Badiru, A. B., "Expert Systems Applications in Engineering and Manufacturing," Prentice Hall, 1992.

Bakshi, B. R., and Stephanopoulos, G., Wave-Net: a multi-resolution, hierarchical neural network with localized learning, *AIChE J.* **39**(1), 57–81 (1993).

Bakshi, B. R., and Utojo, U., Unification of neural and statistical methods that combine inputs by linear projection, *Comput. Chem. Eng.* **22**(12), 1859–1878 (1998).

Ball, G. H., and Hall, D. J., "Isodata, a novel method of data analysis and pattern classification," NTIS Report AD699616 (1965).

Basseville, M., and Benveniste, A., "Detection of Abrupt Changes in Signals and Dynamical Systems. Springer-Verlag, New York, 1986.

Breiman, L., Friedman, J. H., Olshen, R. A., and Stone, C. J., "Classification and Regression Tres." Wadsworth, 1984.

Carpenter, G. A., and Grossberg, S., ART2: self-organization of stable category recognition codes for analog input patterns, *Appl. Opt.* **26**, 4919 (1987a).

Carpenter, G. A., and Grossberg, S., A massively parallel architecture for a self-organizing neural pattern recognition machine, *Comput. Vis. Graphics Image Process* **37**, 54 (1987b).

Carpenter, G. A., and Grossberg, S., The ART of adaptive pattern recognition by a self-organizing neural network, *Computer*, **21**, 77 (1988).

Carpenter, G. A., Grossberg, S., and Rosen, D. B., ART2-A: An adaptive resonance algorithm for rapid category learning and recognition, *Neural Network* **4**, 493 (1990).

Catino, C. A., and Ungar, L. H., Model-based approach to automated hazard identification of chemical plants, *AIChE J.* **41**(1), 97–109 (1995).

Chandrasekaran, B., Models versus rules, deep versus compiled, content versus form: Some distinctions in knowledge systems research, *IEEE Expert* **6**(2), 75–79 (1991).

Chen, L. W., and Modarres, M., Hierarchical decision process for fault administration, *Comput. Chem. Eng.* **16**(5), 425–448 (1992).

Chen, S., Cowan, C. F. N., and Grant, P. M., Orthogonal least squares learning algorithm for radial basis function networks, *IEEE Trans. Neur. Net.* **2**(2), 302–309 (1991).

Coifman, R. R., and Wickerhauser, M. V., Entropy-based algorithms for best basis selection, *IEEE Trans. Inform. Theory* **38**(2), 713–718 (1992).

Cybenko, G., Continuous valued neural networks with two hidden layer are sufficient, Technical Report, Department of Computer Science, Tufts University (1988).

Daubechies, I., Orthonormal bases of compactly supported wavelets, *Comm. Pure Applied Math.* **XLI**, 909–996 (1988).

Davis, J. F., On-line knowledge-based systems in process operations: The critical importance of structure for integration, *Proc. IFAC Symp. on Advanced Control of Chemical Processes*, ADCHEM '94, Kyoto, Japan (1994).

Davis, J. F., and Wang, C. M., Pattern-based interpretation of on-line process data, *in* "Neural Networks for Chemical Engineers" (A. Bulsari, ed.), Elsevier Science, 1995, pp. 443–470.

Davis, J. F., Bakshi, B., Kosanovich, K. A., and Piovoso, M. I., Process monitoring, data analysis and data interpretation, "Proceedings, Intelligent Systems in Process Engineering," AIChE Symposium Series, **92**(312), (1996a).

Davis, J. F., Stephanopoulos, G., and Venkatasubramanian, V., eds., "Proceedings, intelligent systems in process engineering, 1995," AIChE Symposium Series, 92 (312), (1996b).

De Veaux, R. D., Psichogios, D.C., and Ungar, L. H., A comparison of two nonparametric estimation schemes: MARS and neural networks, *Comput. Chem. Eng.* **17**(8), 819–837 (1993).

Dong, D., and McAvoy, T. J., Nonlinear principal component analysis—based on principal curves and neural networks, *Comput. Chem. Eng.* **20**(1), 65 (1996).
Donoho, D. L., Johnstone, I. M., Kerkyacharian, G., and Picard, D., Wavelet shrinkage: asymptopia? *J. R. Stat. Soc. B.* **57**(2), 301 (1995).
Duda, R. O., and Hart, P. E., "Pattern Classification and Scene Analysis." Wiley-Interscience, New York, 1973.
Finch, F. E., and Kramer, M. A., Narrowing diagnostic focus using functional decomposition, *AIChE J.* **34**(1), 25–36 (1988).
Frank, I. E, A nonlinear PLS model, *Chemom. Intell. Lab. Sys.* **8**, 109–119 (1990).
Friedman, J. H., Multivariate adaptive regression splines, *Ann. Statistics* **19**(1), 1–141 (1991).
Friedman, J. H., and Stuetzle, W., Projection pursuit regression, *J. Amer. Stat. Assoc.* **76**, 817–823 (1981).
Gallagher, N. C., Jr., and Wise, G. L., A theoretical analysis of the properties of median filters, *IEEE Trans. Acoustics, Speech, & Signal Proc.* **29**, 1136 (1981).
Geladi, P., and Kowalski, B., Partial least-squares regression: a tutorial, *Anal. Chim. Acta* **185**, 1–17 (1986).
Grossberg, S., "Studies of Mind and Brain: Neural Principles of Learning, Perception, Development, Cognition, and Motor Control," Reidel, Boston, 1982.
Grossberg, S., "The Adaptive Brain, I: Cognition, Learning, Reinforcement and Rhythm." Elsevier/North Holland, Amsterdam, 1987a.
Grossberg, S., "The Adaptive Brain, II: Vision, Speech, Language and Motor Control." Elsevier/North Holland, Amsterdam, 1987b.
Grossberg, S., "Neural Networks and Natural Intelligence," MIT Press, Cambridge, MA, 1988.
Hastie, T. J., and Tibshirani, R. J., "Generalized Additive Models." Chapman and Hall, New York, 1990.
Heinonen, P., and Neuvo, Y., FIR-median hybrid filters, *IEEE Trans. Acoustics and Signal Processing* **35**(6), 832 (1987).
Henrion, M., Breese, J. S., and Horvitz, E. J., Decision analysis and expert systems, *AI Mag.* **12**(4), 64–91 (1991).
Hornik, K., Stinchcombe, M., and White, H., Multilayer feedforward networks are universal approximators, *Neural Networks* **2**(5), 359–36 (1989).
Hoskins, J. C., and Himmelblau, D. M., Artificial neural network models of knowledge representation in chemical engineering, *Comput. Chem. Eng.* **12**, 881–890 (1988).
Hoskins, J. C., Kaliyur, K. M., and Himmelblau, D. M., Fault diagnosis in complex chemical plants using artificial neural networks, *AIChE J.* **37**, 137–141 (1991).
Hwang, J. N., Lay, M., Martin, R. D., and Schimert, J., Regression modeling in back-propagation and projection pursuit learning, *IEEE Trans. Neur. Networks* 5 (1994).
Iserman, R., Process fault detection based on modeling and estimation methods—a survey, *Automatica* **20**, 387–404 (1984).
Jain, A. K., and Dubes, R. C., "Algorithms for Clustering Data." Prentice-Hall, Englewood Cliffs, NJ, 1988.
Jackson, J. E., "A User's Guide to Principal Components." Wiley, New York, 1992.
Jolliffe, I. T., "Principal Component Analysis." Springer-Verlag, New York, 1986.
Josephson, J., and Josephson, S., eds., "Abductive Inference: Computation, Philosophy, Technology." Cambridge University Press, 1994.
Kavuri, S. N., and Venkatasubramanian, V., Using fuzzy clustering with ellipsoidal units in neural networks for robust fault classification, *Comput. Chem. Eng.* **17**(8), 765 (1993).
Kelly, B. E., and Lees, F. P., The propagation of faults in process plants: 1, modeling of fault propagation, *Reliability Eng. and Sys. Safety* **16**, 3 (1986).
Kohonen, T., "Self-Organizing and Associative Memory." Springer-Verlag, New York, 1989.

Kosanovich, K. A., Piovoso, M. J., and Dahl, K. S., Multi-way principal component analysis applied to an industrial batch process, *Ind. & Eng. Chem. Res.* **6,** 739 (1995).

Kramer, M. A., Malfunction diagnosis using quantitative models with non-Boolean reasoning in expert systems, *AIChE J.* **33,** 1130–140 (1987).

Kramer, M. A., Nonlinear principal component analysis using autoassociative neural networks, *AIChE J.* **37**(2), 233–243 (1991).

Kramer, M. A., and Mah, R. S. H., Model-based monitoring, *in* "Proc. Second Int. Conf. on Foundations of Computer Aided Process Operations." (D. Rippin, J. Hale, and J. Davis, eds.). CACHE, 1994.

Kramer, M. A., and Palowitch, B. L., Rule-based approach to fault diagnosis using the signed directed graph, *AIChE J.* **33**(7), 1067–1078 (1987).

Kresta, J. V., and MacGregor, J. F., Multivariate statistical monitoring of process operating performance, *Can. J. Chem. Eng.* **69,** 35–47 (1991).

Kuipers, B., "Qualitative Reasoning: Modeling and Simulation with Incomplete Knowledge." MIT Press, Boston, 1994.

Leonard, J., and Kramer, M. A., Improvement of the back propagation algorithm for training neural networks, *Comput. Chem. Eng.* **14,** 337–341 (1990).

Leonard, J., and Kramer, M. A., Radial basis function networks for classifying process faults, *IEEE Control Systems,* April, p. 31 (1991).

MacGregor, J. F., and Kourti, T., Statistical process control of multivariate processes, *Cont. Eng. Prac.* **3,** 404–414 (1995).

MacGregor, J. F., Jaeckle, C., Kiparissides, C., and Koutoudi, M., Process monitoring and diagnosis by multiblock PLS method, *AIChE J.* **40**(5), 826–838 (1994).

MacQueen, J., Some methods for classification and analysis of multivariate data, *Proc. 5th Berkeley Symp. on Probability and Statistics,* Berkeley, CA (1967).

Mallat, S. G., A theory for multiresolution signal decomposition: The wavelet representation, *IEEE Transactions on Pattern Analysis and Machine Intelligence* **11**(7), 674–693 (1989).

Maren, A., Harston, C., and Rap, R., "Handbook of Neural Computing Applications." Academic Press, New York, (1990).

Mavrovouniotis, M. L., ed., "Artificial Intelligence in Process Engineering." Academic Press, Boston, 1990.

McDowell, J. K., and Davis, J. F., Diagnostically focused simulation: managing qualitative simulation, *AIChE J.* **37**(4), 569–580 (1991).

McDowell, J. K., and Davis, J. F., Problem solver integration based on generic task architectures, *in* "Intelligent Modeling, Diagnosis and Control of Manufacturing Processes" (B. B. Chu and S. Chen, eds.). World Scientific, Singapore, 1992, pp. 61–82.

McVey, S. R., Davis, J. F., and Venkatasubramanian, V., Intelligent systems in process operations, design and safety, *in* "A Perspective on Computers in Chemical Engineering." CACHE Corp., 1997.

Miller, W. T., III, Sutton, R. S., and Werbos, P. J., eds., "Neural Networks for Control." MIT Press, Boston, 1990.

Minsky, M., and Papert, S., "Perceptions: An Introduction to Computational Geometry." MIT Press, Cambridge, MA, 1969.

Mitra, S. K., and Kaiser, J. F., "Digital Signal Processing." Wiley, New York, 1993.

Mohindra, S., and Clark, P. A., A distributed fault diagnosis method based on digraph models: Steady-state analysis, *Comput. Chem. Eng.* **17,** 193 (1993).

Moody, J., "Fast learning in multi-resolution hierarchies," Research Report, Yale University, YALEU/DCS/RR-681 (1989).

Moody, J., and Darken, C. J., Fast learning in networks of locally-tuned processing units, *Neural Computation* **1,** 281–294 (1989).

Mullhi, J. S., Ang, M. L., and Lees, F. P., *Reliability Engr. and Sys. Safety* **23,** 31 (1988).
Myers, D. R., Hurley, C. E. and Davis, J. F. A diagnostic expert system for a sequential, PLC controlled operation, *in* "Artificial Intelligence Applications in Process Engineering." (M. Mavrovoumolis, (ed.). Academic Press, New York, 1990.
Naidu, S., Zafiriou, E., and McAvoy, T. J., Application of neural networks on the detection of sensor failure during the operation of a control system, *Proc. Amer. Control Conf.,* Pittsburgh, PA (1989).
Nason, G. P., Wavelet shrinkage using cross-validation, *J. R. Statist. Soc. B* **58**(2), 463 (1996).
Nomikos, P., and MacGregor, J. F., Monitoring of batch processes using multi-way PCA, *AIChE J.* **40**(5), 1361–1375 (1994).
Nounou, M. N., and Bakshi, B. R., On-line multiscale fillering of random and gross errors without process models, *AIChE J.,* **45**(6), 1041 (1999).
Oja, E., Unsupervised neural learning, *in* "Neural Networks for Chemical Engineers" (A. B. Bulsari, (ed.). Elsevier, Amsterdam, 1995, p. 21.
Pearl, J., "Probabilistic Reasoning in Intelligent Systems," Morgan Kaufmann, San Mateo, CA, 1988.
Piovoso, M. J., and Kosanovich, K. A., Applications of multivariate statistical methods to process monitoring and controller design, *Int. J. Control* **59**(3), 743–765 (1994).
Piovoso, M. J., Kosanovich, K. A., and Yuk, J. P., Process data chemometrics, *IEEE Trans. Instrumentation and Measurements* **41**(2), 262–268 (1992a).
Piovoso, M. J., Kosanovich, K. A., and Pearson, R. K., Monitoring process performance in real-time, *Proc. Amer. Control Conf.,* Chicago, IL, Vol. 3, pp. 2359–2363 (1992b).
Prasad, P. R., and Davis, J. F., A framework for knowledge-based diagnosis in process operations, *in* "An Introduction to Intelligent and Autonomous Control" (P. J. Antsaklis and K. M. Passino, eds.). Kluwer Academic Publishers, Boston, 1992, pp. 401–422.
Prasad, P. R., Davis, J. F., Jirapinyo, Y., Bhalodia, M., and Josephson, J. R., Structuring diagnostic knowledge for large-scale process systems, *Comput. Chem. Eng.,* **22,** 1897–1905 (1998).
Qin, S. J., and McAvoy, T. J., Nonlinear PLS modeling using neural networks, *Comput. Chem. Eng.* **16**(4), 379 (1992).
Quantrille, T. E., and Liu, Y. A., "Artificial Intelligence in Chemical Engineering." Academic Press, San Diego, 1991.
Ragheb, M., and Gvillo, D., Development of model-based fault identification systems on microcomputers, *International Society for Optical Engineering,* pp. 268–275. SPIE, Bellingham, WA, 1993.
Ramesh, T. S., Shum, S. K., and Davis, J. F., A structured framework for efficient problem solving in diagnostic expert systems, *Comput. Chem. Eng.* **12,** 891 (1988).
Ramesh, T. S., Davis, J. F., and Schwenzer, G. M., "Knowledge-based diagnostic systems for continuous process operations based upon the task framework," *Comput. Chem. Eng.* **16**(2), 109–127 (1992).
Rekalitis, G. V., and Spriggs, H. D., *Proceedings of the First International Conference on Foundations of Computer-Aided Operations,* Elsevier Science Publishers, New York, 1987.
Rich, E., and Knight, K., *Artificial Intelligence, 2nd edition,* McGraw-Hill, New York, 1991.
Ripley, B. D., "Neural networks and related methods for classification," *J. Royal Stat. Soc.* **56**(3), 409–456 (1994).
Rippin, D., Hale, J., and Davis, J. F., eds., *Proceedings Foundations of Computer-Aided Process Operations,* CACHE, 1994.
Roosen, C. B., and Hastie, T. J., Automatic smoothing spline projection pursuit. *J. Comput. Graph. Stat.,* **3,** 235 (1994).

Rosenblatt, F., *Principles of Neurodynamics: Perceptions and the Theory of Brain Mechanisms,"* Spartan, NY, 1962.
Roth, E. M., Woods, D. D., and Pople, H. E., "Cognitive simulation as a tool for cognitive task analysis," *Ergonomics* **35**(10), 1163–1198 (1992).
Rumelhart, D. E., Hinton, G. E., and Williams, R. J., "Learning internal representation by error propagation," *in* "Parallel Distributed Processing, Vol. 1., Foundations" (D. E. Rumelhart and J. L. McClelland, eds.). MIT Press 1986.
Shapiro, S. C., Eckroth, D., et al., eds., *"Encyclopedia of Artificial Intelligence."* John Wiley & Sons, New York, 1987.
Simon, J. C., *"Patterns and Operators: The Foundations of Data Representation."* North Oxford Academic Publishers Ltd., London, 1986.
Sravana, K. K., "Diagnostic Knowledge-Based systems for batch chemical processes: hypothesis queuing and evaluation," PhD Thesis, The Ohio State University (1994).
Stephanopoulos, G., ed., *"Knowledge-based systems in process engineering."* CACHE Case Studies Series-I, CATDEX, CACHE Corp., Austin, TX, 1988.
Stephanopoulos, G., and Davis, J. F., eds., CACHE monograph series, *in* "Artificial Intelligence in Process Systems Engineering." Vols. I, II and III, CACHE (1990).
Stephanopoulos, G., and Davis, J. F., eds., CACHE monograph series, *in* "Artificial Intelligence in Process Systems Engineering." Vol. IV, CACHE (1993).
Stephanopoulos, G., and Han, C., Intelligent systems in process engineering. *Proc. PSE 94,* Korea (1994).
Strang, G., Wavelets and dilation equations: a brief introduction. *SIAM Review,* **31**(4), 614–627 (1989).
Tan, S., and Mavrovouniotis, M. L., Reducing data dimensionality through optimizing neural networks inputs, *AIChE J.* **41**(6), 1471–1480 (1995).
Tou, J. T., and Gonzalez, R. C., *"Pattern Recognition Principles."* Addison-Wesley, Reading, MA, 1974.
Ulerich, N. H., and Powers, G. J., On-Line hazard aversion and fault detection: Multiple loop control example. *AIChE Fall National Meeting,* New York, 1987.
Ungar, L. H., Powell, B. A., and Kamens, S. N., Adaptive networks for fault diagnosis and process control, *Comput. Chem. Eng.* **14,** 561–572 (1990).
Vedam, H., Venkatasubramanian, V., and Bhalodia, M., A B-spline base method for data compression, process monitoring and diagnosis. *Comput. Chem. Eng.* **22**(13), S827–S830 (1998).
Venkatasubramanian, V., and Rich, S., An object-oriented two-tier architecture for integrating compiled and deep level knowledge for process diagnosis. *Comput. Chem. Eng.* **12**(9), 903 (1988).
Venkatasubramanian, V., and Vaidhyanathan, R., A knowledge-based framework for automating HAZOP analysis, *AIChE J.* **40**(3), 496 (1994).
Whiteley, J. R., and Davis, J. F., Qualitative interpretation of sensor patterns. *IEEE Expert,* **8**(2) 54–63 (1992a).
Whiteley, J. R., and Davis, J. F., Knowledge-based interpretation of sensor patterns, special issue of *Comput. Chem. Eng.* **16**(4), 329–346 (1992b).
Whiteley, J. R., and Davis, J. F., A similarity-based approach to interpretation of sensor data using adaptive resonance theory. *Comput. Chem. Eng.* **18**(7), 637–661 (1994).
Whiteley, J. R., Davis, J. F., Ahmet, M., and Ahalt, S. C., Application of adaptive Resonance theory for qualitative interpretation of sensor data, *Man, Machine and Cybernetics* **26**(7), July (1996).
Widrow, B., and Hoff, M. E., Adaptive switching circuits, *1960 WESCON Convention Record, Part IV,* 96–104 (1960).

Wilcox, N. A., and Himmelblau, D. M., The possible cause and effect graphs (PCEG) model for fault diagnosis—I methodology, *Comput. Chem. Eng.* **18**(2) 103–116 (1994).

Wold, H., "Soft Modeling: The Basic Design and Some Extensions, Systems Under Indirect Observations" (K. J. Horeskog and H. Wold, eds.). Elsevier Science, North Holland, Amsterdam, 1982.

Wold, S., Ruhe, A., Wold, H., and Dunn, W., The collinearity problem in linear regression. The partial least squares (PLS) approach to generalized inverses, *SIAM J. Sci. and Statistical Computing* **5**, 735–743 (1984).

Wold, S., Esbensen, K., and Geladi, P., Principal component analysis, *Chemometrics and Intelligent Laboratory Systems* **2**, 37–52 (1987a).

Wold, S., Geladi, P., and Ohman, J., Multi-way principal components and PLS analysis, *J. Chemometrics* **1**, 41–56 (1987b).

Wold, S., Kettaneh-Wold N., and Skagerberg, B., Nonlinear PLS modeling, *Chemom. Intell. Lab. Sys.* **7**, 53–65 (1989).

Wold, S., Nonlinear PLS modeling II: spline inner relation, *Chemom. Intell. Lab. Sys.*, **14**, 71–84 (1992).

Wold, S., Kettaneh, N., and Tjessem, K., "Hierarchical multiblock PLS and PC models for easier model interpretation and as an alternative to variable selection," *J. Chemometrics* **10**, 463 (1996).

MIXING AND DISPERSION OF VISCOUS LIQUIDS AND POWDERED SOLIDS*

J. M. Ottino, P. DeRoussel, S. Hansen, and D. V. Khakhar

Department of Chemical Engineering Northwestern University Evanston, IL 60208

I.	Preliminaries	105
II.	Mixing and Dispersion of Immiscible Fluids	124
	A. Breakup	130
	B. Coalescence	151
	C. Breakup and Coalescence in Complex Flows	155
III.	Fragmentation and Aggregation of Solids	159
	A. Fragmentation	163
	B. Aggregation	180
IV.	Concluding Remarks	194
	References	198

I. Preliminaries

Mixing and dispersion of viscous fluids—blending in the polymer processing literature—is the result of complex interaction between flow and events occurring at drop length-scales: breakup, coalescence, and hydrodynamic interactions. Similarly, mixing and dispersion of powdered solids in viscous liquids is the result of complex interaction between flow and

---

* Note on notation: Relations from breakup, coalescence, fragmentation, and aggregation are based on either actual experiments or numerical simulations, the latter commonly referred to as "computer experiments." Computer experiments are often based on crude simplifying assumptions and actual experiments are always subject to errors; the strict use of the equality sign in many of the final results may therefore be misleading. In order to accurately represent the uncertainty associated with the results, the following notation is adopted:

$\approx$	Used in relations when the error is generally less than 25%
$\sim$	Used to denote proportionality and order of magnitude
$\gtrsim, \lesssim$	Used in conditional statements when the limiting value given is correct within a factor of 2
$=, >, <$	Used in expressions related to modeling and standard mathematical manipulations

events—erosion, fragmentation and aggregation—occurring at agglomerate length scales. Important applications of these processes include the compounding of molten polymers and the dispersion of fine particles in polymer melts. Reynolds numbers in both cases are small, even more so given the small length scales that dominate the processes.

There are similarities and, undoubtedly, substantial differences between these two processes. The following analogies are apparent (Fig. 1):

$$\text{Breakup} \leftrightarrow \text{Fragmentation}$$

$$\text{Coalescence} \leftrightarrow \text{Aggregation}$$

The ratio of deforming viscous forces to resisting interfacial tension forces in the case of droplets is the *capillary number,* Ca. Similarly, the ratio of viscous to cohesive forces in agglomerates is the *fragmentation number,* Fa.

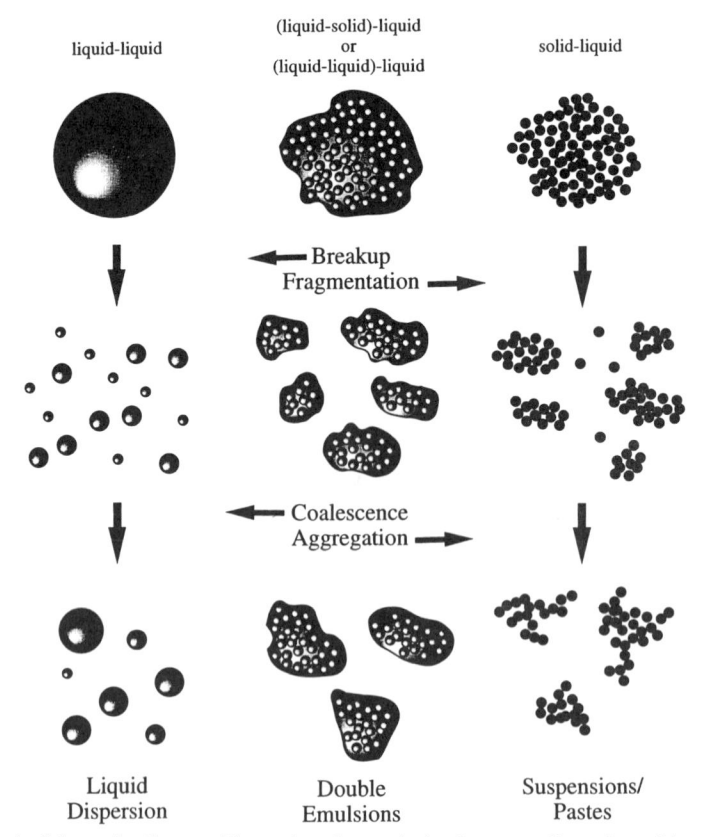

FIG. 1. Schematic diagram illustrating the analogies between dispersion of immiscible liquids and dispersed solids.

Thus, $Ca < O(1)$ and $Fa < O(1)$ determine conditions where no breakup or fragmentation is possible. Both processes, breakup and coalescence for drops, and fragmentation and aggregation for solids, lead to time-varying distributions of drops and cluster sizes that become time-invariant when scaled in suitable ways. There are also parallels between aggregation and fragmentation that underscore mathematical similarity: Both *Smoluchowski's equation* and the *fragmentation equation* are amenable to scaling solutions.

These similarities notwithstanding, it may be argued that it is differences that have hindered understanding. This state of affairs has not been helped by proliferation of terminology: alloying, compounding, and blending all appear in the polymer processing literature; fragmentation and rupture in the solids dispersion literature; breakup, rupture, and burst in the fluid mechanics literature. Both agglomerates and flocs refer to particle clusters. Undoubtedly, droplets are easier to deal with than agglomerates whose structure is complex and can only be known in a statistical sense. Thus, breakup and coalescence, being placed squarely in the realm of classical fluid mechanics, are on more sure footing than fragmentation and aggregation, which demand knowledge of physical chemistry and colloid science, and have been studied to a lesser extent.

Realistic mixing problems are inherently difficult owing to the complexity of the *flow* fields, to the fact that the *fluids* themselves are rheologically complex, and to the coupling of length scales. For this very reason, mixing problems have been attacked traditionally on a case-by-case basis. Modeling becomes intractable if one wants to incorporate all details at once. Nevertheless it appears important to focus on common features and to take a broad view.

Two reviews, published in the same book, may be considered to be the launching basis for the material presented here: Meijer and Janssen (1994) and Manas-Zloczower (1994). In both reviews the fundamentals of the processes are considered at a small length scale (drops, agglomerates), focusing on the effects of flow. Meijer and Janssen (1994) review fundamental studies of droplet breakup and coalescence in the context of the analysis of polymer blending. Manas-Zloczower (1994), on the other hand, focuses on dispersion of fine particles in polymers. The primary objective of the present work is to present both topics in a unified format together with the basics of particle aggregation. The unified presentation serves to highlight the analogies between the processes and consequently to increase understanding. A related goal is to introduce potentially useful recent advances in fundamentals that have not yet been applied to the analysis of practical processing and structuring problems (Villadsen, 1997).

There have been substantial advances in the understanding of *viscous mixing* of single fluids. This has been driven primarily by theoretical developments based on chaos theory and increased computational resources, as well as by advances in fluid mechanics and a host of new experimental results. Such an understanding forms a fabric for the evolution of breakup, coalescence, fragmentation, and aggregation. These processes can in fact be viewed as a population of "microstructures" whose behavior is driven by a chaotic flow; microstructures break, diffuse, and aggregate, causing the population to evolve in space and time. *Self-similarity* is common to all these problems; examples arise in the context of the distribution of stretchings within chaotic flows, in the asymptotic evolution of fragmentation processes, and in the equilibrium distribution of drop sizes generated upon mixing of immiscible fluids.

It may be useful at this juncture to draw the distinction between *theory* and *computations* (or numerical experiments), as the material presented here is somewhat tilted toward theory. Computations are not theory, but theory often requires computations. To the extent that it is not possible to put all elements of a problem into a complete picture, assumptions are necessary—often entailing mechanistic views of the behavior of the system. Nontrivial assumptions leading to nontrivial consequences lead to significant theories. Assuming that a flow field can be imagined as an assembly of weak regions (where stretching of passive elements is linear) and strong regions (where stretching of passive elements is exponential) could form the basis of a theory; real flows are manifestly more complex, but this is clearly a useful approximation. In fact, G. I. Taylor's (1932) pioneering work in drop dynamics can be traced back to this crucial element. A binary breakup assumption, on the other hand, may not form as strong a basis, especially if more precise knowledge can be incorporated with little or no difficulty into the picture: Drops and fluid filaments break, often producing, in single events, a distribution of sizes. Thus, a "theory" based on binary breakup could be revisited and may be successfully augmented. Similar comments apply to fragmentation of agglomerates.

This paper is divided into two main, interconnected parts—breakup and coalescence of immiscible fluids, and aggregation and fragmentation of solids in viscous liquids—preceded by a brief introduction to mixing, this being focused primarily on stretching and self-similarity.

The treatment of mixing of immiscible fluids starts with a description of breakup and coalescence in homogeneous flows. Classical concepts are briefly reviewed and special attention is given to recent advances—satellite formation and self-similarity. A general model, capable of handling breakup and coalescence while taking into account stretching distributions and satellite formation, is described.

The treatment of aggregation and fragmentation processes parallels that of breakup and coalescence. Classical concepts are briefly reviewed; special attention is given to fragmentation theory as well as to flow-driven processes in nonhomogeneous (chaotic) flows. The Péclet number in all instances is taken to be much greater than unity, so that diffusion effects are unimportant. In many examples, hydrodynamic interactions between clusters are neglected to highlight the effects of advection on the evolution of the cluster size distribution and the formation of fractal structures.

The paper is structured to be read at three levels. The main thread of the text is a review of fundamentals and previous studies. *Illustrations* focusing on specific systems or more detailed elaboration of concepts are interspersed in the text. Many of these include new results; they form a second level that can be read as independent subunits. Finally, the conclusions of each section, especially those with significance for practical applications, are summarized as ⟦*heuristics*⟧.

1. Stretching and Chaotic Mixing

Fluid advection—be it regular or chaotic—forms a template for the evolution of breakup, coalescence, fragmentation, and aggregation processes. Let $\mathbf{v}(\mathbf{x}, t)$ represent the Eulerian velocity field (typically we assume that $\nabla \cdot \mathbf{v} = 0$). The solution of

$$\left(\frac{d\mathbf{x}}{dt}\right)_{\mathbf{X}} = \mathbf{v}(\mathbf{x}, t), \tag{1}$$

with $\mathbf{X}$ representing the initial coordinates of material particles located at $\mathbf{x}$ at time t, gives the motion

$$\mathbf{x} = \Phi_t(\mathbf{X}) \text{ with } \mathbf{X} = \Phi_{t=0}(\mathbf{X}), \tag{2}$$

i.e., the particle $\mathbf{X}$ is mapped to the position $\mathbf{x}$ after a time t. This is formally the solution to the so-called *advection* problem (Aref, 1984). The foundations of this area are now on sure footing and reviews are presented in Ottino (1989, 1990). The quantity of interest is stretching of fluid elements.

The stretches of a material filament $d\mathbf{x}$, λ, and material surface element $d\mathbf{a}$, η, are defined as

$$\lambda \equiv \lim_{|d\mathbf{X}| \to 0} \frac{|d\mathbf{x}|}{|d\mathbf{X}|}, \quad \eta \equiv \lim_{|d\mathbf{A}| \to 0} \frac{|d\mathbf{a}|}{|d\mathbf{A}|}, \tag{3}$$

where $d\mathbf{X}$ and $d\mathbf{A}$ represent the initial conditions of $d\mathbf{x}$ and $d\mathbf{a}$, respectively. The fundamental equations for the rate of stretch are

$$\frac{d(\ln \lambda)}{dt} = \mathbf{D} : \mathbf{mm}, \quad \frac{d(\ln \eta)}{dt} = \nabla \cdot \mathbf{v} - \mathbf{D} : \mathbf{nn}, \tag{4}$$

where $\mathbf{D} \equiv [\nabla\mathbf{v} + (\nabla\mathbf{v})^T]/2$ is the stretching tensor, and $\mathbf{m}$ and $\mathbf{n}$ are the instantaneous orientations ($\mathbf{m} = d\mathbf{x}/|d\mathbf{x}|$, $\mathbf{n} = d\mathbf{a}/|d\mathbf{a}|$). The Lagrangian histories $d(\ln\lambda)/dt = \alpha_\lambda(\mathbf{X}, \mathbf{M}, t)$ and $d(\ln\eta)/dt = \alpha_\eta(\mathbf{X}, \mathbf{N}, t)$ are called *stretching functions*. Flows can be compared in terms of their stretching efficiencies. The stretching efficiency $e_\lambda = e_\lambda(\mathbf{X}, \mathbf{M}, t)$ of the material element $d\mathbf{X}$ and the stretching efficiency $e_\eta = e_\eta(\mathbf{X}, \mathbf{N}, t)$ of the area element $d\mathbf{A}$ are defined as

$$e_\lambda \equiv \frac{1}{(\mathbf{D}:\mathbf{D})^{1/2}} \frac{d(\ln \lambda)}{dt} < 1, \qquad e_\eta \equiv \frac{1}{(\mathbf{D}:\mathbf{D})^{1/2}} \frac{d(\ln \eta)}{dt} < 1. \qquad (5)$$

If $\nabla \cdot \mathbf{v} = 0$, $e_i \leq (1/2)^{1/2}$ in two-dimensional (2D) flows and $(2/3)^{1/2}$ in three-dimensional (3D) flows, where $i = \lambda, \eta$. The efficiency can be thought of as the specific rate of stretching of material elements normalized by a factor proportional to the square root of the energy dissipated locally.

The key to effective mixing lies in producing stretching and folding, an operation that is referred to in the mathematics literature as a *horseshoe map*. Horseshoe maps, in turn, imply chaos. The 2D case is the simplest. The equations of motion for a two-dimensional area preserving flow can be written as

$$v_x = \frac{dx}{dt} = \frac{\partial \psi}{\partial y}, \qquad v_y = \frac{dy}{dt} = -\frac{\partial \psi}{\partial x}, \qquad (6)$$

where ψ is the stream function. If the velocity field is steady (i.e., ψ is independent of time), then it is integrable and the system cannot be chaotic. The mixing is thus poor: Stretching for long times is linear, the stretching function decays as $1/t$, and the efficiency decays to zero. On the other hand, if the velocity field, or equivalently ψ, is time-periodic, there is a good chance that the system will be chaotic (Aref, 1984). It is relatively straightforward to produce flow fields that can generate chaos. A necessary condition for chaos is the "crossing" of streamlines: Two successive streamline portraits, say at t and $(t+\Delta T)$ for time periodic flows or at z and $(z+\Delta z)$ for spatially periodic flows, when superimposed, should show intersecting streamlines. In 2D systems this can be achieved by time modulation of the flow field, for example by motions of boundaries or time periodic changes in geometry. Figures 2 and 3 show typical examples of mixing in such flows: the vortex mixing flow (VMF) and the cavity flow, respectively. The vortex mixing flow is generated by alternately rotating one of the cylinders for a fixed period of time; the cavity flow is generated by alternately moving the upper and lower walls of the cavity for a fixed time period. It is clear from the geometry of the systems that both achieve the required "crossing" of streamlines by the time modulation.

Numerous experimental studies have revealed the degree of order and

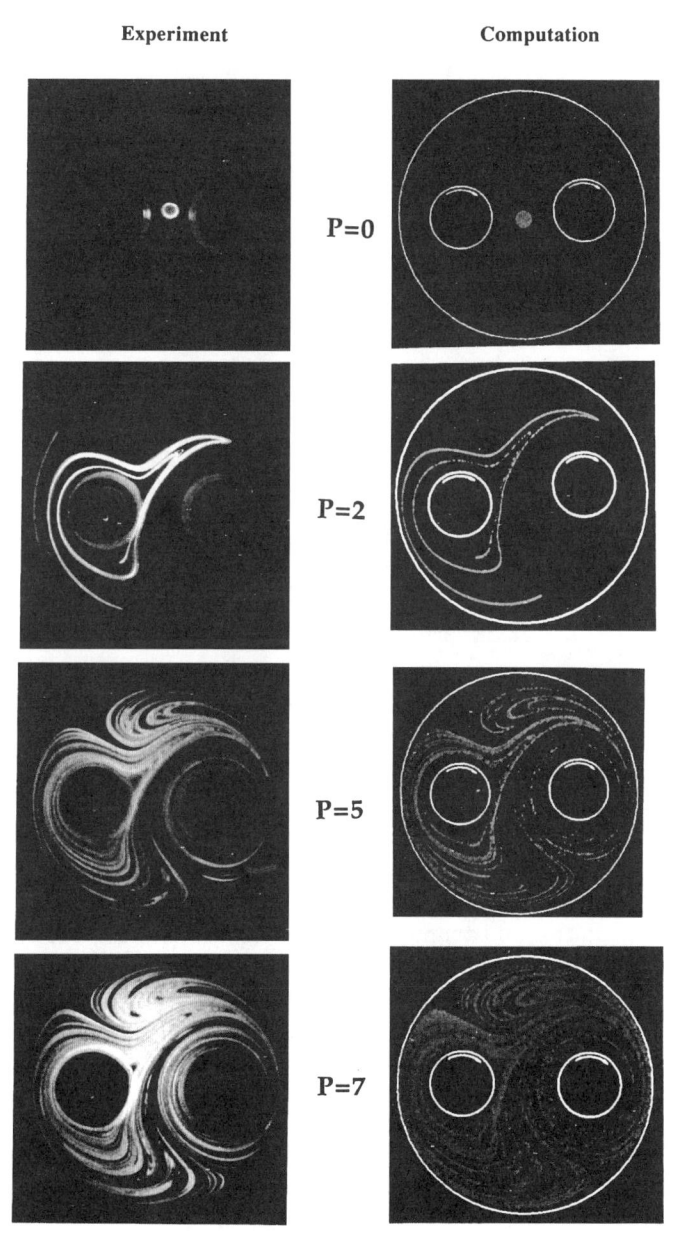

FIG. 2. Mixing in the vortex mixing flow with increasing periods of flow (P). The flow is time periodic with each cylinder rotating alternately for a fixed time period (Jana, Metcalfe and Ottino, 1994).

experiment

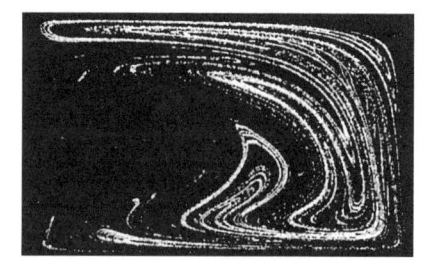

computation

FIG. 3. Mixing in the cavity flow. The flow is time periodic with the upper and lower walls moving alternately for a fixed time period (Leong and Ottino, 1989).

disorder compatible with chaos in fluid flows. Experiments conducted in carefully controlled two-dimensional, time-periodic flows and spatially periodic flows reveal the complexity associated with chaotic motions. The most studied flows are the flow between two rotating eccentric cylinders, several classes of corotating and counterrotating cavity flows, and spatially periodic flows. Even within this theme, many variations are possible. For example, Leong (1990) considered the effect of cylindrical obstructions placed in a cavity flow; Schepens (1996) took this a step further and carried out computations and experiments for the case when the pin itself is allowed to rotate. The 2D time-periodic case is especially illustrative.

Dye structures of passive tracers placed in time-periodic chaotic flows evolve in an iterative fashion; an entire structure is mapped into a new structure with persistent large-scale features, but finer and finer scale features are revealed at each period of the flow. After a few periods, strategically placed blobs of passive tracer reveal patterns that serve as templates for subsequent stretching and folding. Repeated action by the flow generates a lamellar structure consisting of stretched and folded striations, with thicknesses $s(t)$, characterized by a probability density function, $f(s,t)$, whose mean, on the average, decreases with time [$f(s,t)ds$ is the number of striations with striation thicknesses between s and $s + ds$]. The striated pattern quickly develops into a time-evolving complex morphology of poorly mixed regions of fluid (islands) and of well-mixed or chaotic regions (see Fig. 3). Computations capture the evolution of the structure reasonably well for short mixing times (Figs. 2 and 3) and are useful for analysis. Islands translate, stretch, and contract periodically and undergo a net rotation, preserving their identity returning to their original locations after multiples of the period of the flow, symmetry being a common feature of many flows (see Fig. 4). Stretch in islands, on the average, grows linearly and much slower than in chaotic regions, in which the stretch increases exponentially with time. Moreover, since islands do not exchange matter with the rest of the fluid they represent an obstacle to efficient mixing. An important parameter of the flow is the sense of rotation in the alternate periods—corotation tends to produce more uniform mixing.

Duct flows, like steady two-dimensional flows, are poor mixers. This class of flows is defined by the velocity field

$$v_x = \frac{\partial \psi}{\partial y}, \qquad v_y = -\frac{\partial \psi}{\partial x}, \qquad v_z = f(x, y), \tag{7}$$

which is composed of a two-dimensional cross-sectional flow augmented by a unidirectional axial flow; fluid is mixed in the cross-section while it is

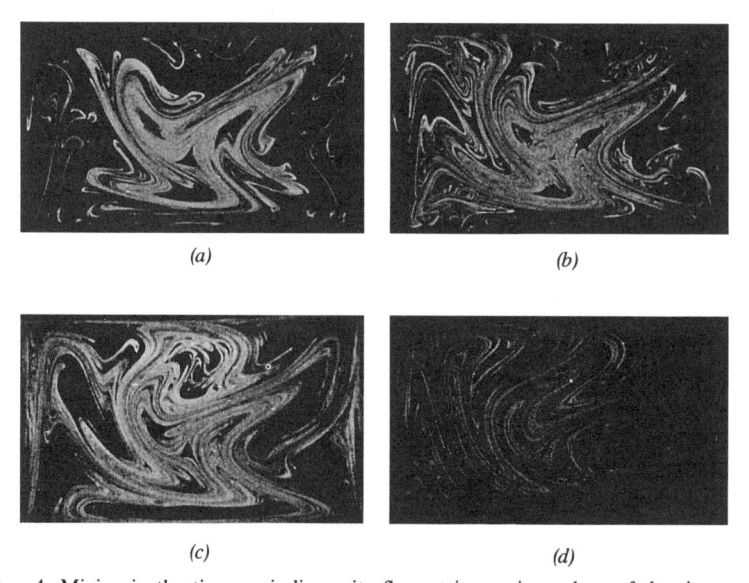

(a)   *(b)*

(c)   *(d)*

FIG. 4. Mixing in the time periodic cavity flow at increasing values of the time period. The mixed state is shown at long mixing times (Leong and Ottino, 1989).

(a)

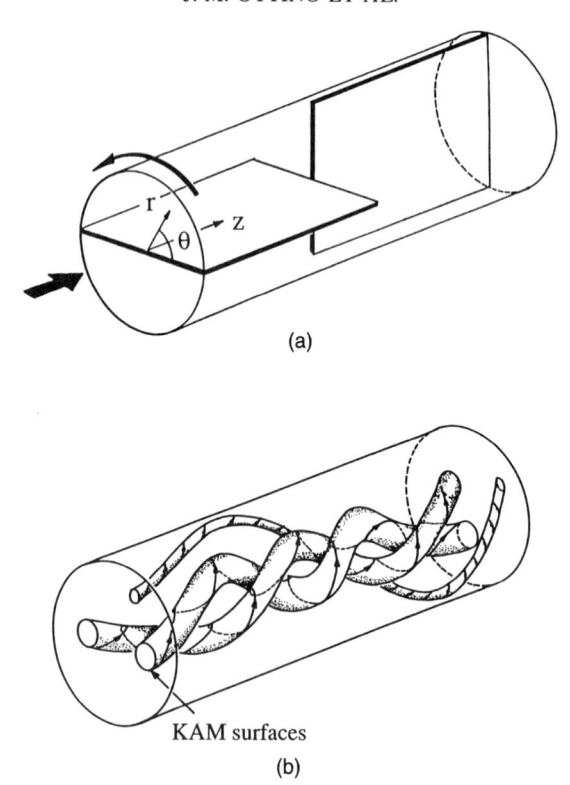

(b)

Fig. 5. Mixing in the partitioned pipe mixer. (a) Schematic view of the mixer geometry; one element of the mixer is shown, and the mixer comprises several such elements joined together. (b) KAM surfaces in flow bounding regions of regular flow. (c) A and B show examples of experimentally obtained streaklines: undeformed streaklines pass through KAM tubes, whereas streaklines in the chaotic region are well mixed. The white arrows indicate the location of a companion streak-tube. (Khakhar, Franjione, and Ottino, 1987; Kusch and Ottino, 1992).

simultaneously transported down the duct axis. In a duct flow, the cross-sectional and axial flows are independent of both time and distance along the duct axis, and material lines stretch linearly in time. A single screw extruder, for example, belongs to this class.

Duct flows can be converted into efficient mixing flows (i.e., flows with an exponential stretch of material lines with time) by time- modulation or by spatial changes along the duct axis. One example of the spatially periodic class is the partitioned-pipe mixer (PPM). This flow consists of a pipe partitioned with a sequence of orthogonally placed rectangular plates (Fig. 5a). The cross-sectional motion is induced through rotation of the pipe with respect to the assembly of plates, whereas the axial flow is caused by

A B

(c)

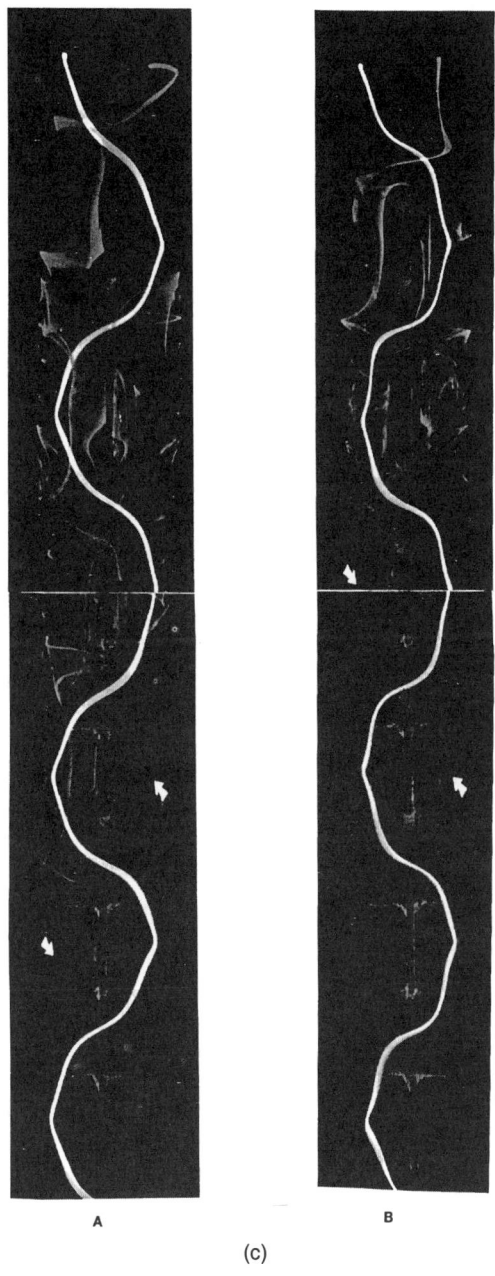

FIG. 5 (*continued*)

a pressure gradient; the behavior of the system is characterized by the ratio of cross-sectional twist to axial stretching, β (Khakhar, Franjione, and Ottino, 1987; Kusch and Ottino, 1992). The flow is regular for no cross-sectional twist ($\beta = 0$) and becomes chaotic with increasing values of β. The KAM (Kolmogorov–Arnold–Moser) surfaces (tubes) bound regions of regular flow and correspond to the islands in 2D systems (Fig. 5b). Experiments reveal the intermingled regular and chaotic regions: A streakline starting in a KAM tube passes through the mixer with little deformation, whereas streaklines in the chaotic regions are mixed well (Fig. 5c a,b).

Illustration: Improved mixing in single screw extruders. The concept of improved mixing by reorienting the flow is often used in various types of single screw extruders. The extruder is divided into a small number of zones with a mixing section between each zone. Each mixing section consists of pins or blades that protrude into the flow and cause random reorientations of fluid elements in the flow; a random reorientation is better than no reorientation at all. The mixing efficiency climbs and mixing is dramatically improved. Such an approach was investigated by Chella and Ottino (1985).

Mixing in such systems can also be significantly improved by means of time-dependent changes in geometry. This idea can be readily implemented in the context of duct flows by adding a secondary baffle. Such a concept has obvious applications in polymer processing; for example, single screw extruders can be imagined as a channel with a moving lid (not shown) as in Fig. 6a. A two-dimensional analog of the extruder channel with a baffle is a cavity flow with steady motion of the top wall and periodic motion of the lower wall with a rectangular block (Fig. 6b). The mixing in this case is greatly improved as compared to a steady cavity flow. Similar designs have been arrived at empirically in engineering practice.

Illustration: Mixing of viscoelastic fluids. Driving a system faster, for example by moving boundaries at a higher speed under a *fixed* protocol, does not imply better mixing: Islands survive and do not go away. Yet another instance where faster action may actually lead to *worse* mixing is provided by the case of viscoelastic fluids. Niederkorn and Ottino (1993) studied, experimentally and computationally, the mixing of Boger fluids— viscoelastic fluids with a constant shear viscosity. The system considered was the flow between two eccentric cylinders, in which the inner and outer cylinders are rotated alternately. The flow is referred to as the "journal bearing flow" and is a classical system in chaotic mixing studies. In the limit of slow flow, a Boger fluid behaves as Newtonian; faster flows lead to viscoelastic effects quantified in terms of Weissenberg number (We), the ratio of the relaxation time of the fluid to a time scale of the flow, e.g., the inverse of the shear rate. Spectacular effects occur at moderate We

(a)

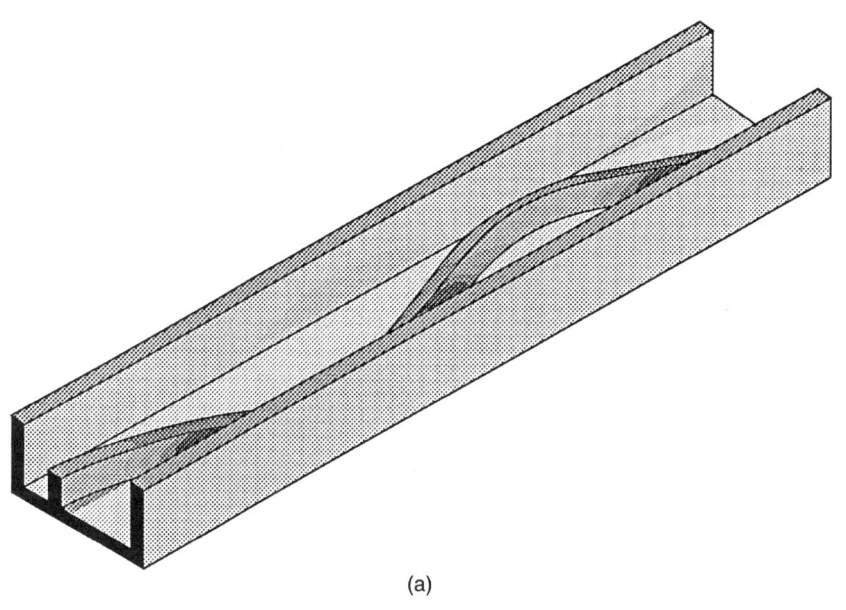

FIG. 6. (a) Schematic view of an extruder channel with an undulating baffle. (b) B, C, D: Steady flow streamlines with and without baffles for initial condition A. E, F: Mixing in a cavity flow with an oscillating baffle. The upper plate moves with a steady velocity while the lower plate with the baffle undergoes linear oscillatory motion (Jana, Tjahjadi, and Ottino, 1994).

(Niederkorn and Ottino, 1993); Fig. 7 shows the contrast between the Newtonian (We≈0) and the non-Newtonian case (We≈0.06). In general, it appears that viscoelastic fluids mix more slowly than the corresponding Newtonian fluid; in most instances also, in the long time limit, the region occupied by regular islands is larger for the viscoelastic case, i.e., the mixing is poorer. This, however, is not always the case, and there are experimentally documented instances where the long-time degree of mixing in the viscoelastic case is better than in the Newtonian case. Shear thinning effects, on the other hand, appear to be milder than viscoelastic effects and relatively high degrees of shear thinning are required to produce substantial effects (Niederkorn and Ottino, 1994).

Kumar and Homsy (1996) carried out a theoretical analysis of slightly viscoelastic flow and mixing in the journal bearing system described earlier. The effect of elasticity on the flow is to shift the stagnation streamline, which affects the size of the mixed region (Kaper and Wiggins, 1993). Depending on the mixing protocol, Kumar and Homsy (1996) found that the size of unmixed islands may increase or decrease with elasticity as in the experiments of Niederkorn and Ottino (1994). Many other comparisons

118   J. M. OTTINO *ET AL.*

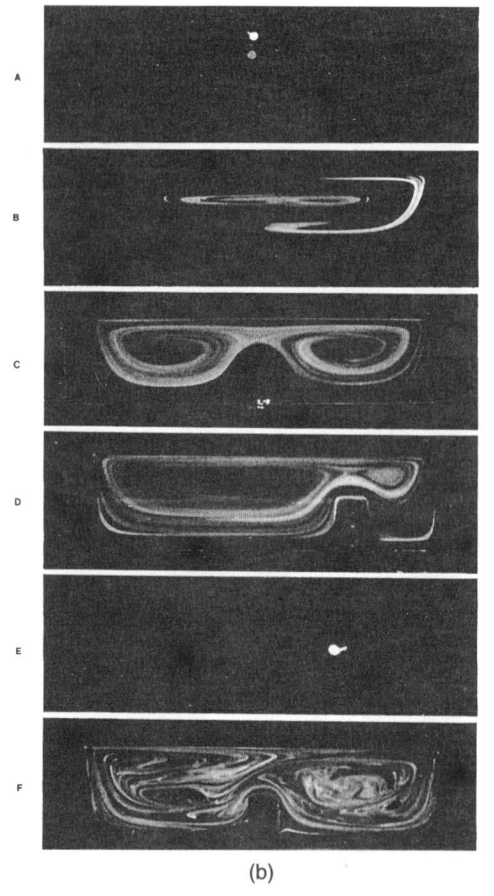

(b)

FIG. 6 (*continued*)

between chaotic advection of Newtonian and viscoelastic fluids are possible. For example, the critical modulation frequency of the inner cylinder at which the largest island disappears is lower for the viscoelastic fluid. Conclusions, however, are specific and dependent on mixing protocols, and little can be said in general about the effects of viscoelasticity. More work should be carried out in this area.

2. Stretching Distributions

Filaments in chaotic flows experience complex time-varying stretching histories. Computational studies indicate that within chaotic regions, the distribution of stretches, λ, becomes self-similar, achieving a scaling limit.

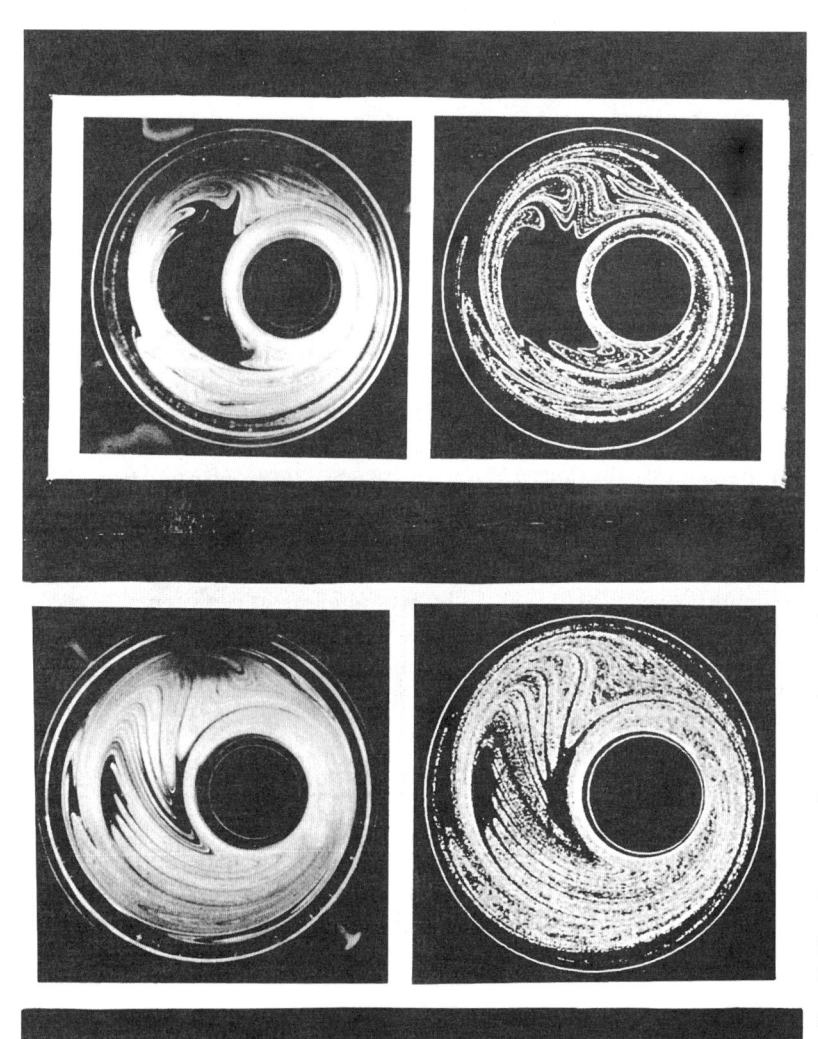

FIG. 7. Experiments and computations on the advection of a dye blob in an eccentric cylinder apparatus (from Niederkorn and Ottino, 1993). *Left*: Top, experiment using Newtonian fluid; bottom, numerical simulation of same situation. *Right*: Top, identical experiment as in upper left, but using viscoelastic fluid (We ≈ 0.06); bottom, numerical simulation at We = 0.04.

The distribution of stretches can be quantified in terms of the probability density function $F_n(\lambda) \equiv dN(\lambda)/d\lambda$, where $dN(\lambda)$ is the number of points that have values of stretching between λ and $(\lambda+d\lambda)$ at the end of period n. Another possibility is to focus on the distribution of $\log \lambda$. In this case we define the measure $H_n(\log\lambda) \equiv dN(\log\lambda)/d(\log\lambda)$.

Muzzio, Swanson, and Ottino (1991a) demonstrated that the distribution of stretching values in a globally chaotic flow approaches a log-normal distribution at large n: A log-log graph of the computed distribution approaches a parabolic shape (Fig. 8a) as required for a log–normal distribution. Furthermore, as n increases, an increasing portion of the curves in the figure (Fig. 8b) overlap when the distribution is rescaled as

$$\mathcal{H}(z) = [\log \lambda]^2 \frac{H_n(\log \lambda)}{m_{\log \lambda}(n)}, \qquad (8)$$

where $m_{\log \lambda}(n) = \Sigma \log \lambda H_n(\log \lambda)$ is the first moment of $H_n(\log \lambda)$ and $z = \log \lambda/\log \Lambda_g$, with $\log \Lambda_g = [\Sigma \log \lambda H_n(\log \lambda)]/[\Sigma H_n(\log \lambda)]$, and the sums are over a large number of fluid elements in the flow. The rescaled distribution is thus independent of the number of periods of the flow (n) at large n, and the distributions $H_n(\log \lambda)$ are said to be self-similar.

The explanation for the approach to a log-normal distribution is as follows. Consider, for simplicity, flows as those considered by Muzzio, Swanson, and Ottino (1991a): two-dimensional time periodic flows. Let $\lambda_{n,k}$ denote the length stretch experienced by a fluid element between periods n and k. The total stretching after m periods of the flow, $\lambda_{0,m}$, can be written as the product of the stretchings from each individual period:

$$\lambda_{0,m} = \lambda_{0,1}\lambda_{1,2}\lambda_{2,3} \ldots \lambda_{m-1,m}. \qquad (9)$$

The amount of stretching between successive periods (i.e., $\lambda_{1,2}$ and $\lambda_{2,3}$) is strongly correlated; however, the correlation in stretching between nonconsecutive periods (e.g., $\lambda_{0,1}$ and $\lambda_{4,5}$) grows weaker as the separation between periods increases as a result of chaos (the presence of islands in the flow complicates the picture, but these issues are not considered here). Thus, $\lambda_{0,m}$ is essentially the product of random numbers, which when rewritten as

$$\log \lambda_{0,m} = \log \lambda_{0,1} + \log \lambda_{1,2} + \log \lambda_{2,3} + \cdots + \log \lambda_{m-1,m} \qquad (10)$$

gives a sum of random numbers. According to the central limit theorem, any collection of sums of random numbers will converge to a Gaussian. So, when all material elements are considered, the distribution of $\lambda_{0,m}$ should be log–normal (Fig. 8a). This conjecture has been verified by numerous computations (Muzzio et al., 1991a).

Illustration: Optimum strain per period in shear flows with periodic reorientation. Many practical mixing flows (e.g., single screw extruder with mixing

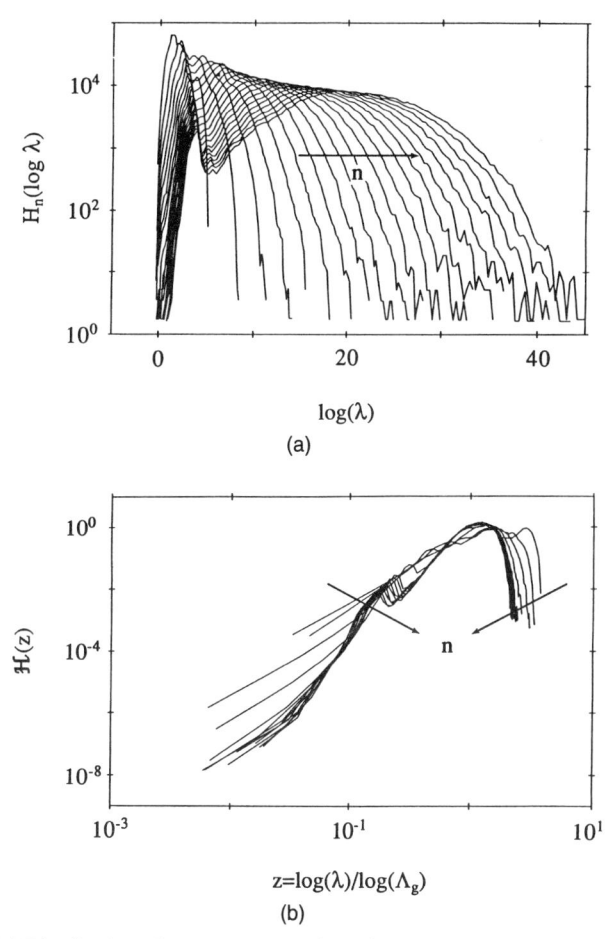

FIG. 8. (a) Distribution of length stretches (log λ) for increasing periods of flow (n) in a journal bearing flow. The distribution approaches a log–normal form (parabolic shape) at large n. (b) Rescaled distribution of length stretches. Data in (a) collapse to a single curve, implying self-similarity (Muzzio et al., 1991a).

zones), as well as the chaotic cavity flows discussed earlier, are composed of a sequence of shear flows with periodic random reorientation of material elements relative to the flow streamlines. In all cases, the effect of the reorientation is an exponential stretching of material elements. The interval between two successive reorientations is an important parameter of such systems, and the following argument demonstrates that an optimum interval must exist at which the total length stretch is maximum for a fixed time of mixing: In the limit of very small periods, material elements are stretched and compressed at random, and hence the average length stretch is small;

in the limit of very large time periods, the flow approaches a steady shear flow and again the length stretch is small. Khakhar and Ottino (1986a) showed a maximum in the average stretching efficiency when the strain per period was between 4 and 5 for a simple shear flow and a vortical flow with random periodic reorientation (Fig. 9). Since the average efficiency is simply the length stretch per period normalized by $(\mathbf{D}:\mathbf{D})^{1/2}$, which is proportional to the square root of the viscous dissipation, the optimum corresponds to the maximum stretching for a fixed energy input. Similar results are obtained when there is a distribution of shear rates.

Numerical analysis provides an important tool for quantitative estimates of stretching in practical mixers. The flow in most such mixers is three-dimensional and time dependent, and numerical analysis is required to obtain the velocity field; stretching is then obtained computationally from this. Avalosse and Crochet (1997a,b) obtained the velocity field for a rotating cam mixer (a 2D time periodic flow) and a Kenics static mixer (a spatially periodic duct flow) using the finite element method (Avalosse, 1993). The 2D mixer comprises two triangular cams rotating in a cavity and is similar in design to the Banbury mixer (see Fig. 34). The Kenics static mixer comprises a pipe with inserts—helically twisted tape sections joined at right angles to each other—and the mixing action is similar to the partitioned pipe mixer. In both cases, an exponential increase in length of material line is obtained. As may be expected based on studies using idealized systems (Khakhar et al., 1986) corotating cams are found to give much better mixing than counterrotating cams in the 2D mixer. Good agreement is obtained between the computed mixing patterns and experimental results for both flows. More recently, Hobbs and Muzzio (1997) carried out an analysis of the Kenics static mixer and showed that the stretching distributions produced are self-similar.

HEURISTICS

- 2D time periodic flows and spatially periodic duct flows can produce chaos. A necessary condition for chaos is crossing of streamlines.

FIG. 9. Average efficiency of stretching of material elements (e_∞) in a simple shear flow with random reorientation after an average length stretch γ_m. ρ gives the with of the distribution of length stretch about the mean value (γ_m). Results for a random distribution (top) and a normal distribution (bottom) of length stretch are shown. The maximum in the efficiency corresponds to the maximum length stretch for a fixed amount of energy dissipated and occurs at an average stretch of about 5 per period (Khakhar and Ottino, 1986a).

Importance of reorientation

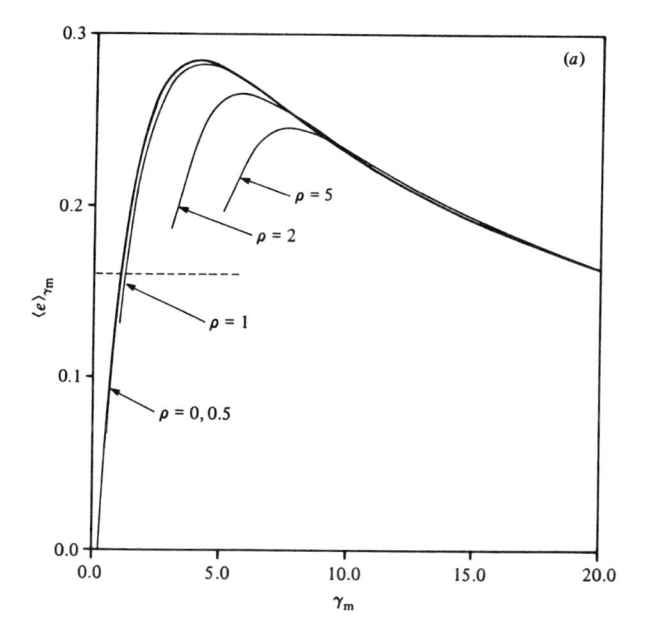

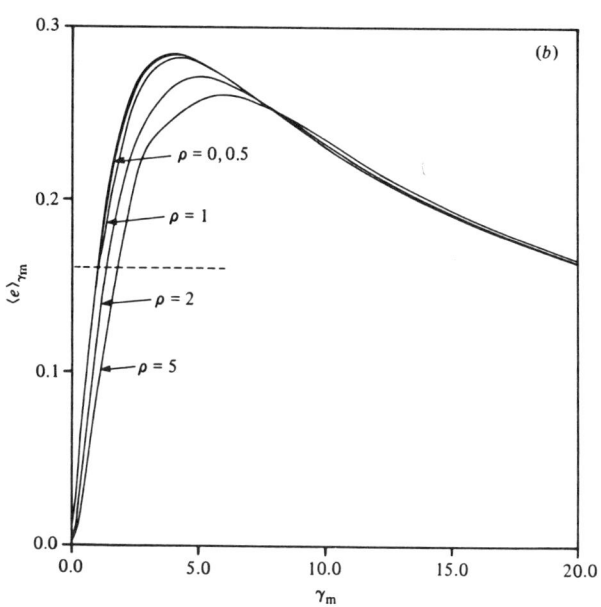

(a)

$\langle e \rangle_{\gamma_m}$ vs γ_m, curves labeled $\rho = 5$, $\rho = 2$, $\rho = 1$, $\rho = 0, 0.5$

(b)

$\langle e \rangle_{\gamma_m}$ vs γ_m, curves labeled $\rho = 0, 0.5$, $\rho = 1$, $\rho = 2$, $\rho = 5$

- In general, time periodic (or spatially periodic) flows generate islands (or tubes). Stretching in islands and tubes is linear. Stretching in chaotic regions is exponential.
- Stretchings of the order of 10^4 can be obtained in about six cycles (or reorientations). A stretch of about 4–5 per cycle seems to be optimal.
- Corotational flows distribute material better than counterrotational flows.
- Viscoelasticity (We as low as 0.06) results in substantially different patterns of mixing—larger islands and slower rates of mixing may result.
- Shear thinning typically has little effect on mixing.

II. Mixing and Dispersion of Immiscible Fluids

1. Physical Picture

The importance of viscous mixing can be justified on purely business grounds. Consider the case of polymers. The world production (by volume) of plastics has surpassed that of metals, and new polymers with extraordinary properties are constantly being produced in laboratories around the world. Less than 2% of all new polymers, however, ever find a route to commercial application. There are two reasons for this low figure. The first is the inherent high cost of producing new materials. The second is that often new properties can be obtained by compounding, blending and alloying (all synonyms for mixing) existing polymers together with additives, to produce "tailored" materials with the desired properties. Mixing provides the best route to commercial competitiveness while optimizing properties-to-price ratio. Similar arguments can be made in the consumer products industry, where imparting the "right structure" is crucial to the value of the product (Villadsen, 1997).

Quenched, possibly metastable structures are ultimately responsible for the properties observed, so the key linkage is between *mixing* and *morphology*. Thus, for example, the properties of a polymer blend—e.g., permeability, mechanical properties—are a strong function of the mixing achieved, and this in turn depends strongly on the equipment used. Thin, ribbon-like anisotropic structures are produced in single screw extruders; fine drop dispersions in static mixers and twin screw extruders. Knowledge about the mixing process can be used to produce targeted properties (see, for example, Liu and Zumbrunnen, 1997).

Experiments reveal mechanisms at work. Figure 10 shows a 45/55 blend of PS and HDPE; the processing temperature is such that the viscosities are nearly matched. Long ribbons of PS in the process of breakup by capillary instabilities, as well as coalesced regions, are apparent in this figure (Meijer *et al.*, 1988). In simple shear flow, the phenomena are somewhat different and pellets may be stretched into sheets that break by the formation and growth of holes (Sundaraj, Dori, and Macosko, 1995).

It may appear surprising that large length reductions—initial pellet sizes being on the order of a few millimeters, final drop scales on the order of microns—can be achieved in short residence times and relatively low shear rates. For example, in a typical extruder there are high-shear zones in which residence times (t_{res}) are short ($\dot{\gamma} \sim 100$ s^{-1}, $t_{res} \sim 0.1$ s), and low-shear zones with longer residence times ($\dot{\gamma} \sim 3$ s^{-1}, $t_{res} \sim 10$ s), with material elements visiting each zone several times (Janssen and Meijer, 1995). This leads to a strain of about 10 in high shear zones and 30 in low shear zones, which is considerably less than the length stretches shown in Fig. 10. The key to efficient mixing of immiscible liquids, as in the case of single fluids, lies in reorientations; six reorientations, with a stretch per reorientation of about 5, generate a stretch of 1.5×10^4.

A precondition for stretching of initially spherical drops is that hydrodynamic stresses acting on the drop be large enough to overcome surface tension that tends to return the drop to a spherical shape; stretched drops eventually break by surface-tension-driven instabilities. Complex flow in mixers results also in collisions between dispersed phase drops and, eventually, coalescence if the film between the colliding drops breaks. The dynamic balance between breakup and coalescence, both driven by the flow, determines the distribution of drop sizes and morphology in the blend. Fundamental studies of single-drop breakup and coalescence of pairs of drops, mainly for Newtonian fluids, provide a basis for the analysis of the physical processes. The non-Newtonian rheology of blends and the effects of high loading of the dispersed phase (e.g., increase in effective viscosity and phase inversion) complicate the analysis, and much remains to be done at a basic level in this regard. Also, little will be said here about the mid-column in Fig. 1. The reader interested in this topic will find leads in the papers by Kao and Mason (1975) for cohesionless aggregates, and Ulbrecht *et al.* (1982), Stroeve and Varanasi (1984), Srinivasan and Strove (1986), and Varanasi *et al.* (1994) for the case of double emulsions.

2. Mixing: From Large to Small Scales

Mixing involves a reduction of length scales. Let us now consider a typical mixing process as it progresses from large to small scales, as illustrated in

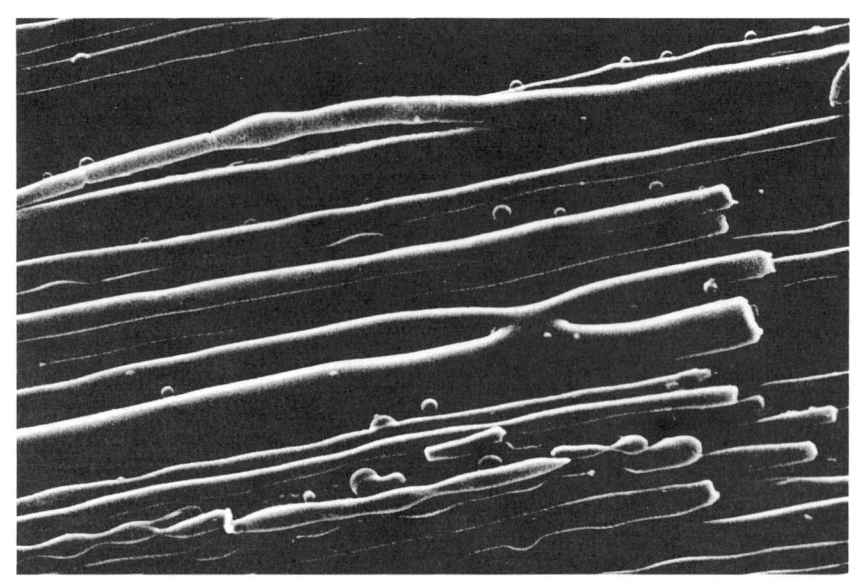

Fig. 10. Scanning electron micrograph of a fracture surface parallel to the direction of extrusion of an extrudate of a 45:55 PS–HDPE blend with a viscosity ratio $p \approx 1$. Fibrous PS is shown at different stages of breakup; the diameter of the largest fiber is about 1 μm (Meijer *et al.,* 1988).

Fig. 11. The initial condition corresponds to a large blob of the dispersed phase (d), suspended in the continuous phase (c). At the beginning of the mixing process, the capillary number, which is the ratio of the viscous forces to the interfacial forces, is large and interfacial tension is unimportant. A description of mixing essentially amounts to a description of the evolution of the interface between the two large masses of fluid: an initially designated material region of fluid (Fig. 11, top) stretches and folds throughout space. An exact description of mixing is thus given by the location of the interfaces as a function of space and time. This level of description is, however, rare because the velocity fields usually found in mixing processes are complex, and the deformation of the blobs is related in a complicated way to the velocity field. Moreover, relatively simple velocity fields can produce exponential area growth due to stretching and folding, and numerical tracking becomes impossible. Realistic problems can take years of computer time with megaflop machines (Franjione and Ottino, 1987).

The problem of following the interface for Newtonian fluids can be described by the Stokes equations,

$$\mu_i \nabla^2 \mathbf{v}_i = \nabla P_i, \quad \nabla \cdot \mathbf{v}_i = 0 \quad (i = \text{c, d}), \tag{11}$$

where c denotes the continuous phase and d denotes the dispersed phase. The boundary conditions at the interface come from a jump in the normal

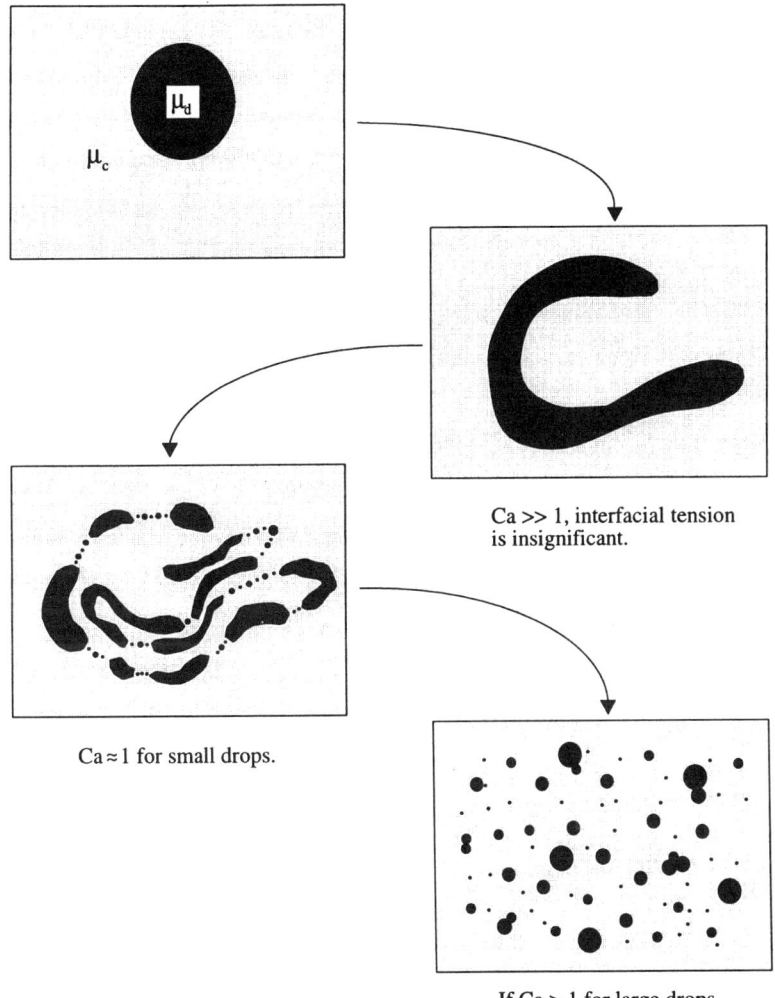

FIG. 11. Schematic view of the stages in the mixing of two immiscible viscous liquids. The large drop of the dispersed phase is stretched out and folded by the flow and breaks up into smaller droplets. Smaller drops may collide and coalesce to form larger drops.

stress due to the interfacial tension, σ, between the two fluids and the kinematic condition

$$\mathbf{n} \cdot (\mathbf{T}_c - \mathbf{T}_d) = \sigma \mathbf{n}(\boldsymbol{\nabla}_s \cdot \mathbf{n}) \quad \text{for } \mathbf{x} = \mathbf{x}_s \qquad (12)$$

$$\frac{d\mathbf{x}}{dt} = (\mathbf{v}_i \cdot \mathbf{n})\mathbf{n} \quad (i = c, d) \quad \text{for } \mathbf{x} = \mathbf{x}_s. \qquad (13)$$

In addition, the velocity field is continuous across the interface

$$\mathbf{v}_c = \mathbf{v}_d, \tag{14}$$

and boundary conditions at the system boundaries and the initial condition must be specified. Points in both the dispersed and continuous phases are denoted by the position vector $\mathbf{x}$ and points at the interface are given by $\mathbf{x}_s$. The stress tensors $\mathbf{T}_c$ and $\mathbf{T}_d$ are given by $\mathbf{T}_i = -P_i\mathbf{I} + 2\mu_i\mathbf{D}_i$, where $\mathbf{D}_i = [\nabla \mathbf{v}_i + (\nabla \mathbf{v}_i)^T]/2$. The mean curvature of the interface is given by $\nabla_s \cdot \mathbf{n}$, where the local normal $\mathbf{n}$ is directed from the dispersed phase to the continuous phase and ∇_s denotes the surface gradient.

If the length scales associated with changes in velocity are normalized by ∇_v (characteristic length scale for Stokes flow), length scales associated with changes in curvature are normalized by δ_s (typical striation thickness) and velocities normalized by V (a characteristic velocity), then the normal stress condition becomes,

$$\mathbf{n} \cdot (\mathbf{T}'_c - p\mathbf{T}'_d) = \frac{1}{\mathrm{Ca}} \mathbf{n}(\nabla'_s \cdot \mathbf{n}), \tag{15}$$

where p is the viscosity ratio, μ_d/μ_c, primed quantities denote dimensionless variables, and $\mathrm{Ca} = \mu_c V \delta_s / (\delta_v \sigma)$ may be interpreted as the ratio of viscous forces, $\mu_c V/\delta_v$ (V/δ_v is the characteristic shear rate), to capillary forces, σ/δ_s. This ratio is the so-called *capillary number*. During the initial stages of mixing Ca is relatively large because of large striation thicknesses, δ_s, although interfacial tension effects may be noticeable in regions of high curvature, such as folds. However, as the mixing process proceeds, δ_s is reduced and interfacial tension starts to play a larger role. The coupling between the flow field and interfacial tension occurs at length scales of order $\sigma/\dot{\gamma}\mu$, where $\dot{\gamma} = V/\delta_v$ is a characteristic shear rate [i.e., Ca = $O(1)$].

Illustration: Initial stages of mixing. Bigg and Middleman (1974) and Chakravarthy and Ottino (1996) approached the problem depicted in Fig. 12 by solving the governing equations (11)–(14) using a modified finite difference technique. Using this method the evolution of the interface can be followed and the length stretch of the interface and striation thicknesses can be obtained. However, this technique eventually breaks down as the interface becomes highly convoluted. Numerical difficulties quickly arise because of the large number of points that must be used to follow the interface. Alterantive methods for the numerical simulation of the early stages of mixing are presented by Chella and Viñals (1996) and Zumbrunnen *et al.* (e.g., Zhang and Zumbrunnen, 1996a,b) based on the spatial evolution of an *order parameter* (which is related to the mass fraction of the phases) due to convection and diffusion, together with fluid flow. The

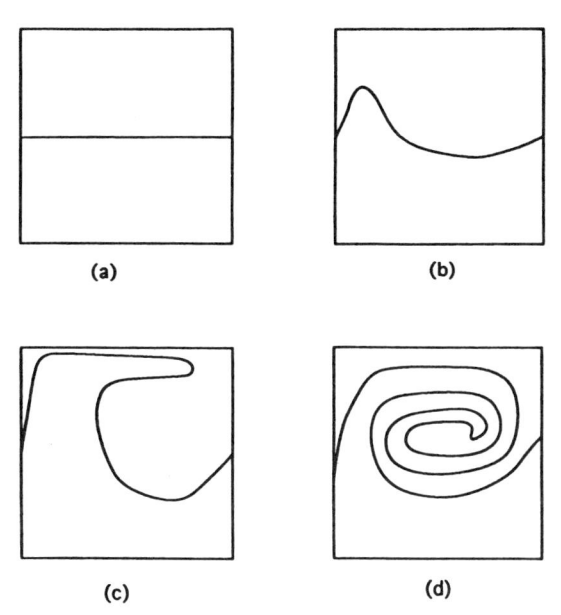

FIG. 12. Deformation of the interface between two immiscible viscous liquids in a steady cavity flow with increasing time of flow. The viscosity ratio is $p = 0.8$, and the upper fluid is of lower viscosity (Chakravarthy and Ottino, 1996).

interface between the phases is diffuse and is determined by the gradient of the order parameter. Interfacial tension forces are calculated from the order parameter field and are included in the fluid flow equations.

Illustration: Typical mixing in polymer processing. In order to illustrate the changing scales in mixing, consider a typical example in polymer processing where viscosities are on the order of 100 Pa · s and interfacial tension 0.005 N/m. Consider a typical screw with a diameter of 64 mm and a channel that can be approximated by the barrel wall moving over a cavity with dimensions 3 mm by 30 mm. If the screw rotates at 60 rpm, the relative speed of the upper wall to the screw is 0.2 m/s. A characteristic length scale for changes in velocity is around 3 mm and a typical initial value for δ_s is 1 mm. Using these typical values gives an initial capillary number of Ca ~ 1300, which clearly demonstrates that interfacial tension is unimportant in the initial stages of mixing. Interfacial tension stress does not become significant, Ca = $O(1)$, until the striation thickness is reduced to about 10^{-5}m (10 μm).

Illustration: Role of dispersed phases on flow structure. Dispersed fluid phases that have properties different from those of the continuous phase can disrupt the structure of chaotic and regular regions obtained for single phase mixing. Zhang and Zumbrunnen (1996a) studied the patterns of mixing of a tracer in a time periodic cavity flow with and without a dispersed phase blob. The capillary number for the flow with the blob is small enough so that the blob does not deform significantly in the flow. The presence of the blob disturbs the flow in a random fashion when it is placed in the chaotic region, and thus eliminates islands by breaking KAM surfaces. From a macroscopic viewpoint, dispersed phases improve mixing. Put in another way, once some breakup has been achieved, further mixing becomes more efficient.

A. Breakup

1. Small Scales

As the mixing proceeds, the capillary number decreases. At $Ca = O(1)$, interfacial tension stresses become of the same order of magnitude as viscous stresses, and the extended thread breaks into many smaller drops. Large drops, corresponding to $Ca > 1$, may stretch and break again, while smaller drops begin to collide with each other and coalesce into larger drops, which may in turn break again.

To a first approximation, the velocity field with respect to a frame fixed on the drop's center of mass, denoted $\mathbf{X}$, and far away from it, denoted by the superscript ∞, can be approximated by

$$\mathbf{v}^\infty = \mathbf{x} \cdot \mathbf{L} + \textit{higher order terms}, \quad (16)$$

where $\mathbf{L} = \mathbf{L}(\mathbf{X}, t) = (\mathbf{D}+\mathbf{\Omega})$ is a function of the fluid mechanical path of the drop, and $\mathbf{D}$ and $\mathbf{\Omega}$ are defined as $\mathbf{D} \equiv [\nabla \mathbf{v}^\infty + (\nabla \mathbf{v}^\infty)^T]/2$, and $\mathbf{\Omega} \equiv [\nabla \mathbf{v}^\infty - (\nabla \mathbf{v}^\infty)^T]/2$, respectively. The central point is to investigate the role of $\mathbf{L}$ in the stretching and breakup of the drop.

As before, the problem is governed by the creeping flow equations and boundary conditions given earlier [Eqs. (11)–(14)]. The far field boundary condition in this case is

$$\mathbf{v} \to \mathbf{v}^\infty = \mathbf{x} \cdot \mathbf{L} \quad \text{as} \quad |\mathbf{x}| \to \infty. \quad (17)$$

The tensor $\mathbf{L}$ defines the character of the flow. The capillary number for the drop deformation and breakup problem is

$$Ca = \frac{\mu_c \dot{\gamma} R}{\sigma}, \quad (18)$$

where R is the radius of the initially spherical drop, and $\dot{\gamma} = \sqrt{2\mathbf{D}:\mathbf{D}}$ is the shear rate.

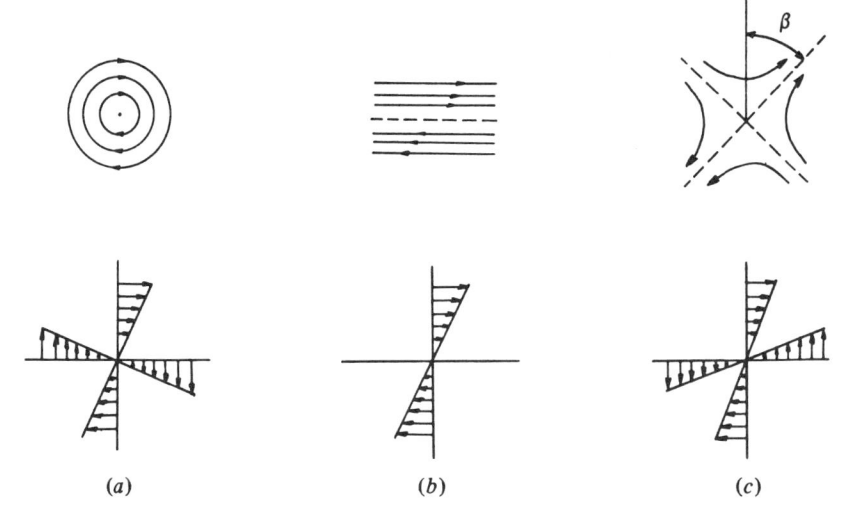

FIG. 13. Streamlines and velocity profiles for two-dimensional linear flows with varying vorticity. (a) $K = -1$: pure rotation, (b) $K = 0$: simple shear flow, (c) $K = 1$: hyperbolic extensional flow.

Illustration: Common flow types. Experimental studies of drop breakup have been mainly confined to linear, planar flows. All linear flows in 2D are encapsulated by the general velocity field equations

$$v_x = Gy, \qquad v_y = KGx, \tag{19}$$

where K determines the vorticity of the flow and G is a measure of the shear rate. Planar extensional flow corresponds to $K = 1$ (zero vorticity) and simple shear flow corresponds to $K = 0$. Streamlines for the flows with different values of K are shown in Fig. 13. Uniaxial extensional flow is another common flow type encountered, and is defined by

$$v_z = \dot{\varepsilon}z, \qquad v_x = -\frac{1}{2}\dot{\varepsilon}x, \qquad v_y = -\frac{1}{2}\dot{\varepsilon}y. \tag{20}$$

For comparison of the different flows, it is necessary to define a consistent shear rate for all the flows. A natural choice is

$$\dot{\gamma} = \sqrt{2\mathbf{D}:\mathbf{D}}. \tag{21}$$

This definition gives $\dot{\gamma} = G$ when applied to a simple shear flow (i.e., $K = 0$).

Flows may be classified as strong or weak (Giesekus, 1962; Tanner, 1976) based on their ability to stretch material elements after long times of stretching, and this characteristic can be inferred from the velocity field.

Flows that produce an exponential increase in length with time are referred to as strong flows, and this behavior results if the symmetric part of the velocity gradient tensor (**D**) has at least one positive eigenvalue. For example, 2D flows with $K > 0$ and uniaxial extensional flow are strong flows; simple shear flow ($K = 0$) and all 2D flows with $K < 0$ are weak flows.

2. Critical Capillary Number

The degree of deformation and whether or not a drop breaks is completely determined by Ca, p, the flow type, and the initial drop shape and orientation. If Ca is less than a critical value, Ca_{crit}, the initially spherical drop is deformed into a stable ellipsoid. If Ca is greater than Ca_{crit}, a stable drop shape does not exist, so the drop will be continually stretched until it breaks. For linear, steady flows, the critical capillary number, Ca_{crit}, is a function of the flow type and p. Figure 14 shows the dependence of Ca_{crit} on p for flows between elongational flow and simple shear flow. Bentley and Leal (1986) have shown that for flows with vorticity between simple shear flow and planar elongational flow, Ca_{crit} lies between the two curves in Fig. 14. The important points to be noted from Fig. 14 are these:

- For $p > 4$, stretching of a drop into a thread is impossible for shear flow.
- It is easiest to stretch drops when $p \approx 1$.
- Elongational flow is more effective than simple shear flow for a given viscosity ratio.

It is also important to note Ca_{crit} says nothing about the drop sizes produced upon breakup: The value of Ca_{crit} only gives the *maximum* drop size that can survive in a given flow in the absence of coalescence. This result may appear to suggest that the most effective dispersion—leading to the finest drop sizes—occurs when viscosities are nearly matched. As we shall see later on, this perception turns out to be incorrect. Nevertheless, an understanding of Fig. 14 constitutes the minimum level of knowledge needed to rationalize dispersion processes in complex flows.

3. Affine Deformation

For $Ca > Ca_{crit}$ a drop continually stretches until it breaks. If $Ca > \kappa Ca_{crit}$, where κ is about 2 for simple shear flow and 5 for elongational flow (Janssen, 1993), the drop undergoes *affine deformation*, i.e., the drop acts as a material element, and it is stretched into an extended cylindrical thread with length L and radius R according to

$$\text{(Simple shear flow)} \quad \frac{L}{R_o} \sim \dot{\gamma}t, \quad \frac{R}{R_o} \sim (\dot{\gamma}t)^{-1/2} \qquad (22)$$

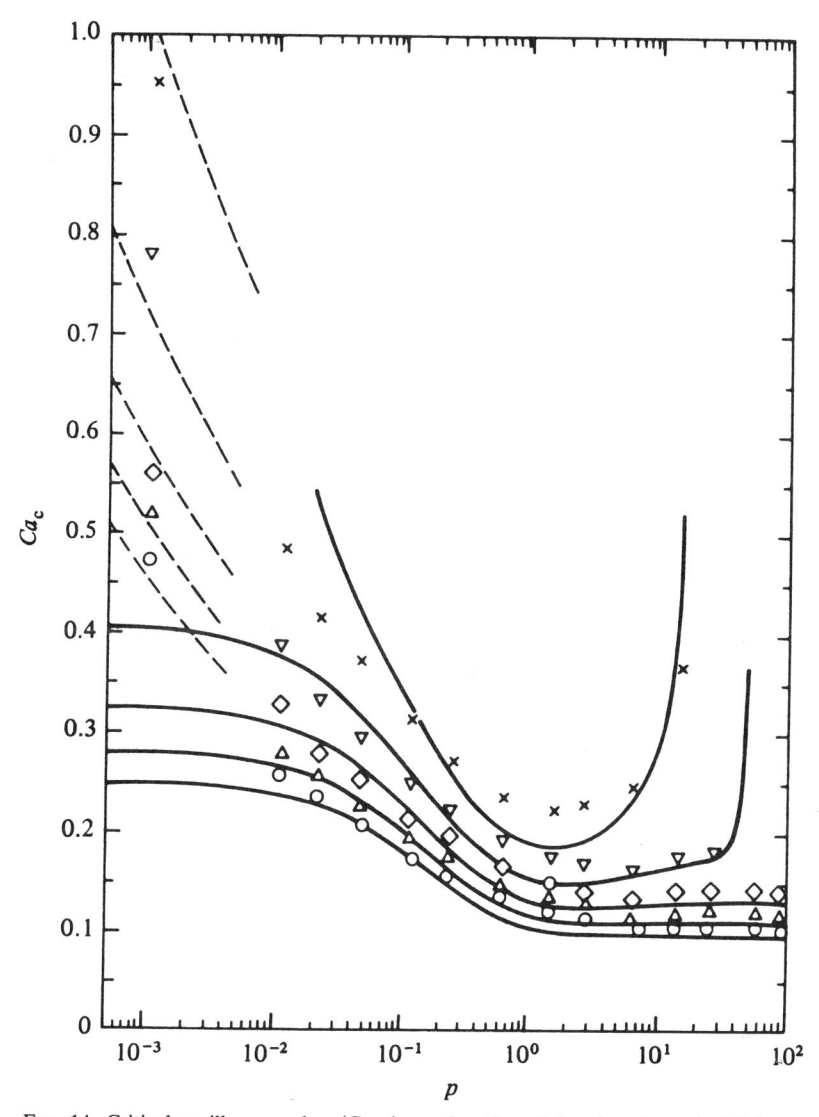

FIG. 14. Critical capillary number (Ca$_{crit}$) as a function of the viscosity ratio (p) for two-dimensional linear flows with varying vorticity (Bentley and Leal, 1986).

$$\text{(Extensional flow)} \quad \frac{L}{R_o} \sim \exp(\dot{\gamma}t), \quad \frac{R}{R_o} \sim \exp(-\dot{\gamma}t/2). \tag{23}$$

These expressions approach an exact equality as $\dot{\gamma}t$ becomes large. Figure 15 illustrates the point at which a highly stretched drop can be treated as

134   J. M. OTTINO *ET AL.*

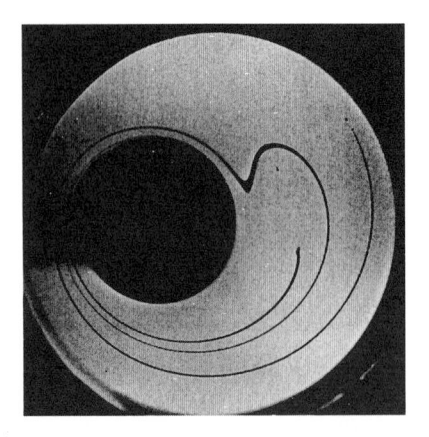

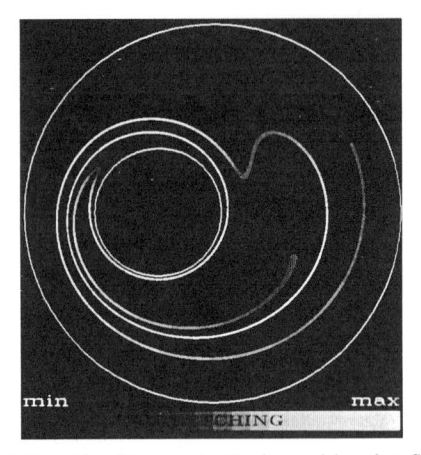

Fig. 15. Affine stretching of a filament in the journal bearing flow. Experiments (top) agree well with computations (bottom) carried out assuming that the filament deforms as the suspending fluid (*i.e.*, affine deformation) (Tjahjadi and Ottino, 1991).

a material element (i.e., it deforms affinely). This figure shows computations and experiments done by Tjahjadi and Ottino (1991) where a drop of fluid is dispersed in a second fluid in the journal bearing flow (Swanson and Ottino 1990, Chaiken et al. 1986). In the computations, the drop was treated as a material element, and as can be seen, the agreement between the computations and experiments is quite good. A necessary condition for stretching is that Ca must surpass Ca_{crit} as illustrated in Fig. 16: One drop does not reach the critical Ca and remains underformed; the other breaks into thousands of drops.

Illustration: Importance of reorientations. The stretching rate of long filaments in shear flow can be improved from being linear to exponential

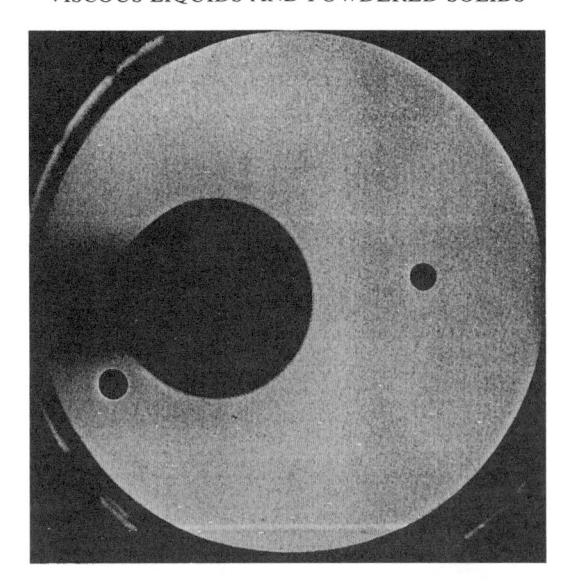

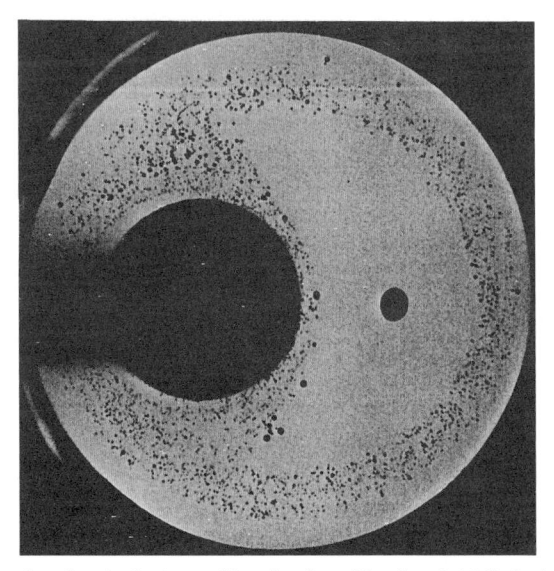

FIG. 16. Drop breakup in the journal bearing flow. The drop initially in the chaotic region of the flow deforms into a thin filament that breaks to produce a fine dispersion of droplets. The drop initially in the regular region of the flow (island) remains undeformed (Tjahjadi and Ottino, 1991).

by incorporating periodic reorientations in the flow, as seen earlier for material lines. The basic idea is instead of using one long shear flow to divide the flow into shorter sections with reorientations between each section. When this is done the amount of stretching is given by (Erwin, 1978)

$$\frac{L}{R_o} \sim \left(\frac{\gamma_{tot}}{n+1}\right)^{(n+1)}, \quad \frac{R}{R_o} \sim \left(\frac{\gamma_{tot}}{n+1}\right)^{-(n+1)/2}, \tag{24}$$

where γ_{tot} is the *total* shear in the mixer. The improvement with reorientations is illustrated in Fig. 17, where R_o/R is plotted vs time for the cases

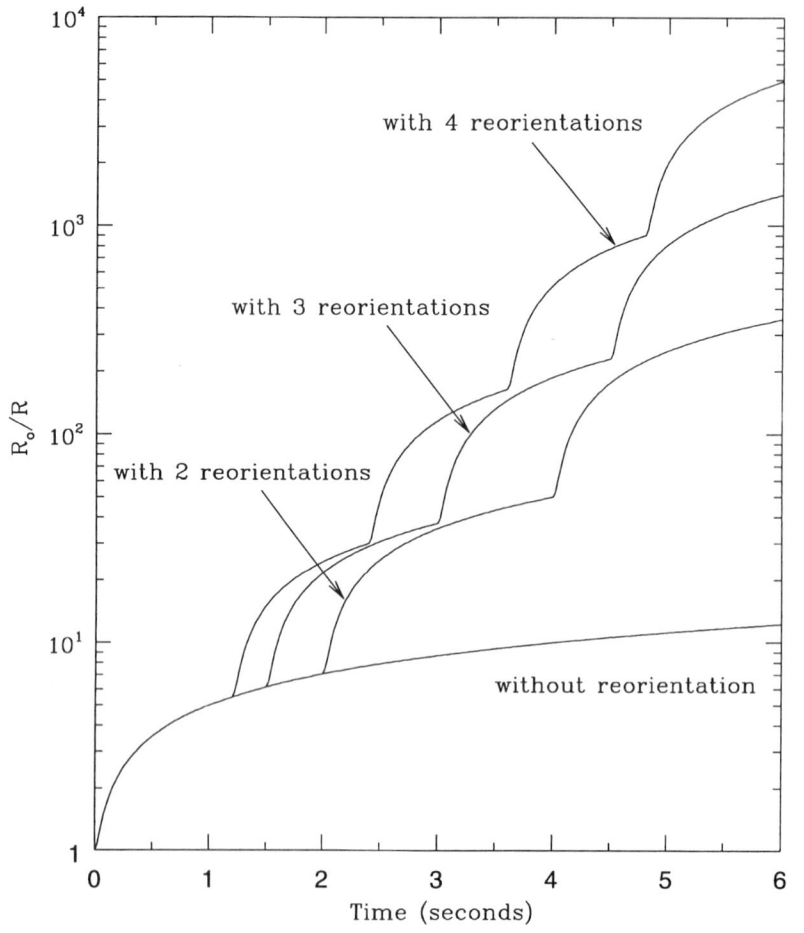

FIG. 17. Stretching of a filament in a simple shear flow with random reorientation at different time intervals. The length stretch is shown vs time for $\dot{\gamma} = 50 \text{ s}^{-1}$.

of 0, 2, 3, and 4 reorientations. After a total shear of 300 ($\dot\gamma = 50$ s^{-1} for 6 s), the length scale is reduced from 10^{-3} m to 6×10^{-5} m. If a reorientation is added after every 2 s, the length scale is reduced from 10^{-3} to 10^{-6} m (a typical length scale reduction in polymer processing). To get this same reduction without reorientation would take approximately 5.5 h. Four or five reorientations are typically enough.

Although the preceding equations illustrate clearly the role of reorientation in stretching, it should be noted that the equations are valid only when the strain per period is large ($\gamma_{tot}/(n+1) > 1$). The length of the filament after n reorientations is in fact given by

$$\frac{L}{R_o} = \prod_{i=1}^{n+1}\left\{\left(\cos\theta_i + \frac{\gamma_{tot}}{n+1}\right)^2 + \sin^2\theta_i\right\}^{1/2}, \quad (25)$$

where θ_i is the angle between the filament and the streamlines at the start of the ith period (Khakhar and Ottino, 1986a). For large $\gamma_{tot}/(n+1)$ Eq. (25) reduces to Eq. (24). However, when this condition is not satisfied, actual strains may be less than or greater than unity depending on the orientation of the material element (θ_i). Consequently, Eq. (24) would not give an accurate estimate of the optimum number of reorientations when the length stretch is maximum.

Illustration: Stretching of low-viscosity-ratio elongated drops. For the case $p \ll 1$ and Ca/Ca$_{crit}$ = $O(1)$, the dynamics of a nearly axisymmetric drop with pointed ends, characterized by an orientation **m** ($|\mathbf{m}| = 1$) and a length $L(t)$, is given by (Khakhar and Ottino, 1986b,c)

$$\frac{1}{L}\frac{dL}{dt} = \mathbf{D}:\mathbf{mm} - \underline{\left(\frac{\sigma}{2(5)^{1/2}\mu_c R}\right)\left(\frac{(L(t)/R)^{1/2}}{1 + 0.8p(L(t)/R)^3}\right)} \quad (26a)$$

$$\frac{d\mathbf{m}}{dt} = (\mathbf{GD} + \mathbf{\Omega})^T \cdot \mathbf{m} - (\mathbf{GD}:\mathbf{mm})\mathbf{m}, \quad (26b)$$

where $G(t) = (1 + 12.5R^3/L(t)^3)/(1 - 2.5R^3/L(t)^3)$. The underlined term in the first equation acts as a resistance to the deformation [contrast Eq. (26a) with Eq. (4) for stretching of a material element]. A very long drop, $[L(t)/a] \to \infty$, $G \to 1$, rotates and stretches as a passive element since the resistance to stretching becomes negligible. Note also that since $G > 1$ the droplet "feels" a flow that is slightly more extensional than the actual flow. The preceding equation is a special case of the linear vector model (Olbricht et al., 1982), which describes the dynamics of deformation of an arbitrary microstructure that is specified by its length and orientation. We shall show later in this paper that the fragmentation and separation of agglomerates is also described by an equation very similar to the preceding equation.

In the context of the preceding model, a drop is said to break when it undergoes infinite extension and surface tension forces are unable to balance the viscous stresses. Consider breakup in flows with **D : mm** constant in time (for example, an axisymmetric extensional flow with the drop axis initially coincident with the maximum direction of stretching). Rearranging Eq. (26) and defining a characteristic length $R/p^{1/3}$, we obtain the condition, for a drop in equilibrium,

$$\frac{\mathbf{D:mm}}{(\mathbf{D:D})^{1/2}} = e_\lambda = \left(\frac{1}{2(5)^{1/2}E}\right)\left(\frac{L_s^{1/2}}{1 + 0.8L_s^3}\right), \tag{27}$$

where L_s denotes the steady-state length and $E = p^{1/6}Ca$. A graphical interpretation of the roots L_s is given in Fig. 18. The horizontal line represents the asymptotic value of the efficiency (i.e., corresponding to $d\mathbf{m}/dt = 0$), which in three dimensions is $(2/3)^{1/2}$, and the value of the resistance is a function of the drop length for various values of the dimensionless strain rate E. For $E < E_c$ there are two steady states: one stable and the other unstable. For $E > E_c$ there are no steady states and the drop extends indefinitely.

4. Breakup Modes

Once a drop is subjected to a flow for which $Ca > Ca_{crit}$, it stretches and breaks depending on the degree of deformation, the viscosity ratio

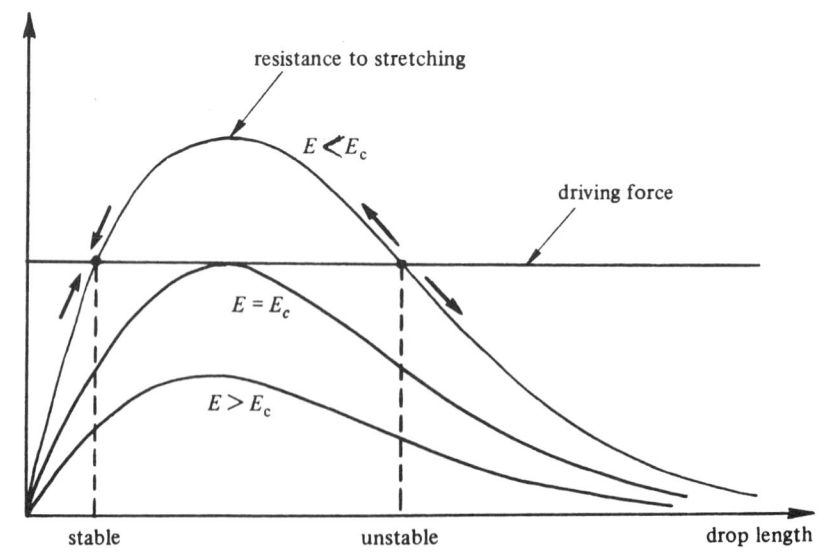

FIG. 18. Graphical interpretation of the criterion for breakup of pointed drops in an extensional flow (Khakhar and Ottino, 1986c).

and the flow type (see, for example, Fig. 16). When a drop breaks, it does so by one of the four mechanisms illustrated in Fig. 19 (Stone, 1994). These four mechanisms will be briefly discussed here. For more details, see reviews by Eggers (1997), Stone (1994), Rallison (1984), and Acrivos (1983).

Moderately extended drops ($L/R_0 \lesssim 15$, where R_0 is the radius of a spherical drop of the same volume) break by a *necking* mechanism (Rumscheidt and Mason, 1961). In this type of breakup, the two ends of the drop form bulbous ends and a neck develops between them. The neck continuously thins until it breaks, leaving behind a few smaller drops between two large drops formed from the bulbous ends. This necking mechanism generally occurs during a sustained flow where Ca is relatively close to Ca_{crit}. Little has appeared in the literature on the number and size of drops formed upon breakage by necking. One general observation is that the number of drops produced is less than 10 (Grace, 1982).

The necking mechanism has also been investigated using theoretical and numerical techniques. The theoretical approach, based on small deformation analysis (Barthès-Biesel and Acrivos, 1973) for the case of low Ca or high p shows the formation of lobes on the drop for $Ca > Ca_{crit}$. Numerical techniques (Rallison, 1981) for $p = 1$ give similar results. The general conclusion is confirmation of the experimentally determined curve for Ca_{crit}; the drops in this case may break up rather than extend indefinitely.

Tipstreaming, in which small drops break off from the tips of moderately extended, pointed drops, is another mechanism for drop breakup, though not of much significance to the dispersion process. Tipstreaming is generally attributed to gradients in interfacial tension along the surface of the drop (de Bruijn, 1993), but the exact conditions at which tipstreaming occurs are not well known.

Relaxation of a moderately extended drop under the influence of surface tension forces when the shear rate is low may lead to breakup by the *end-pinching* mechanism (Stone *et al.*, 1986; Stone and Leal, 1989). For example, this type of breakup occurs if a drop is deformed past a critical elongation ratio, $(L/R_o)_{crit}$, at Ca close to Ca_{crit} and then the flow is stopped abruptly. The critical elongation ratio necessary for breakup once the flow is stopped is dependent on p with a minimum deformation around $p \approx 0.2$–2, but is nearly independent of flow type (Stone *et al.*, 1986). Smaller-than-critical elongations result in relaxation of the drop to a spherical shape. The relaxation of the drop in this case is driven by the surface tension stress generated by the ends of the extended drop. The high viscosities and low surface tensions encountered in polymer processing lead to high relaxation times for the drop; hence, breakup by end pinching may not occur to any significant effect in such systems. However, this mechanism would be the dominant mode of breakup in the emulsification of low-viscosity liquids.

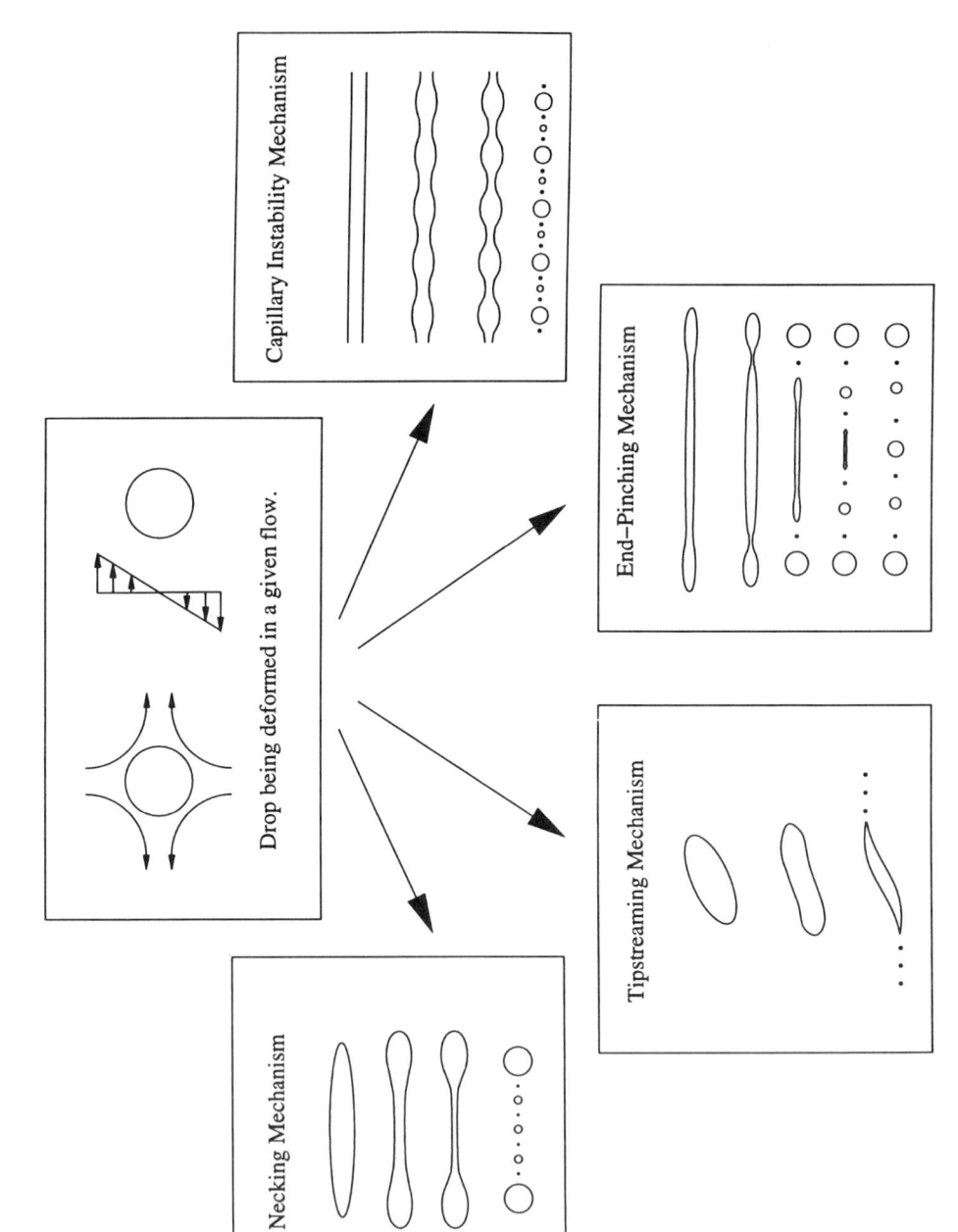

The three breakup mechanisms previously discussed occur for moderately extended drops; however, when Ca ≳ κCa$_{crit}$, the drop is stretched affinely and becomes a highly extended thread. The extended thread is unstable to minor disturbances and will eventually disintegrate into a number of large drops with satellite drops between the larger mother drops (see Fig. 19). The driving force behind this process is provided by interfacial tension minimizing the surface area: All sinusoidal disturbances cause a decrease in surface area. However, only disturbances with a wavelength greater than the filament perimeter produce pressure variations along the filament (due to the normal stress boundary condition) that magnify the disturbance and lead to breakup. The analysis for how these disturbances grow depends on whether the thread is at rest or being stretched. Each of these two cases is considered next.

For the case of a thread at rest, the initial growth of a disturbance can be relatively well characterized by linear stability theory. In the initial stages, the deformation of the thread follows the growth of the fastest growing disturbance (Tomotika, 1935). Eventually the interfacial tension driven flow becomes nonlinear, leading to the formation of the smaller satellite drops (Tjahjadi *et al.*, 1992).

Although linear stability theory does not predict the correct number and size of drops, the time for breakup is reasonably estimated by the time for the amplitude of the fastest growing disturbance to become equal to the average radius (Tomotika, 1935):

$$t_{\text{break}} \approx \frac{2\mu_c R_o}{\sigma \Omega_m} \ln\left(\frac{(2/3)^{1/2} R_o}{\alpha_o}\right). \tag{28}$$

The nondimensional growth rate, Ω_m, is a unique function of the wavenumber and p. Kuhn (1953) estimated the magnitude of the initial amplitude of the disturbances (α_o) to be 10^{-9} m based on thermal fluctuations. Mikami *et al.* (1975) gave a higher estimate of 10^{-8} to 10^{-7} m.

For the case of a thread breaking during flow, the analysis is complicated because the wavelength of each disturbance is stretched along with the thread. This causes the dominant disturbance to change over time, which results in a delay of actual breakup. Tomotika (1936) and Mikami *et al.* (1975) analyzed breakup of threads during flow for 3D extensional flow, and Khakhar and Ottino (1987) extended the analysis to general linear flows. Each of these works uses a perturbation analysis to describe an equation for the evolution of a disturbance.

In general, disturbances will damp, then grow, and then damp again as the wavelength of a particular disturbance increases due to stretching of the thread. However, disturbances cannot damp below the initial amplitude,

142   J. M. OTTINO ET AL.

α_o, caused by thermal fluctuations. Hence, the initial damping stage is omitted and once the thread reaches a critical radius, R_{crit}, the disturbance starts to grow from α_o. If the amplitude of a disturbance reaches the average size of the thread, disintegration into drops occurs. Disturbances grow and damp at different rates depending on their initial wavenumber x_o. The disturbance that reaches the amplitude of the average thread radius first is the dominant disturbance and causes breakup.

Mikami et al. (1975) and Khakhar and Ottino (1987) presented a numerical scheme for determining t_{grow}, which is the time for the dominant disturbance to grow from α_o to an amplitude equal to the average thread radius. The total breakup time, t_{break}, is the sum of t_{grow} and t_{crit}, where t_{crit} is the time to reach R_{crit} from R_o. The value of R_{crit} is also obtained from the numerical scheme for calculating t_{grow}. Tjahjadi and Ottino (1991) used this numerical scheme and fitted the results to the expression

$$R_{crit} \approx (37.8 \pm 3.8)10^{-4} e^{-0.89} p^{-0.44} \left(\frac{\sigma}{\mu_c G}\right)^\chi, \tag{29}$$

where $0.84 < \chi < 0.92$ for $10^{-3} \leq p \leq 10^2$ and R_{crit} is in cm. Janssen and Meijer (1993) took the approach of Tjahjadi and Ottino (1991) one step further and reduced all the results to a graphical representation of R_{drops}, R_{crit}, and t_{grow} that only depends on the dimensionless parameters p and $\mu_c \epsilon \alpha_0/\sigma$. The Illustration following this section indicates how to calculate t_{break} using these graphs.

Although the area of breakup can be divided into four distinct categories—necking, tipstreaming, end-pinching, and capillary instabilities—more than one mechanism may be present in a given flow. Stone et al. (1986) and Stone and Leal (1989) demonstrated that if a drop is stretched enough, both end-pinching and capillary instabilities will be present. The end-pinching mechanism dominates the breakup close to the ends and the capillary instabilities dominate the breakup of the drop toward the middle. An examination of the pictures of Tjahjadi and Ottino (1991) shows the presence of three breakup mechanisms, necking, end-pinching, and capillary instabilities, on different portions of the same extended thread; however, capillary instabilities dominate the breakup.

Illustration: Breakup time for threads during flow. Consider a thread being deformed in a 2D extensional flow, with the following material and process parameters:

$$\dot{\varepsilon} = 30\,\text{s}^{-1}, \qquad \sigma = 5 \times 10^{-3}\,\text{N/m}$$
$$\mu_c = 100\,\text{Pa}\cdot\text{s}, \qquad L_o = 1 \times 10^{-2}\,\text{m}$$
$$\mu_d = 100\,\text{Pa}\cdot\text{s}, \qquad R_o = 1 \times 10^{-5}\,\text{m}.$$

Janssen and Meijer (1993) have shown the calculation of the time for breakup of a thread during flow can be reduced to the graphs in Fig. 20, which only depend on the viscosity ratio, p = 1, and the parameter $\mu_c \dot{\varepsilon} \alpha_o / \sigma = 6 \times 10^{-4}$ using $\alpha_o = 1 \times 10^{-9}$m (Kuhn, 1953). Using these two parameters and Fig. 20, the following values are obtained:

$$R_{crit}/\alpha_o = 50, \quad R_{drops/\alpha o} = 50, \quad e_f t^*_{grow} = 1.5.$$

Here, $e_f = \sqrt{2}\dot{\varepsilon}/\dot{\gamma}$ and $t_{grow}* = t_{grow} \dot{\gamma}$. R_{crit} is the radius of the thread at which the critical disturbance begins to grow, R_{drops} is the size of the drops produced once the thread breaks, and t_{grow} is the time for the disturbance to grow from its initial amplitude to half the average radius at which breakup occurs. For $\alpha_o = 1 \times 10^{-9}$m, we obtain $R_{crit} = 5 \times 10^{-8}$m, $R_{drops} = 5 \times 10^{-8}$m, and $t_{grow} = 0.04$s. The time to reach R_{crit}, t_{crit}, is found from the equation for stretching in a 2D exponential flow:

$$\frac{R_{crit}}{R_o} = \exp(-\dot{\varepsilon}t) \rightarrow t_{crit} = 0.18 \text{ s}$$

$$t_{break} = t_{crit} + t_{grow} = 0.22 \text{ s}.$$

Illustration: Satellite formation in capillary breakup. The distribution of drops produced upon disintegration of a thread at rest is a unique function of the viscosity ratio. Tjahjadi et al. (1992) showed through inspection of experiments and numerical simulations that up to 19 satellite drops between the two larger mother drops could be formed. The number of satellite drops decreased as the viscosity ratio was increased. In low-viscosity systems [$p < O(0.1)$] the breakup mechanism is self-repeating: Every pinch-off results in the formation of a rounded surface and a conical one; the conical surface then becomes bulbous and a neck forms near the end, which again pinches off and the process repeats (Fig. 21). There is excellent agreement between numerical simulations and the experimental results (Fig. 21).

Illustration: Comparison between necking and capillary breakup. Consider drops of different sizes in a mixture exposed to a 2D extensional flow. The mode of breakup depends on the drop sizes. Large drops ($R \gg \text{Ca}_{crit}\sigma/\mu_c\dot{\gamma}$) are stretched into long threads by the flow and undergo capillary breakup, while smaller drops ($R \approx \text{Ca}_{crit}\sigma/\mu_c\dot{\gamma}$) experience breakup by necking. As a limit case, we consider necking to result in binary breakup, i.e., two daughter droplets and no satellite droplets are produced on breakup. The drop size of the daughter droplets is then

$$R \approx \frac{\text{Ca}_{crit}\sigma}{\mu_c \dot{\gamma}} 2^{-1/3}. \tag{30}$$

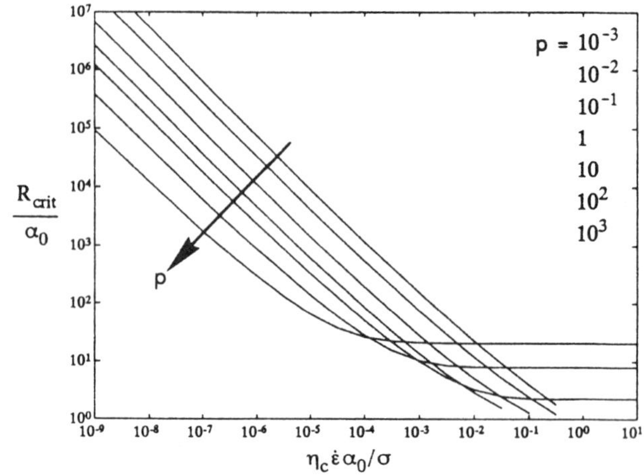

$p = 10^{-3}$
10^{-2}
10^{-1}
1
10
10^2
10^3
R_{crit}/α_0
p
$\eta_c \dot{\varepsilon} \alpha_0/\sigma$

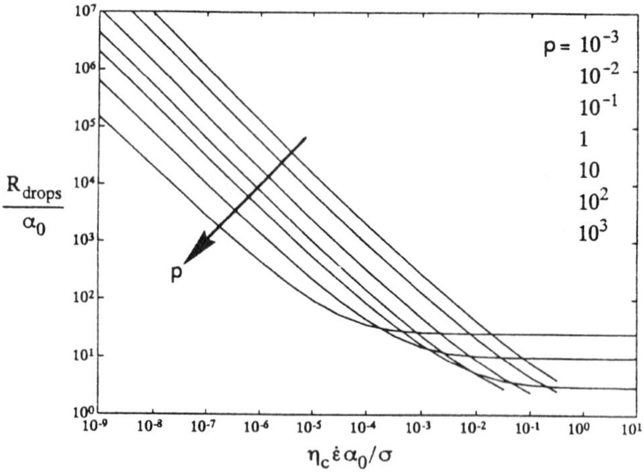

$p = 10^{-3}$
10^{-2}
10^{-1}
1
10
10^2
10^3
R_{drops}/α_0
p
$\eta_c \dot{\varepsilon} \alpha_0/\sigma$

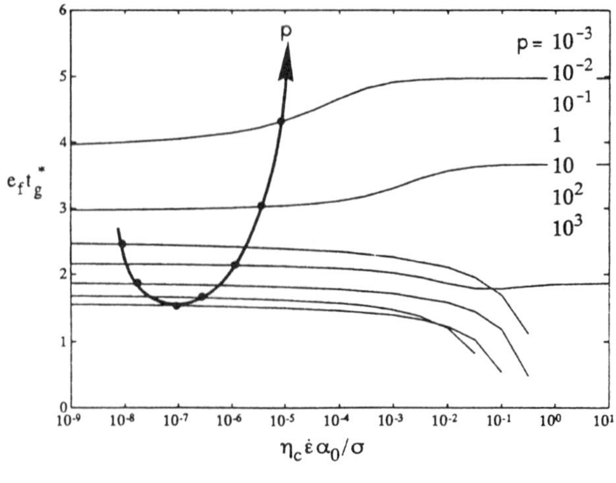

$p = 10^{-3}$
10^{-2}
10^{-1}
1
10
10^2
10^3
$e_f t_g^*$
p
$\eta_c \dot{\varepsilon} \alpha_0/\sigma$

The radius of drops produced by capillary breakup is independent of the initial drop size, and is determined essentially by the viscosity ratio. Figure 22 shows a comparison of the drop size produced on breakup by the two different mechanisms (Janssen and Meijer, 1993). The size of the daughter drops produced by capillary breakup is significantly smaller than that for binary breakup for the case of high viscosity ratios. However, in practical situations in which coalescence and breakup occur during mixing, a range of drop sizes would exist, and thus the binary breakup radius gives an upper limit for the drop size.

Illustration: Drop size distributions produced by chaotic flows. Affinely deformed drops generate long filaments with a stretching distribution based on the log–normal distribution. The amount of stretching (λ) determines the radius of the filament locally as

$$R \sim \frac{1}{\lambda^{1/2}}. \tag{31}$$

Upon breakup, the filament breaks into a set of *primary* or mother drops whose sizes are, to a first approximation, proportional to R. The size of drops produced when the filament breaks can then be obtained from the distribution of R. Each mother drop produced upon breakup carries a distribution of satellites of diminishing size; for example, each mother drop of radius r has associated with it one large satellite of radius $r^{(1)}$, two smaller satellites of radius $r^{(2)}$, four satellites of radius $r^{(3)}$, and so on. For breakup at rest, the distribution of smaller drops is a unique function of the viscosity ratio.

If we assume that drops break only once, the drop-size distribution can be predicted based on the log–normal stretching distributions typical of mixing flows. Assuming a mean stretch of 10^4 gives a range of stretching from 1 to 10^{10} (Muzzio *et al.*, 1991a). Therefore, the corresponding distribution of mother drops is log–normal with a range from R_o to $10^{-5}R_o$, where R_o is the radius of the initial drop. The distribution for each satellite drop will be the same as that of the mother drop, except the position of the distribution will be shifted by a factor of r/r^i and the amplitude multiplied by the number of satellite drops per mother drop. When all of the individual

---

FIG. 20. Radius of drops produced on capillary breakup in hyperbolic extensional flow (R_{drops}), radius of the thread at which the disturbance that causes breakup begins to grow (R_{crit}), and the time for growth of the disturbance (t_{grow}) for different values of the dimensionless parameters p and $\mu_c \varepsilon \alpha_0 / \sigma$. The time for capillary breakup of the extending thread (t_{break}) can be obtained from these graphs (see Illustration for sample calculations) (Janssen and Meijer, 1993).

146 J. M. OTTINO ET AL.

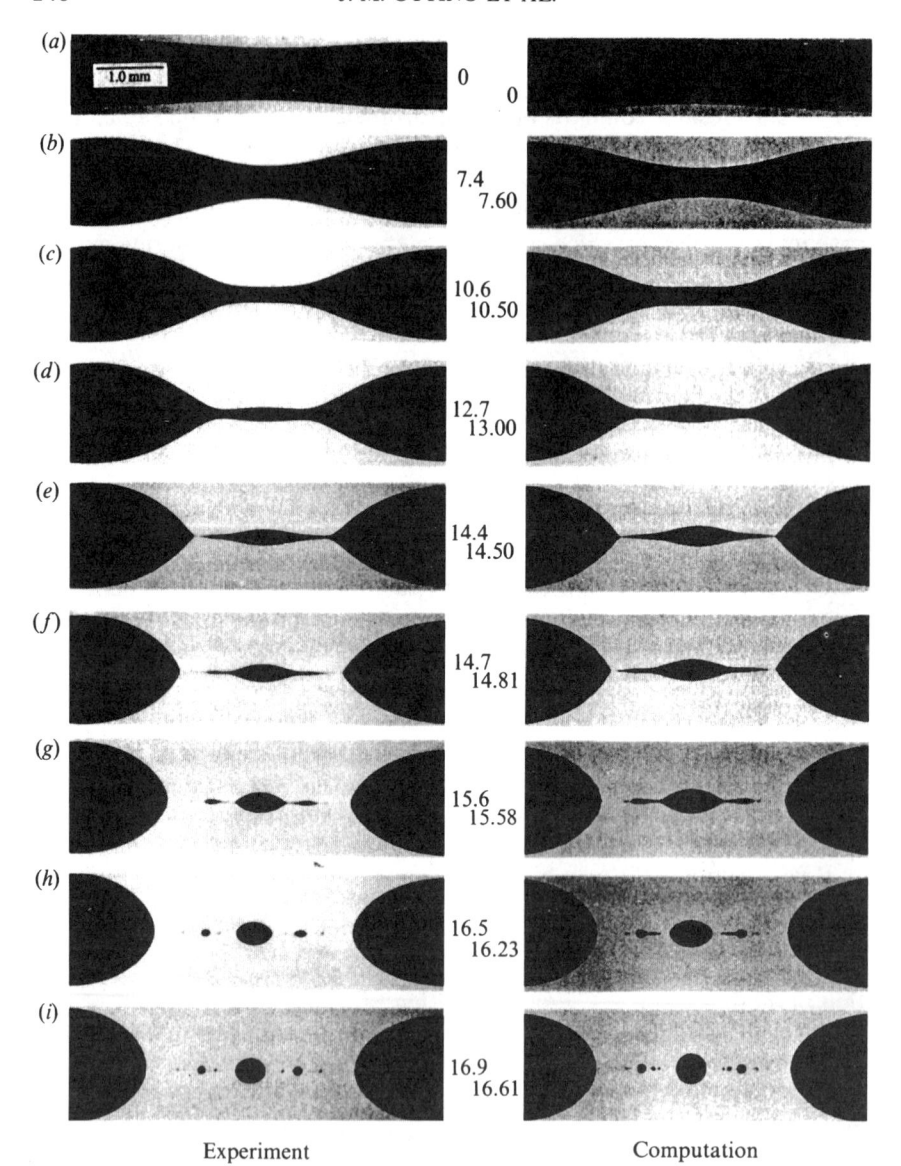

FIG. 21. Formation of satellite drops during the breakup of a filament at rest. A comparison between computations and experimental results is shown (Tjahjadi, Stone, and Ottino, 1992). Numbers refer to dimensionless times with $t = 0$ corresponding to Fig (a).

VISCOUS LIQUIDS AND POWDERED SOLIDS 147

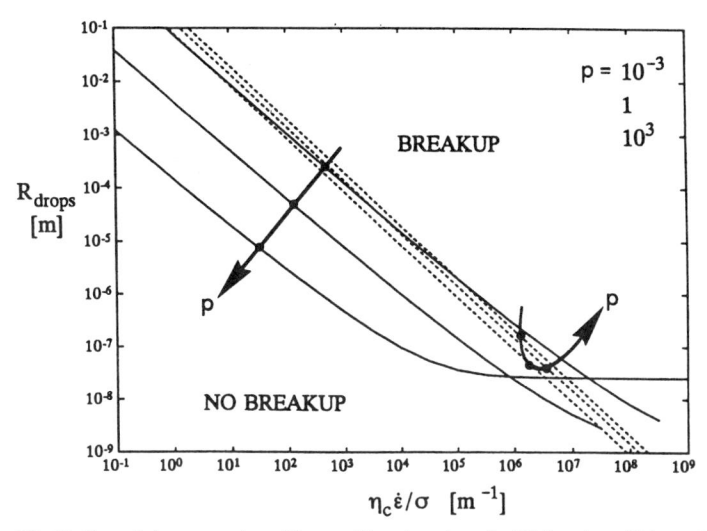

FIG. 22. Radius of drops produced by capillary breakup (solid lines) and binary breakup (dotted lines) in a hyperbolic extensional flow for different viscosity ratios (p) and scaled shear rate ($\mu_c \dot{\gamma}/\sigma$) (Janssen and Meijer, 1993). The initial amplitude of the surface disturbances is $\alpha_0 = 10^{-9}$ m. Note that significantly smaller drops are produced by capillary breakup for high viscosity ratios.

distributions are added, to give the overall drop distribution, a log–normal distribution is approached, as illustrated in Fig. 23.

The preceding example, as well as the repetitive nature of stretching and breakup, suggests that the experimentally determined equilibrium drop size distribution $f(V, p)$ might be self-similar [$f(V, p)dV$ is the number of drops with sizes between V and $V + dV$]. Figure 24 shows the distribution of drops obtained experimentally in the journal bearing flow resulting from repeated stretching and breakup. Figure 25 shows that rescaled experimental drop-size distributions $V^2 f(V, p)$ vs $V/\langle V \rangle$, where $\langle V \rangle$ is the arithmetic mean of volume sizes, collapses all data into a single curve. (Muzzio et al. 1991b).

Illustration: Effect of viscosity ratio on drop size distributions. Experiments show that the equilibrium size distributions corresponding to high viscosity ratio drops are more nonuniform than those corresponding to low viscosity ratios, and that, in general, the mean drop size decreases as the viscosity ratio p increases (Tjahjadi and Ottino, 1991). The experiments pertain to the case of low number densities of drops when coalescence is negligible. There seem to be two distinct mechanisms: one-time breakup and repeated breakup. Low viscosity drops stretch passively but extend relatively little before they break, resulting in the formation of large droplets; these droplets undergo subsequent stretching, folding, and breakup.

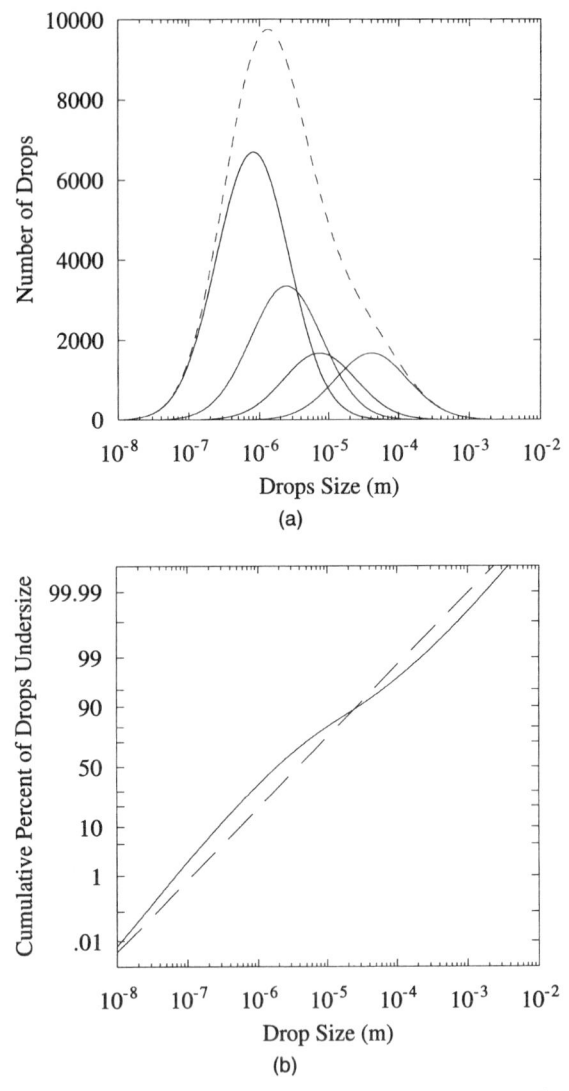

FIG. 23. (a) Distribution of drop sizes for mother droplets and satellite droplets (solid lines) produced during the breakup of a filament (average size = 2×10^{-5} m) in a chaotic flow. The total distribution is also shown (dashed line). A log-normal distribution of stretching with a mean stretch of 10^{-4} was used. (b) The cumulative distribution of mother droplets and satellite droplets (solid line) approaches a log-normal distribution (dashed line).

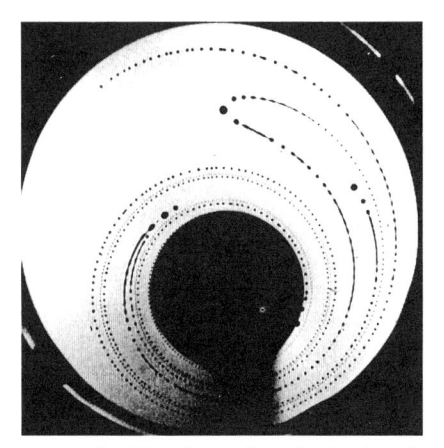

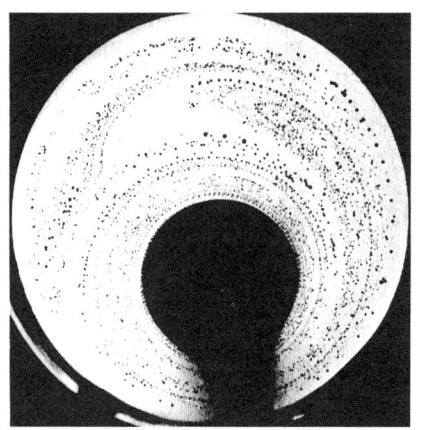

FIG. 24. Experimental results for drop breakup in a journal bearing flow. The Figure on the right shows repeated stretching and breakup, which is observed for low-viscosity drops (Tjahjadi and Ottino, 1991).

Under identical conditions, high-viscosity-ratio drops stretch substantially, $O(10^3-10^4)$, before they break producing very small fragments; these small fragments rarely break again. It is apparent, contrary to a long-held belief, that the finest dispersion does not correspond to $p = 1$.

HEURISTICS

- *Necking:*

Breakup by necking occurs in sustained flows when Ca is close to Ca_{crit}.
The number of drops produced upon breakup by necking is generally less than 10.
The time for breakup by necking increases with p.

- *End-Pinching:*

Breakup by end-pinching occurs when a drop is deformed at Ca close to Ca_{crit}, and the flow is stopped abruptly.
Breakup by end-pinching is most difficult for very viscous or inviscid drops.

- *Capillary Instabilities:*

Breakup due to capillary instabilities dominates when the length of the filament is more than 15 times the initial radius of the drop.
The size of daughter droplets produced is independent of the initial drop size.

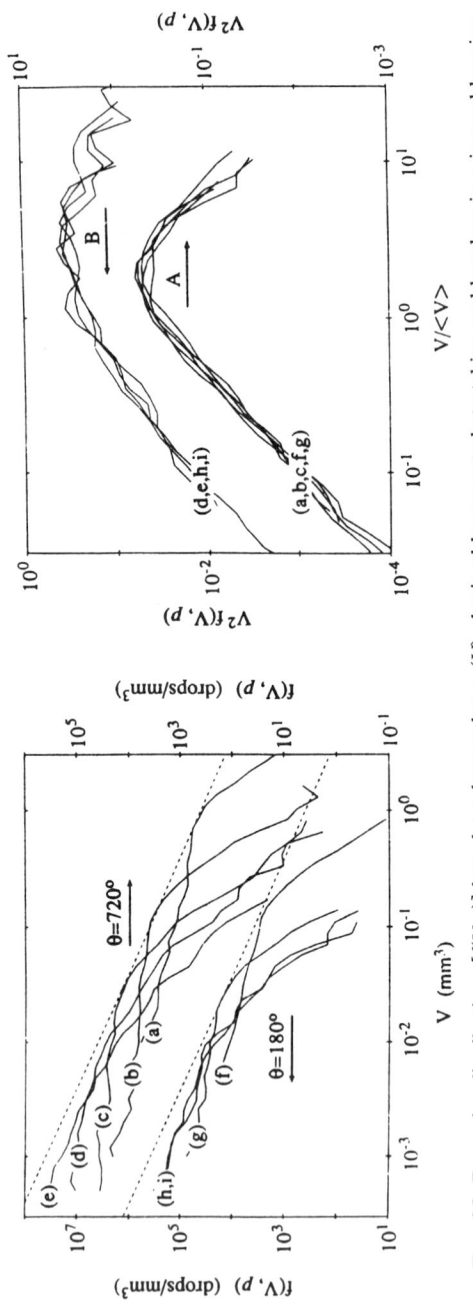

FIG. 25. Drop size distributions [$f(V,p)$] based on drop volume (V) obtained by repeated stretching and breakup in a journal bearing flow for different viscosity ratios (p) (left). The curves for the different distributions overlap when the distribution is rescaled (right) (Muzzio, Tjahjadi, and Ottino, 1991).

The number of satellite drops produced upon breakup by capillary instabilities decreases as p increases (minimum of 3 to maximum of 16).

Breakup during flow is delayed relative to breakup at rest, so in general the easiest way to break an extended thread is to stop the flow.

Viscosity ratio $p = 1$ does not produce the finest dispersion; average drop sizes decrease with viscosity ratio.

B. Coalescence

1. Collisions

As dispersion proceeds drops come into close contact with each other and may coalesce. Coalescence is commonly divided into three sequential steps (Chesters, 1991): "collision" or close approach of two droplets, drainage of the liquid between the two drops, and rupture of the film (see Fig. 26).

The collision frequency between drops may be estimated by means of Smoluchowski's theory (see, for example, Levich, 1962). The collision frequency ($\omega \equiv$ number of collisions per unit time per unit volume) for randomly distributed rigid equal-size spheres, occupying a volume fraction ϕ, is given by

$$\omega \approx \frac{4\dot{\gamma}\phi n}{\pi}, \tag{32}$$

where n is the number density of drops (number per unit volume). Because hydrodynamic interactions are neglected, this equation gives at best an order-of-magnitude estimate for the collision frequency. However, it is important to note that the result is independent of flow type if $\dot{\gamma}$ is interpreted as $\sqrt{2\mathbf{D}:\mathbf{D}}$ (see following Illustration). The collision rate for small drops considering hydrodynamic interactions is given in Wang, Zinchenko, and Davis (1994). A thorough analysis of coagulation in the presence of hydrodynamic interactions, interparticle forces, and Brownian diffusion in random velocity fields is given by Brunk *et al.* (1997).

Illustration: Effect of flow type on shear induced collisions in homogenous linear flows. The collision frequency for a general linear flow [Eq. (15)] is obtained following Smoluchowski's (1917) approach as (Bidkar and Khakhar, 1990)

$$\omega = 2d^3 n \int_0^{2\pi} d\phi \int_0^{\pi} \sin\theta \, d\theta |\mathbf{D}:\mathbf{mm}| \tag{33}$$

in spherical coordinates (r, θ, ϕ), where $d = 2R$ is the collision radius and $\mathbf{m}$ is the unit normal to the collision surface. If the coordinate axes are

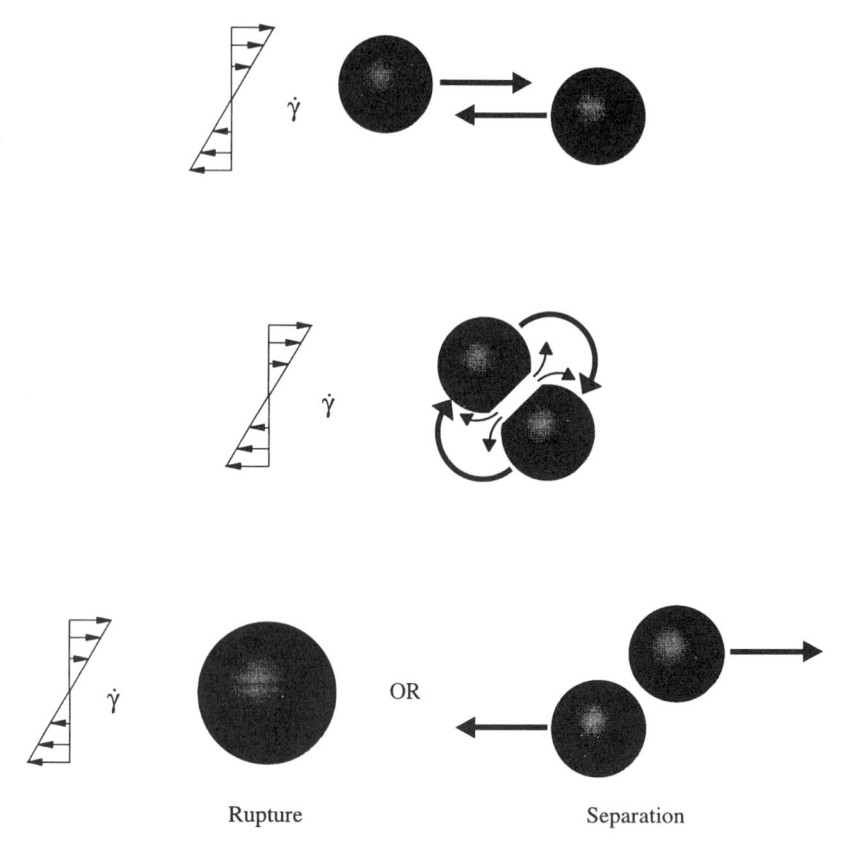

Fig. 26. Schematic representation of the three basic steps of coalescence.

chosen to be the principal axes, the tensor **D** is diagonal, with the diagonal elements given by the roots of

$$\lambda^3 - \frac{1}{4}\lambda + \det(\tilde{\mathbf{D}}) = 0, \tag{34}$$

where $\tilde{\mathbf{D}} = \mathbf{D}/\dot{\gamma}$ and $\dot{\gamma} = \sqrt{2\mathbf{D}:\mathbf{D}}$. Thus, in the context of Smoluchowski's theory, the only parameter that specifies the flow type is det (**D**), which has a range of possible values $(-1/12\sqrt{3}, 1/12\sqrt{3})$. Clearly, vorticity plays no role in the process. For 2D flows, det (**D**) = 0, so that the collision frequency is independent of flow type and is given by

$$\omega_{2D} = \frac{4}{\pi}\dot{\gamma}n\phi. \tag{35}$$

For general linear flows, the limit case of axisymmetric flows [$|\det(\tilde{\mathbf{D}})|$ =

1/12√3] gives

$$\omega_{\text{axs}} = \frac{16}{9\pi}\dot\gamma n\phi = \frac{\pi}{3}\omega_{\text{2D}}. \tag{36}$$

These results show that the Smoluchowski collision frequency is independent of flow type for planar flows and that the maximum collision frequency (obtained for axisymmetric flows) is only 5% larger than that for planar flows.

2. Film Drainage

Once a "collision" occurs, the liquid between the drops is squeezed, forming a film. As the drops are continually squeezed by the external flow field, the drops rotate as a dumbbell and the film drains. At some distance h_o, the drops begin to influence each other and their rate of approach, dh/dt, decreases and is now governed by the rate of film drainage.

The rate of approach, dh/dt, is determined by the different boundary conditions of the interface, which characterize the *mobility* and *rigidity* of the interface. The mobility of the interface is essentially determined by the viscosity ratio and determines the type of flow occurring during film drainage. The rigidity of the interface is determined by the interfacial tension and determines the degree of flattening of the drop. These boundary conditions, along with the different expressions for dh/dt, are displayed in Fig. 27 (Chesters, 1991).

The required time for complete film drainage is given by integrating the equations for dh/dt from h_o to the critical film thickness at which film drainage ends and rupture begins. If the driving force for film drainage is taken, as a first approximation, to be the Stokes drag force acting on the drops,

$$F \approx 6\pi\mu_c\dot\gamma R_o, \tag{37}$$

integrating dh/dt is straightforward. This is clearly a simplified picture, but consistent with the assumptions used for calculation of the collision rate [Eq. (32)]. A more accurate estimate of the force may be obtained from Wang et al. (1994). The film drainage times for the different boundary conditions are given in Fig. 27. The expressions differ significantly, particularly with regard to the dependence on the applied force (F) and initial separation (h_o). In all cases, except for the perfectly mobile interface, the drainage time is directly proportional to the continuous phase viscosity (μ_c) and interfacial tension is important only in those cases where the drops are deformed. The equations in Fig. 27 are based on the formation of a flat circular film between the drops of radius a, and this dimension can be calculated from the applied force.

	Drainage Rate	Drainage Time	Criteria
Rigid Drops	$-\dfrac{dh}{dt} \approx \dfrac{2hF}{3\pi\mu_c R_o^2}$	$t_{drain} \approx \dfrac{3\pi\mu_c R_o^2}{2F} \ln\!\left(\dfrac{h_o}{h_{crit}}\right)$	$\dfrac{a}{R} < 2\!\left(\dfrac{h}{R}\right)^{1/2}$
Immobile Interfaces	$-\dfrac{dh}{dt} \approx \dfrac{8\pi\sigma^2 h^3}{3\mu_c R_o^2 F}$	$t_{drain} \approx \dfrac{3\mu_c R_o^2 F}{16\pi\sigma^2}\!\left(\dfrac{1}{h_{crit}^2} - \dfrac{1}{h_o^2}\right)$	$p > \dfrac{3a}{h}$
Partially Mobile Interfaces	$-\dfrac{dh}{dt} \approx \dfrac{2(2\pi\sigma/R_o)^{3/2} h^2}{\pi\mu_d F^{1/2}}$	$t_{drain} \approx \dfrac{\pi\mu_d F^{1/2}}{2(2\pi\sigma/R_o)^{3/2}}\!\left(\dfrac{1}{h_{crit}} - \dfrac{1}{h_o}\right)$	$\dfrac{6h}{a} < p < \dfrac{3a}{h}$
Fully Mobile Interfaces	$-\dfrac{dh}{dt} \approx \dfrac{2\sigma h}{3\mu_c R_o}$	$t_{drain} \approx \dfrac{3\mu_c R_o}{2\sigma} \ln\!\left(\dfrac{h_o}{h_{crit}}\right)$	$p < \dfrac{6h}{a}$

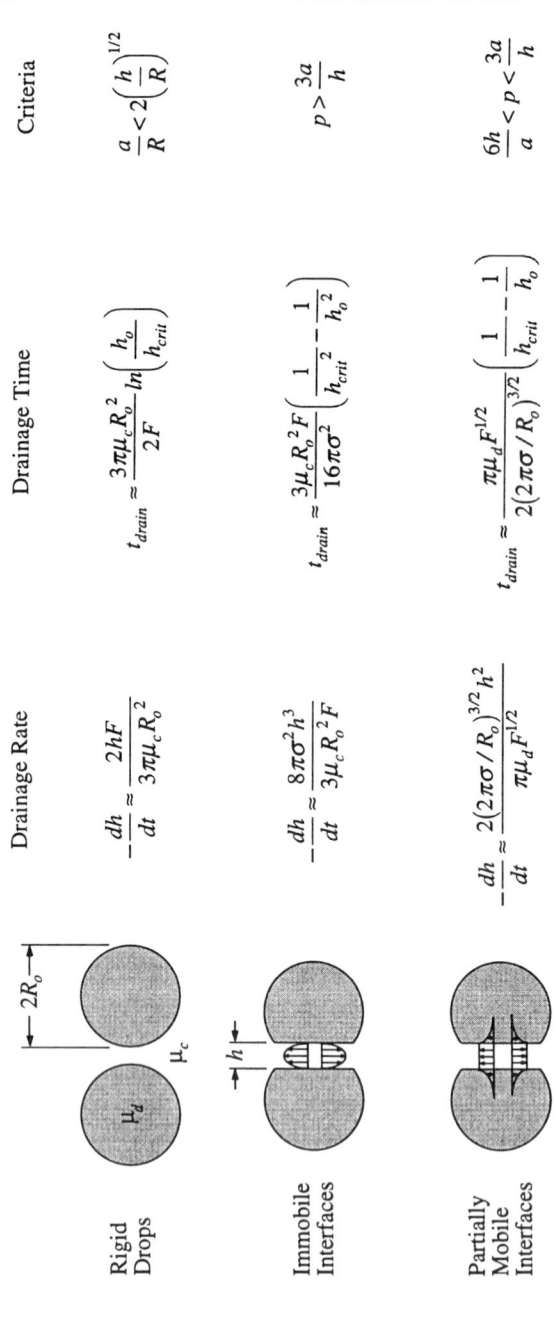

FIG. 27. Various modes of film drainage and the corresponding equations for the rate of film thinning and drainage time are shown. The criteria for specific modes are also indicated.

If a critical film thickness is not reached during film drainage, the drops separate from each other. Conversely, if the critical film thickness is reached, the film ruptures—as a result of van der Waals forces—and the drops coalesce. This generally occurs at thin spots, because van der Waals forces are inversely proportional to h^4 (Verwey and Overbeek, 1948). The value of h_{crit} can be determined by setting the van der Waals forces equal to the driving force for film drainage, giving (Verwey and Overbeek, 1948)

$$h_{crit} \approx \left(\frac{HR_0}{8\pi\sigma}\right)^{1/3}, \tag{38}$$

where H is the Hamaker constant.

A few general conclusions can be drawn from the equations presented in Fig. 27. In general, the drainage time is shortest when the drops are rigid (top equation). Smaller drops are more likely to coalesce than larger ones because the drainage time decreases with drop size. Since the drainage time decreases with force in all cases except the fully mobile interface, coalescence is more likely in lower shear rate zones where the force is lower [Eq. (37)]. Furthermore, surfactants and high-viscosity dispersed phases, both of which reduce the mobility of the interface, result in longer drainage times and thus lower probabilities of coalescence.

HEURISTICS

- Collision frequency is nearly independent of flow type.
- Coalescence is important for dispersed phase volumes fraction (ϕ) greater than about 0.005; the rate of coalescence increases with ϕ.
- Smaller drops are more likely to coalesce after collision than larger drops.
- Coalescence is more likely to take place in regions of low shear rate.
- Coalescence becomes more likely as the mobility of the interface increases.

C. BREAKUP AND COALESCENCE IN COMPLEX FLOWS

There have been several attempts at models incorporating breakup and coalescence. Two concepts underlie many of these models: binary breakup and a flow subdivision into weak and strong flows. These ideas were first used by Manas-Zloczower, Nir, and Tadmor (1982, 1984) in modeling the dispersion of carbon black in an elastomer in a Banbury internal mixer. A similar approach was taken by Janssen and Meijer (1995) to model blending of two polymers in an extruder. In this case the extruder was divided into two types of zones, strong and weak. The strong zones correspond to regions of high shear with short residence times, where stretching of drops into threads and breakup of threads during flow takes place. The weak zones correspond to regions of low shear and long residence times, where breakup of threads at rest and coalescence of drops occurs. Janssen and Meijer (1995) modeled the strong zone as elongational flow and the weak zone as simple shear flow. After each zone, conditions were checked for stretching, capillary breakup, and coalescence according to local coalescence and breakup theory. An initial drop size distribution was passed through a series of alternating strong and weak zones a specific number of times, resulting in a final drop size distribution. Using this model the effects of material properties and process parameters on the final drop size distribution were evaluated. Another model was proposed by Huneault, Shi, and Utracki (1995), again, to model dispersion in extruders. In this case, however, a simplified flow analysis was used to model the flow in the extruder, which gave estimates for the average values of Ca. According to the average Ca in each zone, the average drop diameter was evolved according to a number of rules that included binary breakup. The droplet size distribution in simple shear flow was studied by Patlazhan and Lindt (1996) using a population balance approach based on simple models to predict droplet breakage and coalescence rates.

Let us consider the so-called viscous immiscible liquid mixing (VILM) model (DeRoussel, 1998). This model incorporates most of the important physical processes occurring during drop breakup described in the previous sections, though simplifications are inevitable. The basic approach is similar to that of Janssen and Meijer (1995)—a strong zone modeled by elongational flow in which stretching and breakup by capillary instabilities during flow occur and a weak zone modeled by shear flow in which coalescence and breakup by capillary instabilities at rest occur. The physical aspects of the VILM model that are not included in the model of Janssen and Meijer are summarized as follows:

- Stretching distributions are incorporated into the strong zone.
- Strong and weak zones are divided into subzones that allow for a distribution of shear and elongation rates and residence times.
- Threads breaking by capillary instabilities break into a distribution of drops rather than drop sizes of equal size.
- Coalescence between drops of different size is allowed.
- Variation of the driving force during film drainage is taken into account.

The basic procedure of the VILM model is to send an initial distribution of drops through a specified number of strong and weak zones. With each pass through the strong and weak zones, the evolution of the drop distribution is determined based on the fundamentals of breakup and coalescence.

The dispersed phase is considered to be in the form of either extended threads or spherical drops. In the strong zone, the evolution of drops and threads is essentially the same, except for drops that have $Ca < Ca_{crit}$ and hence remain undeformed. The stretching of drops (if $Ca > 5Ca_{crit}$) and threads is affine and proceeds according to $L/R_o \sim \exp(\dot{\varepsilon}t)$ and $R/R_o \sim \exp(-\dot{\varepsilon}t/2)$. If the residence time in the strong zone is long enough, $\tau_s > t_{crit} + t_{grow}$, the thread breaks into a number of drops with a distribution of sizes. The size of the mother droplets is determined by the wavelength of the fastest-growing disturbance. Corresponding to each mother droplet is a distribution of satellite droplets that depends on the viscosity ratio (see Fig. 21). The distributions obtained by Tjahjadi, Stone, and Ottino (1992) for the breakup of a stationary thread are used in the computations.

The elongation rates for each of the subzones in the strong zone are chosen so that the amount of stretching incurred matches a given stretching distribution. So, if stretching distributions are known for a given mixer, a connection can be made between the relatively simple parameters of the VILM model and the complex flow of the mixer. Techniques for determining the stretching distributions in a mixer are addressed in Muzzio, Swanson, and Ottino (1991a) and Hobbs and Muzzio (1997).

Computations for the weak zone are carried out in discrete time steps with the time step taken to be the average time between collisions, as given by Smoluchowski's theory [Eq. (32)]. At each time step, the following procedure is performed. Two drops are chosen at random and placed in a collision array along with the time the collision occurred and the drainage time for these two particular drops. Each extended thread is checked to see if it has been at rest long enough for breakup due to capillary instabilities to occur. If enough time has elapsed for the thread to break, the corresponding number of drops replaces the thread. The collision array is checked to see if any of the drop pairs have been in contact long enough for the film to drain. When sufficient time has elapsed, the two colliding drops coalesce to form a single larger one. At the end of the time in the weak zone, all drop pairs remaining in the collision array are returned to the drop general population and the next cycle or subzone is started.

The foregoing procedure requires a calculation of the drainage time for each collision. When the number of collisions that occur during a single pass through the weak zone becomes large, $O(10^7)$, the procedure takes too long and is no longer practical. In order to deal with this complication, the number of drops is reduced by a factor that gives a feasible collision frequency. The shape of the distribution is preserved when the number of drops is reduced, so nothing is lost in the simulation.

By using the procedure just discussed, the VILM model allows general trends to be found as process and material parameters are varied. Figure 28

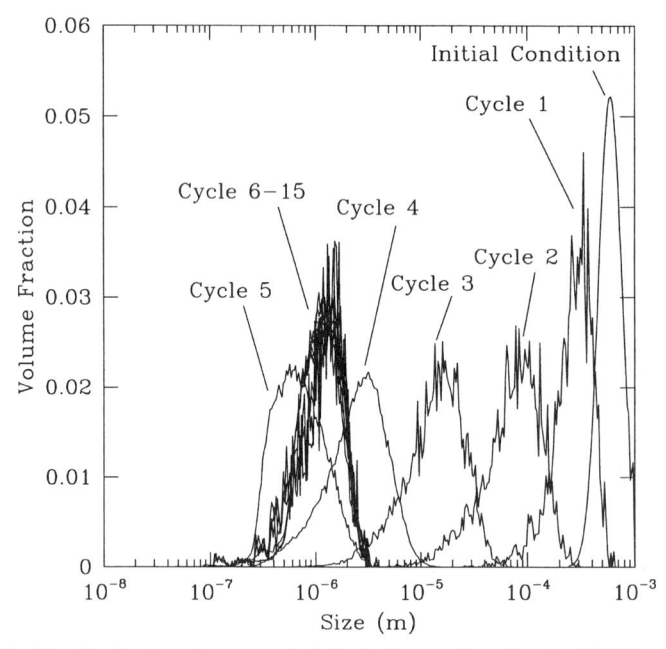

FIG. 28. Size distributions from a typical simulation produced by the VILM model. After six cycles a steady size is reached. Smaller sizes are obtained after five cycles as compared to the final distribution. The conditions for the simulation are $\mu_c = \mu_d = 100$ Pa s, $\phi = 0.2$, and $\sigma = 5 \times 10^{-3}$ N/m.

shows the results of a typical simulation in which an initial drop distribution passes through a number of cycles until a steady-state is reached after six cycles. An important point to note is that after five cycles the resulting distribution is at a smaller size than the final distribution. This phenomena is referred to as overemulsification in the emulsion technology literature (Becher and McCann, 1991), but apparently has not been documented in polymer processing. Overemulsification is a result of the many small satellite drops produced by capillary breakup. The mean drop size then increases with time because the small drops cannot break further and coalescence dominates. Thus, breakup dominates the early part of the process and coalescence dominates the later stages of mixing.

One variable that is commonly used to "classify" the morphology of a blend is the viscosity ratio. Quite often the average drop size is given as a function of the viscosity ratio. Figure 29 gives the results of a number of simulations in which the average size is plotted vs the viscosity ratio for different continuous and dispersed phase viscosities. As can be seen from this figure, for a given viscosity ratio the average size can vary a great deal.

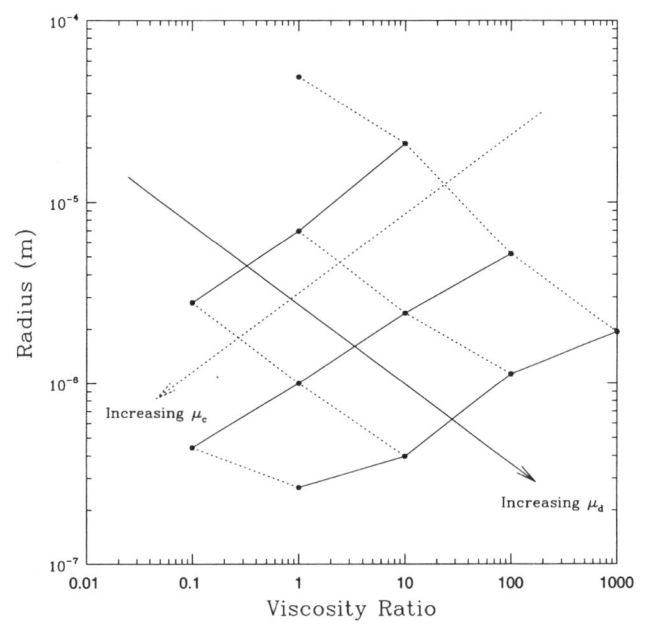

FIG. 29. Average steady-state size of the dispersed phase at different viscosity ratios. The solid and dashed lines represent simulations in which μ_d and μ_c are held constant. Other process parameters are the same as used for Fig. 28 (except $\phi = 0.05$). It is clear that the magnitudes of both viscosities must be considered rather than just the viscosity ratio. The lowest viscosity in each case is 1 Pa·s and the highest 1000 Pa·s. The curves are equally spaced on a logarithmic scale for viscosity.

Hence it is the values of each viscosity, dispersed and continuous, and not just the viscosity ratio that is important in determining the average size. The average size increases with a decrease in either continuous or dispersed phase viscosity for fixed operating conditions.

HEURISTICS
- The average drop size increases with decrease in continuous or dispersed phase viscosity.
- Intermediate mixing times may produce the smallest drops; long mixing times result in an equilibrium size that is larger.

III. Fragmentation and Aggregation of Solids

1. Physical Picture

Powders dispersed in liquids consist of *agglomerates*—a collection of *aggregates*—which in turn are composed of *primary particles.* Agglomerates break because of flow; aggregates do not. Often these particles are of colloidal size, with a size ratio agglomerates/aggregates of about 10^3. In the case of carbon black, for example, aggregates are of the order of 0.1 μm and agglomerates are of the order of 10–100 μm and larger. Thus, the length reduction in solid dispersion is about of the same order of magnitude as in dispersion of liquids. Often we will refer to both aggregates and agglomerates as clusters, a cluster being composed of particles. The *size* of a cluster is given by the *number* of particles composing the cluster.

The objective of mixing—or dispersion—of solids is to break agglomerates to aggregate size, the process giving rise to broad, time-evolving cluster size distributions. The entire process of *dispersion* of a powder into a liquid involves several stages, which may all be occurring with some degree of overlap. Several scenarios have been proposed and, unavoidably, a proliferation of terms has ensued. For instance, Parfitt's (1992) scenario consists of four stages: *Incorporation* is the initial contact of the solid with the medium. *Wetting,* which follows incorporation, may in turn consist of (i) adhesion of the medium to the solid, (ii) immersion of the solid into the fluid, and (iii) spreading of the liquid into the porous solid. *Breakup* (or fragmentation) and *flocculation* (or aggregation) conclude the dispersion process. A much narrower definition of *dispersion* is commonly used in the polymer processing literature: fragmentation of agglomerates into aggregates, and the distribution of the aggregates into the medium.

In fact, the term fragmentation is commonly used in the physics literature to refer to a broader class of processes involving breakup of solids, such as rocks. Much of this literature can be adapted, *mutatis mutandi,* to the dispersion of agglomerates as is of interest here (see Fig. 1). Fragmentation may be in turn divided into two modes of breakup (Redner, 1990): *rupture* and *erosion*—rupture referring to the breakage of a *cluster* into several fragments of comparable size, erosion to the gradual shearing off of small fragments from larger clusters (Fig. 30). The main qualitative difference between these two mechanisms is energy input: low for erosion, high for rupture. Erosion dominates dispersion when stresses are low. Finally, materials may *shatter,* producing a large number of smaller fragments in a single event, as in the case of high-energy fragmentation. While the physical mechanisms may be different, there are similarities between fragmentation of solids and breakup of liquid drops, at least with respect to the size distribution of fragments produced on breakup: Tip streaming is analogous to erosion, necking to rupture, and capillary breakup to shattering.

Aggregation (*flocculation* is the term commonly used in the rubber industry) may be imagined as being the reverse of dispersion. Aggregates come together, interact via hydrodynamic forces and particle potentials, and eventually bind. Two bonding levels are possible: strong and weak. Strongly

Rupture and Erosion

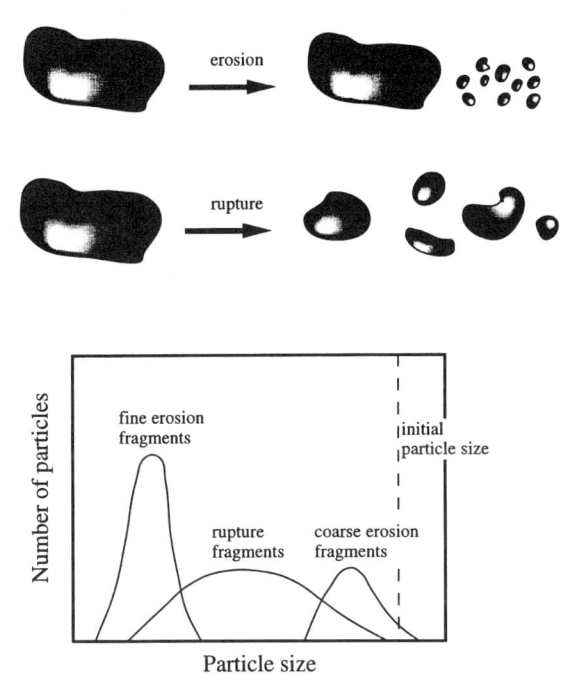

FIG. 30. Schematic view of rupture and erosion of particles and the typical size distribution of fragments obtained.

bound aggregates cannot be redispersed by future stirring; weakly bonded aggregates can be dispersed by stirring. As opposed to coalescence of droplets, structure in this case may be preserved and fractal-like structures (Fig. 31) are common. The mass of such clusters increases with radius according to R^D rather than R^3 for compact agglomerates, where $D < 3$ is the fractal dimension. As we shall see, flows can in fact be manipulated to tailor structures.

2. Small Scales: Particle Interactions

The current level of understanding of how aggregates form and break is not up to par with droplet breakup and coalescence. The reasons for this discrepancy are many: Aggregates involve multibody interactions; shapes may be irregular, potential forces that are imperfectly understood and quite susceptible to contamination effects.

Thus, analyses of how aggregates break have resorted to idealizations inspired by traditional fluid mechanical analysis, two limit cases being per-

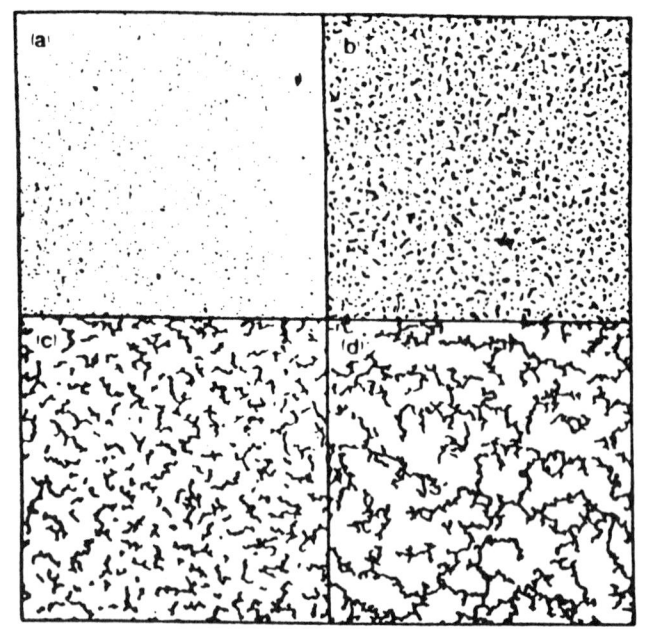

FIG. 31. Fractal structures obtained experimentally at different stages of aggregation of a colloidal monolayer of 1 μm sulfonated polystyrene particles on the surface of an aqueous calcium chloride solution, initially uniformly distributed (Robinson and Earnshaw, 1992).

meable and impermeable spheres. Quite possibly the simplest model is to regard an agglomerate as a pair of bound spheres. As in the case of drop breakup, the analysis considers scales such that the surrounding flow is linear. The velocity of one particle relative to the other, taking into account hydrodynamic and potential interactions between the two particles, is

$$\mathbf{v} = \mathbf{D} \cdot \mathbf{r} + \omega \times \mathbf{r} - \left\{ A(r)\frac{\mathbf{rr}}{r^2} + B(r)\left[\mathbf{I} - \frac{\mathbf{rr}}{r^2}\right] \right\} \cdot \mathbf{D} \cdot \mathbf{r} + \frac{C(r)}{6\pi\mu R}\mathbf{F}_c, \quad (39)$$

where $\mathbf{r}$ is a vector joining the centers of the two spheres, $\mathbf{D}$ is the rate of strain tensor, ω is the vorticity of the driving flow, $A(r)$ and $B(r)$ are known functions (Batchelor and Green, 1972), and $\mathbf{I}$ is the identity matrix. In the last term, $\mathbf{F}_c$ is the physicochemical force between the particles, μ is the fluid viscosity, R is the radius of the particles, and $C(r)$ is a parameter that accounts for the particle proximity effect on drag (Spielman, 1970). The first three terms in this equation give the relative velocity between the spheres in a linear flow field under the influence of hydrodynamic interactions, and the last term gives the relative velocity due to the physicochemical forces between the spheres. Rupture of this idealized aggregate occurs if hydrodynamic forces overcome the binding physicochemical forces as we show later. The equation also applies to the analysis of the aggregation of initially separated particles in a linear flow (Zeichner and Schowalter, 1977).

The physicochemical forces between colloidal particles are described by the DLVO theory (DLVO refers to Derjaguin and Landau, and Verwey and Overbeek). This theory predicts the potential between spherical particles due to attractive London forces and repulsive forces due to electrical double layers. This potential can be attractive, or both repulsive and attractive. Two minima may be observed: The primary minimum characterizes particles that are in close contact and are difficult to disperse, whereas the secondary minimum relates to looser dispersible particles. For more details, see Schowalter (1984). Undoubtedly, real cases may be far more complex: Many particles may be present, particles are not always the same size, and particles are rarely spherical. However, the fundamental physics of the problem is similar. The incorporation of all these aspects into a simulation involving tens of thousands of aggregates is daunting and models have resorted to idealized descriptions.

A. FRAGMENTATION

There is a large body of theoretical work dealing with fragmentation. General aspects, primarily in the context of mathematical aspects of particle size distributions produced on fragmentation, are covered by Redner (1990) and Cheng and Redner (1990), whereas a review of various modeling approaches and experimental results, addressing grinding of solids, is presented by Austin (1971). Of special interest is the distribution of fragments upon rupture. Power law forms for cumulative distributions based on particle radius are commonly obtained and, in many cases, the distribution of fragments produced in a single rupture event is *homogeneous* (i.e., the distribution depends only on the ratio of the mass of the fragment to the mass of the original particle). Erosion produces fragments much smaller than the original particle, and consequently the particle size distribution is bimodal. This body of literature provides a starting point for the understanding of the dispersion of agglomerates in viscous flows (Fig. 30).

Fragmentation of agglomerates is similar to rupture of solids in that both agglomerates and granular solids deform only slightly before breaking. Differences arise mainly from the complex internal structure of agglomerates. In addition, the weaker bonding in agglomerates results in fragmentation at relatively low stresses. Fragmentation may be caused by several mechanisms, for example, application of direct compressive loads and particle–particle and wall–particle impacts. However, here we focus only on fragmentation by hydrodynamic forces, which is of most relevance to polymer processing. By analogy with liquid droplets and the capillary number [Eq. (18)] the dimensionless parameter that characterizes the fragmentation process is the ratio of the viscous shear stress to the strength of the agglomerate. We term this ratio the *Fragmentation number*, Fa:

$$\text{Fa} = \frac{\mu \dot{\gamma}}{T}. \tag{40}$$

The term T denotes the characteristic cohesive strength of the agglomerate and plays a role analogous to the surface tension stress (σ/R) in the definition of the Capillary number for liquid drops. Unlike surface tension, however, the agglomerate strength is not a material property but depends on internal structure, density (degree of compaction), moisture, and many other variables. A similar definition of a dimensionless Fragmentation number appears in previous studies (e.g., Rwei, Manas-Zloczower, and Feke, 1990), though it was not termed as such.

1. Agglomerate Strength

The cohesive strength of an agglomerate owes its origins to interparticle bonds due to electrostatic charges, van der Waals forces, or moisture. Experimental methods for the measurement of the characteristic agglomerate strength include the tensile testing of compacted pellets (Rumpf, 1962; Hartley and Parfitt, 1984), notched bending tests of compacted beams (Kendall, 1988), and compression testing of compacted beds by penetration of a conical tester (Lee, Feke, and Manas-Zloczower, 1993). The three tests measure different properties: the tensile strength, the tensile strength in the presence of flaws, and the cohesivity, respectively. Values obtained from different test methods for the same agglomerate would clearly be different; trends with changes in parameters are, however, similar.

Two idealized models have been reasonably successful in predicting the strength of agglomerates and we review them here. Rumpf (1962) assumed the agglomerates to be spatially uniform and composed of identical spheres of radius a bound to touching neighbors by van der Waals forces. Considering a planar rupture surface, the tensile strength T is

$$T \approx \frac{H}{48 a z_0^2} \frac{\phi}{1-\phi}, \tag{41}$$

where H is the Hamaker constant, ϕ is the solids volume fraction, and z_0 is the equilibrium separation distance between the sphere surfaces. Kendall

(1988), on the other hand, used the Griffith criterion for crack growth to obtain

$$T \approx 15.6 \frac{\phi^4 \Gamma_c^{5/6} \Gamma^{1/6}}{(2ac)^{1/2}}, \tag{42}$$

where Γ is the interfacial energy, Γ_c is the fracture energy, and c is the initial length of the crack (edge notch). The two mechanisms, fracture at a plane and crack growth, give very different expressions, particularly with respect to dependence on the solids volume fraction (ϕ) and the aggregate (primary particle) radius (a). Surprisingly, agglomerates may rupture by either of these mechanisms. For example, experiments using titanium dioxide *agglomerates* carried out by Lee, Feke, and Manas-Zloczower (1993) found the strength of agglomerates ($a \approx 0.16$ μm) to follow Rumpf's model, whereas Kendall (1988) found the agglomerates ($a \approx 0.23$ μm) to follow the crack growth model. A notable difference was the order of magnitude lower strength measured by Lee *et al.* (1993) as compared to the agglomerates of Kendall (1988); however, this may be due to the different test methods used.

Illustration: Effect of flow type on agglomerate separation after rupture. The nature of the local flow significantly affects the separation of fragments produced on rupture of an agglomerate, as shown by Manas-Zloczower and Feke (1988), for equal-sized spherical fragments. Significantly higher shear rates are required for separation in a simple shear flow as compared to extensional flows. As the following analysis demonstrates, the process of separation is very similar to the breakup of slender drops (Khakhar and Ottino, 1986b,c; Ottino, 1989); thus, previous results are useful for generalizing the analysis of fragment separation for all linear flows.

The spherical fragments, initially in close contact, move relative to each other according to Eq. (B.1). The underlying physics is revealed more clearly by recasting equation (B.1) in the following dimensionless form:

$$\frac{1}{r}\frac{dr}{dt} = (1 - A(r))\mathbf{D}:\mathbf{mm} - C(r)\frac{F_c}{\text{Fa}}, \tag{43}$$

$$\frac{d\mathbf{m}}{dt} = [(1 - B(r))\mathbf{D} - \Omega] \cdot \mathbf{m} - (1 - B(r))(\mathbf{D}:\mathbf{mm})\mathbf{m}. \tag{44}$$

Here r is the center-to-center distance, and $\mathbf{m}$ is a unit vector along the line joining the centers of the two fragments, so that $\mathbf{r} = r\mathbf{m}$. Distances are made dimensionless with respect to the radius of a fragment (a), shear rates and time with respect to the characteristic shear rate $\dot{\gamma} = \sqrt{2\mathbf{D}:\mathbf{D}}$, and $F_c = |\mathbf{F}_c|$ with respect to H/a. These equations are identical in form to those for the breakup for a slender drop (Khakhar and Ottino, 1986b) discussed earlier.

Several conclusions result from the preceding equations: They reveal that the dimensionless parameter of the system is the fragmentation number (Fa) with the characteristic strength given by

$$T = \frac{H}{6\pi a^3}. \tag{45}$$

The rate of separation of the fragments depends on the functions $A(r)$, $C(r)$, F_c, and the fragmentation number, while the rate of rotation depends only on the function $B(r)$. Further, it is apparent that the separation between the fragments increases only when the hydrodynamic force exceeds the binding physicochemical force. The pair of fragments rotates as a material element in an apparent flow with an effective velocity gradient tensor

$$\mathbf{M} = \mathbf{D} + \Omega/(1 - B). \tag{46}$$

Since B is positive, the apparent flow appears to have a higher vorticity relative to the strain rate than the imposed flow.

Based on the foregoing discussion, a criterion for the separation of the fragments is easily obtained. If at least one eigenvalue of the tensor $\mathbf{M}$ is positive, the pair orients along the corresponding principal axis and the critical Fragmentation number for separation is given by

$$\mathrm{Fa}_{\mathrm{sep}} = \frac{\sqrt{2}}{e_\lambda} \frac{1}{12 z_0^2}, \tag{47}$$

where $e_\lambda / \sqrt{2}$ is the asymptotic value of $\mathbf{D} : \mathbf{mm}$ (i.e., the largest positive eigenvalue of $\mathbf{M}$) and $z_0 = (r - 2)$ is the equilibrium separation. For $\mathrm{Fa} > \mathrm{Fa}_{\mathrm{sep}}$ the fragments separate indefinitely. In arriving at the preceding equation, the physicochemical force was estimated for a pair of equal-sized spheres as $F_c = 1/12 z_0^2$ (Rumpf, 1962), and the result $(1 - A)/C = 1$ for small separations was used. Special cases of this result were presented by Manas-Zloczower and Feke (1988). For purely extensional flows ($\Omega = \mathbf{0}$), we have $e_\lambda = 1/\sqrt{2}$ for planar flows and $e_\lambda = \sqrt{2/3}$ for axisymmetric flows. In the case of simple shear flow, eigenvalues are imaginary; hence, no asymptotic orientation exists and the separation distance oscillates with time as the fragments rotate in the flow. Separation of fragments in this flow occurs, according to Manas-Zloczower and Feke (1988), when the potential energy due to the van der Waals forces at the maximum separation distance is less than the thermal energy. This criterion gives the critical separation as

$$z_c \approx \frac{Ha}{12 k_B T}, \tag{48}$$

where k_B is the Boltzmann constant. The critical fragmentation number for separation should be at least as large as that for planar extension; numerical simulation is required to obtain Fa$_{sep}$ at which the criterion $z > z_c$ is satisfied.

2. Rupture—Critical Fragmentation Number

Agglomerates in a sheared fluid rupture when the hydrodynamic stress exceeds a critical value; in dimensionless form the criterion for rupture is Fa > Fa$_{crit}$. Rupture occurs within a short time of application of the critical stress, and thus can be distinguished from erosion, which occurs over much longer time scales.

Experimental data for the critical fragmentation number is sparse and certainly much less than what is available for droplets. Rwei, Manas-Zloczower, and Feke (1990) carried out experiments to determine the critical stress for the rupture of compact agglomerates in simple shear flow. Carbon black pellets of different densities (ρ) and silicone oils of different viscosities were used. The tensile strength of the agglomerates was measured and is well correlated by Kendall's (1988) model: $T \approx K\rho^4$ with $K \approx 1.7 \times 10^6$ Pa (g/cm^3)$^{-4}$. The critical shear stress applied at the point of rupture was found to be $\mu\dot{\gamma}_c \approx A\rho^4$ with $A \approx 8.7 \times 10^4$ Pa (g/cm^3)$^{-4}$ for the different density pellets and the fluids of varying viscosity. The reported data thus gives Fa$_{crit} \approx 0.05$ independent of pellet density. The dependence of Fa$_{crit}$ on the agglomerate size was not studied. (Note: The magnitude of Fa$_{crit}$ depends on the method used for measuring the strength of the agglomerate, T.) The size distribution of the fragments formed upon rupture is log–normal. This seems to be in agreement with the theory for random breakage discussed later in this section.

Rupture of fractal (flocculated) aggregates of polystyrene latices in simple shear flow and converging flow was studied by Sonntag and Russel (1986, 1987b). For simple shear flow and low electrolyte concentrations, the critical fragmentation number decreases sharply with agglomerate radius (R) as

$$\text{Fa}_{crit} \approx 1.6 \times 10^4 \left(\frac{R}{a}\right)^{-3}, \tag{49}$$

where the agglomerate strength, $T = 595$ dyn/cm^2, is obtained by fitting a theory (discussed later) to the experimental data. The fractal dimension of the agglomerates is $D \approx 2.5$. Ionic effects at high electrolyte concentrations are well accounted for by assuming T to be proportional to the force between two primary particles calculated using the DLVO theory (the bond strength between the primary particles increases with increasing ionic strength). In the converging flow, the mass of the critical agglomerates decreases linearly with increasing shear rate, like the shear flow, but the radius of gyration is nearly constant. Such behavior could be due to the formation of elongated fragments upon rupture.

Theoretical prediction of the critical fragmentation number requires an estimate of the hydrodynamic stress acting on the agglomerate. For the case of spherical impermeable agglomerates Bagster and Tomi (1974) showed that rupture in simple shear flow occurs on a plane through the center of the agglomerate, and Fa_{crit} is independent of the agglomerate size. In the analysis, the hydrodynamic stress on the surface is given by

$$\mathbf{F} = 5\mu \mathbf{D} \cdot \mathbf{r} \tag{50}$$

(Brenner, 1958). Feke and Manas-Zloczower (1991) extended this analysis to the case of spherical agglomerates with a single flaw considering rupture to occur by crack growth. Again Fa_{crit} is independent of size. A significant result is that rupture is more likely in simple shear flow, as compared to extensional flows, since the agglomerate rotates in a simple shear flow and the rupture surface (flaw) experiences the entire range of stresses possible.

The case of permeable porous agglomerates gives qualitatively different results. Adler and Mills (1979) showed that Fa_{crit} decreases with agglomerate size initially, but approaches the impermeable limit for large agglomerates ($R/k^{1/2} \gg 1$, where μ/k is the permeability). Sonntag and Russel (1987a) analyzed fractal structures using a similar approach. The fractal structure is incorporated into the mean field approach by considering a radially varying solids volume fraction

$$\phi(r) \approx \phi_0 \left(\frac{r}{a}\right)^{D-3}, \tag{51}$$

with the strength given by $T = K\phi^n$ (similar to Kendall's model). This model predicts the experimentally determined dependence of Fa_{crit} on agglomerate size when $n = 4.45$ (Fig. 32). In both the models for porous agglomerates, hydrodynamic stresses within the sphere are calculated using the Brinkman equation with creeping flow in the surrounding fluid. The Mises rupture criterion

$$(\mathbf{T}_D : \mathbf{T}_D)^{1/2} > T \tag{52}$$

is used in both analyses, where $\mathbf{T}_D$ is the deviatoric part of the stress tensor (i.e., $\mathbf{T} = -P\mathbf{I} + \mathbf{T}_D$). Thus, rupture can occur at any point in the sphere.

Most recently, Horwatt, Manas-Zloczower, and Feke (1992a, b) carried out a numerical study of the rupture of fractal clusters generated by different aggregation protocols (more details of these protocols are given in Section

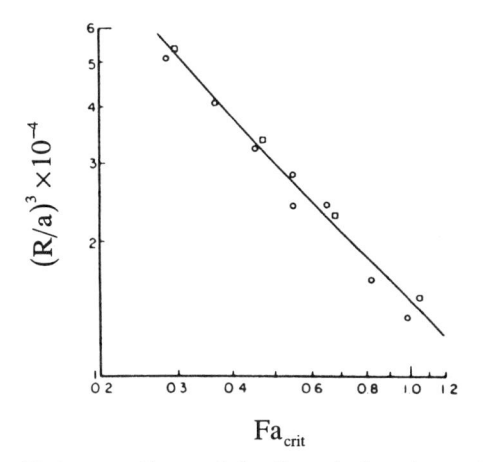

FIG. 32. Relationship between Fa$_{crit}$ and the dimensionless cluster size (R/a) for fractal clusters $(D \approx 2.5)$ of polystyrene latex in a simple shear flow. Data points are experimental results and the solid line is the theoretical prediction (Sonntag and Russel, 1986, 1987a).

III, B on aggregation). The clusters are modeled as impervious spheres and the hydrodynamic force is calculated following the approach of Bagster and Tomi (1974). The cluster ruptures if the resultant stress on the bonds between clusters exceeds a specified bond strength. The critical fragmentation number decreases with agglomerate radius for clusters with low fractal dimension (D) and is nearly independent of cluster size for compact clusters. This is in agreement with previous theories. The magnitude of Fa$_{crit}$, however, does not seem to have a strong correlation with the fractal dimension of the cluster.

3. Erosion

Erosion of agglomerates due to hydrodynamic stresses occurs over long time scales and at low values of the fragmentation number (Fa ≪ Fa$_{crit}$). The fragments eroded are much smaller than the parent particle (volume of fragment <10% of volume of parent). Of primary interest from the viewpoint of dispersion of agglomerates is the rate of erosion and the size distribution of eroded fragments. Only a few studies focusing on these aspects have been carried out, and we summarize the main results here.

Erosion kinetics for compact spherical structures are well described by

$$\frac{R_0 - R(t)}{R_0} \approx k_e \dot{\gamma} t \qquad (53)$$

at short times of erosion, where R is the radius of the parent agglomerate, R_o is the initial radius, and k_e is a dimensionless erosion rate constant. This equation implies that the rate of erosion (volume eroded per unit time) is directly proportional to the surface area of the parent. Thus, k_e is a first-order rate constant, and for long times $R \approx R_o \exp(-k_e \dot{\gamma} t)$.

In the case of compact *cohesionless clusters* (no attractive forces between aggregates forming agglomerate), Powell and Mason (1982) found k_e to be independent of shear rate but dependent on flow type and the ratio (a/R_0), where a is the radius of the primary particle (aggregate). In contrast, Rwei, Manas-Zloczower, and Feke (1991), in a study of compact carbon black agglomerates in simple shear flow, found k_e to depend only on the fragmentation number (Fa) based on experiments carried out at different shear rates and pellet densities (Fig. 33). The critical shear rate below which erosion stops was found to be $\mu \dot{\gamma}_{c,e} \approx 100$ Pa by extrapolating data obtained at different stresses to a zero erosion rate. Using the strength of agglomerates determined in their earlier work (Rwei *et al.*, 1990), the critical fragmentation number for erosion is obtained as $Fa_{c,e} \approx 2 \times 10^{-3}$, which is an order of magnitude lower than Fa_{crit}. The size (radius) distribution of the fragments was found to be Gaussian with a mean and standard deviation proportional to the initial radius of the parent. In a separate study, carried out at significantly higher values of Fa, Rwei, Manas-Zloczower, and Feke (1992) found the radius to decrease exponentially with time.

Physicochemical interaction between the agglomerate and fluid is an additional factor that affects the rate of erosion. For example, in a study of erosion under simple shear flow using four different types of titanium dioxide agglomerates in silicone oil, Lee, Feke, and Manas-Zloczower (1993) found k_e to be proportional to the product FaW_a, where W_a is the work of adhesion of the fluid to the particles (nearly equal to the dispersive component of the fluid interfacial tension). Further, increasing the particle porosity sharply increased the erosion rate. This is a consequence of the faster penetration of the fluid into the more porous agglomerates (initially dry), resulting in the reduction of the cohesive strength of the bonds between the primary particles. There is much that remains to be done in this area. For example, erosion of fractal structures does not seem to have been sufficiently studied.

Illustration: Kinetics of dispersion: the two-zone model. The models for agglomerate rupture when integrated with a flow model are useful for the modeling of dispersion in practical mixers, as was discussed for the case of drop dispersion. Manas-Zloczower, Nir, and Tadmor (1982), in an early study, presented a model for the dispersion of carbon black in rubber in a Banbury mixer (Fig. 34). The model is based on several simplifying assumptions: Fragmentation is assumed to occur by rupture alone, and each rupture produces two equal-sized fragments. Rupture is assumed to occur

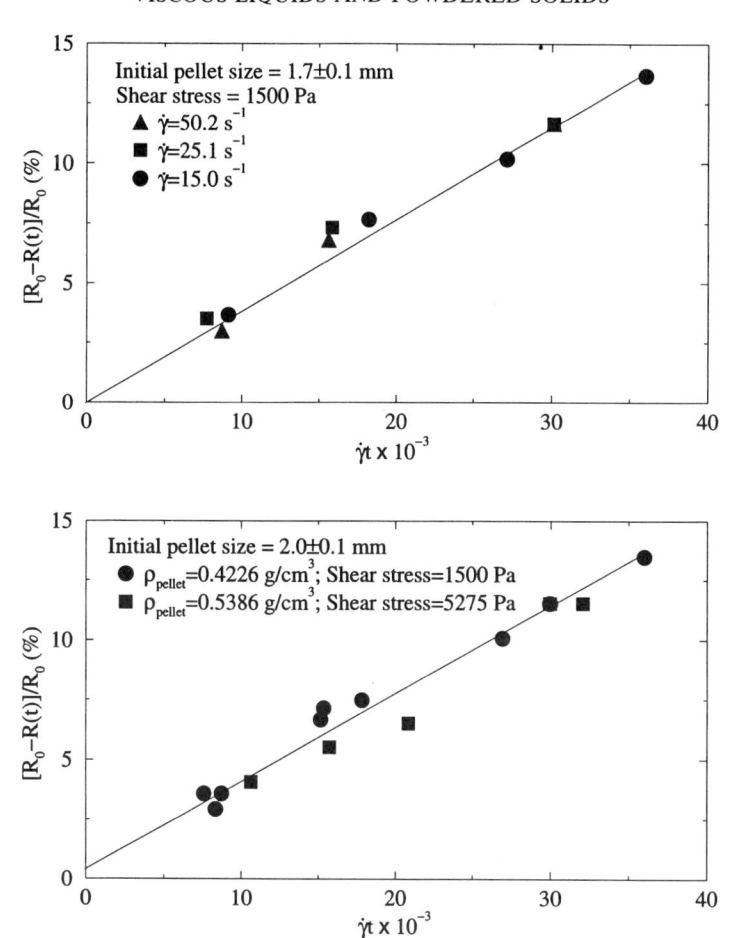

FIG. 33. Variation of dimensionless radius with dimensionless time for compact carbon black particles suspended in silicone oil undergoing a simple shear flow. *Top*: Data for initial radius $R_0 \approx 1.7$ mm for different viscosities and different shear rates, but with Fa ≈ 0.28. *Bottom*: Data for initial radius $R_0 \approx 2.0$ mm for different densities and different shear rates, but with Fa ≈ 0.28. For all cases the erosion rate constant $k_e \approx 0.37$. (See Eq. 53).

along a specific plane in the agglomerate (similar to the model of Kendall, 1988), but the strength of the agglomerate is calculated using Rumpf's (1962) model. Hydrodynamic forces are calculated assuming the agglomerate to be a pair of touching spheres [Eq. (39)]. The flow in the Banbury mixer is taken to be a well-mixed low-shear zone from which a stream is continuously passed through a small high-shear zone and recycled. Only a fraction of the agglomerates passing through the high-shear zone rupture,

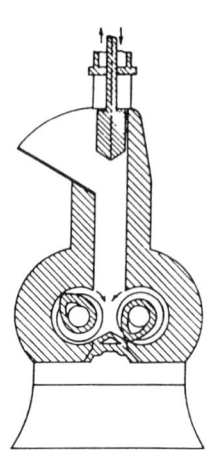

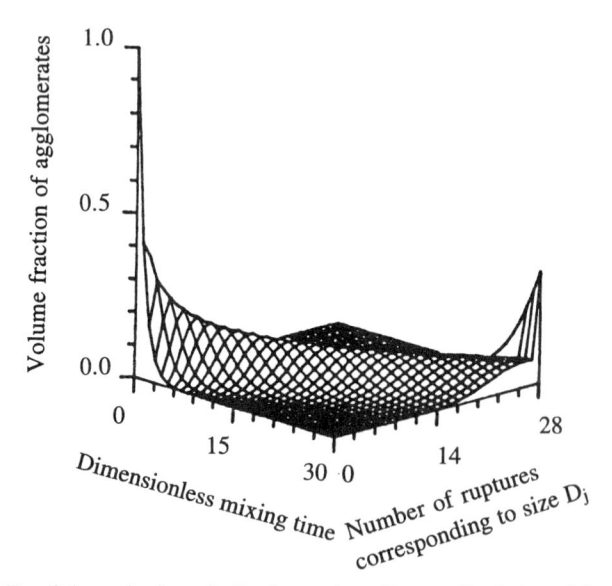

FIG. 34. *Top*: Schematic view of a Banbury mixer. *Bottom*: Evolution of the agglomerate size distribution with time predicted by the two-zone model with parameter values corresponding to the dispersion of carbon black in rubber in a Banbury mixer. The size of an agglomerate after j ruptures is given by $D_j = D_0/2^{j/3}$, where D_0 is the initial size of the agglomerates. The size of the aggregate (primary particle) corresponds to $j = 28$ (Manas-Zloczower, Nir, and Tadmor, 1982).

and this is determined by whether or not the rupture plane is exposed to a sufficiently high force. An agglomerate may break once at most during a single pass through the high-shear zone, and no fragmentation takes place in the low-shear zone. Statistical arguments are used to obtain the distribution for the number of passes made by an agglomerate through the high-shear zone, and to obtain the agglomerate size distribution.

In spite of the simplifying assumptions made in the model, the results give a qualitative insight into the operation of the process. Figure 34 shows the evolution of the agglomerate size distribution with time for a representative case. At the start of the process only large agglomerates (diameter D_0) are present; with increasing mixing time the distribution broadens and the peak shifts to smaller sizes. At the same time a new peak is formed corresponding to the aggregate size; aggregates cannot be further ruptured by the flow.

Illustration: Simultaneous erosion and rupture. Agglomerates when sheared at Fa > Fa_{crit} fragment by both rupture and erosion mechanisms. Experiments of Rwei, Manas-Zloczower, and Feke (1992) for carbon black agglomerates in molten polystyrene show that rupture of the particles occurs within a short time of application of the shear flow, and a log–normal distribution of fragment sizes is produced. Subsequent shearing results only in a slow reduction of the average agglomerate size, and the average radius decreases exponentially with time ($R = R_0 \exp(-k_e \dot{\gamma} t)$) which is characteristic of erosion. Thus rupture and erosion can be considered to occur sequentially in homogeneous flows because of their different time scales. However, in flows with spatially varying shear rates, erosion would occur in all regions of the flow with varying rates depending on the local value of Fa, while rupture would result during the first visit of an agglomerate to a high shear zone where Fa > Fa_{crit}.

4. Global Scales: Particle Size Distributions

Analysis of the particle size distributions produced in fragmentation processes go back to the early work of Kolmogorov (1941). Considering that the size of fragments produced on breakage are random, Kolmogorov (1941) used statistical arguments, similar to those outlined earlier for the stretching distributions, to show that the size distribution is log–normal. A more general starting point for the study of dispersion of powdered solids is provided by the so-called fragmentation theory (Redner, 1990). Irreversible, continuous breakup of solids in a well-mixed batch system can be described by the linear fragmentation equation,

$$\frac{\partial c}{\partial t} = -a(x)c(x, t) + \int_x^\infty a(y)f(x|y)c(y, t)\,dy, \qquad (54)$$

where $c(x, t)$ is the concentration of clusters of mass x at time t, $a(x)$ is the *overall rate* of breakup of clusters of size x, and $f(x|y)$, the *relative rate* of breakup, is the expected number of clusters of size x produced from the breakage of a cluster of size y.

The *breakup kernels*, $a(x)$ and $f(x|y)$, determine the kinetics of the fragmentation process. However, general conclusions can be drawn without exact specification of these kernels by scaling analysis of the fragmentation equations. For example, if small clusters break up like larger clusters, behavior experimentally observed in many systems, the breakage kernels are *homogeneous*. The overall rate of breakup in this case is given by $a(x) \sim x^\lambda$, where λ is known as the homogeneity index. Homogeneity also implies that the distribution of fragments produced on breakup is

$$f(x|y) = \frac{1}{y} b\left(\frac{x}{y}\right). \qquad (55)$$

The average number of particles produced on a single breakup event is then $\int_0^1 b(r)dr$, and $\int_0^1 rb(r)dr = 1$, because of conservation of mass.

The fragmentation equation can be scaled into the following time-invariant form for the case of homogeneous kernels:

$$w\left[2\phi + \eta\frac{d\phi}{d\eta}\right] = -\eta^\lambda \phi + \int_\eta^\infty \xi^{\lambda-1}\phi(\xi)b\left(\frac{\eta}{\xi}\right)d\xi. \qquad (56)$$

Here, both concentration and size are rescaled with respect to the average agglomerate size $s(t)$ as

$$\eta = x/s(t), \qquad \phi(\eta) = c(x, t)/s^{-\theta}. \qquad (57)$$

For conservation of mass it is required that $\theta = 2$. For the scaled equation (56) to be time-invariant, the separation constant (w) given by

$$w = -\frac{ds}{dt}s^{-(1+\lambda)} \qquad (58)$$

must be independent of time. In general, this condition is satisfied at long times as the particle size distribution asymptotically approaches the self-similar distribution, independent of the initial condition.

Filippov (1961) rigorously showed the above scaling to be valid when λ is positive; if $\lambda < 0$ the scaling breaks down as mass is lost to the formation of infinitesimal size fragments (Filippov, 1961). For positive values of λ, Eq. (58) gives $s \sim t^{-1/\lambda}$ at long times.

The moments of order α of the bare and scaled size distribution are defined as

$$M_\alpha(t) \equiv \int_0^\infty x^\alpha c(x, t)\, dx, \quad \text{and} \quad m_\alpha \equiv \int_0^\infty x^\alpha \phi(x)\, dx, \tag{59}$$

respectively, and have the form of Mellin transforms (Morse and Feshbach, 1953). The moments have a physical significance for the process: for example, M_0 is the total number of clusters per unit volume, and M_1 is the total mass per unit volume. The *number* average size is then $s_n(t) = M_1/M_0$, and the *weight* average size is $s_w(t) = M_2/M_1$. Finally, the *polydispersity*, which characterizes the width of the distribution, is

$$P = \frac{s_w}{s_n} = \frac{M_2 M_0}{M_1^2}. \tag{60}$$

The scaled moments for the case of homogeneous kernels are obtained as

$$m_{\alpha+\lambda} = w \frac{1-\alpha}{L_\alpha - 1} m_\alpha, \text{ with } L_\alpha \equiv \int_0^1 r^\alpha b(r)\, dr \tag{61}$$

in the limit of long times when the scaling solution is valid. This recursive relationship between the scaled moments, and use of inverse Mellin transforms, allows the determination of the form of the scaling distribution for the tails of the size distribution (Cheng and Redner, 1990). In the limit of large mass $[x \gg s(t)]$, the size distribution is

$$c(x, t) \sim \exp(-\text{const.}\, tx^\lambda). \tag{62}$$

The cluster size distribution in the limit of small mass $[x \ll s(t)]$ depends on the properties of the agglomerates undergoing fragmentation. If infinitesimal particles are formed on a single breakage, that is, $b(r) \sim r^\nu$, then

$$c(x, t) \sim x^\nu \quad \text{as} \quad x \to 0, \tag{63}$$

and if clusters are broken into small fragments by several steps, then the distribution is log–normal in the limit of small x. This is in agreement with the theory of Kolmogorov (1941). Thus, limiting forms of the cluster size distribution are obtained by invoking only the assumption of homogeneous breakup kernels.

Several works over the past decade or so (Ziff, 1991; Ziff and McGrady, 1986; McGrady and Ziff, 1987, 1988; and Williams, 1990) have addressed the behavior of systems with specified breakup kernels. Certain specific forms for the breakup kernels lead to analytical solutions for the cluster size distribution. For example, Ziff (1991) obtained explicit forms of the size distribution for homogeneous breakup kernels of the form

$$b(r) = \alpha q r^{\alpha-2} + \beta(1-q) r^{\beta-2}, \tag{64}$$

where q, α, and β are adjustable parameters, which allow for fitting of a range of breakup processes. The limiting forms of these solutions are predicted by the preceding analysis.

5. Erosion-Controlled Fragmentation

The dynamics of the particle size distribution produced by fragmentation processes dominated by erosion are qualitatively different from those involving rupture. Although rate of breakage function is homogeneous $[a(x) = kx^\lambda]$ because the rate of erosion is proportional to surface area, the distribution of fragments produced on breakage, in general, is not homogeneous. As seen earlier, erosion produces small fragments with a relatively narrow Gaussian distribution. Thus, each erosion step may be considered as a binary breakup process forming one fragment of size ε and one of size y-ε, with $y \gg \varepsilon$. With these assumptions the relative rate of breakup becomes

$$f(x|y) = \delta(x - (y - \varepsilon)) + \delta(x - \varepsilon), \qquad (65)$$

where y is the mass of the parent agglomerate and ε is the mass of the primary particle (aggregate).

A solution to this problem (Hansen and Ottino, 1996a) reveals that the cluster size distribution is bimodal, as expected, with $c(x,t)$ for large x dependent upon the initial conditions (Fig. 35a). The distribution thus does not approach a self-similar form and the scaling results just given are not valid for this problem. This is a result of the non-homogeneous relative rate of breakup.

Experimental data on erosion processes is limited—but there is some data from the comparable process of attrition. Neil and Bridgwater (1994) determine, by sieving, the cumulative mass fraction of clusters $M(l)$, with a characteristic length less than l. Hansen and Ottino (1996a) assume that $x \approx \pi \rho l^3/6$, where ρ is the density of the solids. Figure 35b shows the fitting to available experimental data. Deviation of the experimental data from the model could be due to inefficiencies in the sieving process, the assumed relationship between x and l, and/or the presence of rupture. Better data are clearly needed before hard conclusions can be drawn (concentrations would be more enlightening than cumulative mass, and the reported size distributions are relatively narrow). It is nevertheless apparent that the relative constancy of $M(l)$ with l indicates the presence of a bimodal distribution.

Advection is important in fragmentation processes, and an initially homogeneous system may evolve spatial variations due to spatially dependent fragmentation rates. For example, Fig. 36 shows the spatial distribution of eroded clusters in the journal bearing flow operating under good mixing

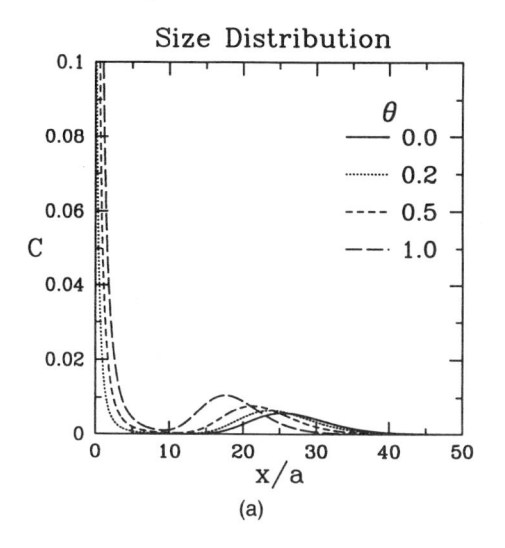

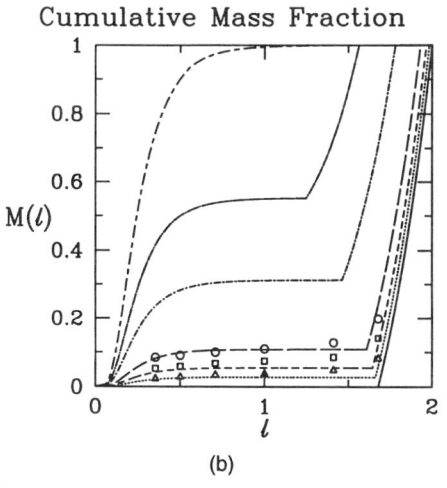

FIG. 35. (a) Size distributions [$c(x, q)$] obtained for erosion-controlled fragmentation for increasing fragmentation time (q). (b) Comparison of theoretically predicted cumulative size distribution for erosion-controlled fragmentation to experimental data (Hansen and Ottino, 1996a).

and poor mixing conditions (Hansen, 1997). Clusters are eroded by hydrodynamic stresses and parent clusters are coded in the figure according to size. The aggregates eroded from the cluster are not shown. The poorly mixed system (Fig. 36a) shows large clusters (clusters eroded to a lesser extent) trapped in a regular island in which the shear rates are low, and

178   J. M. OTTINO *ET AL.*

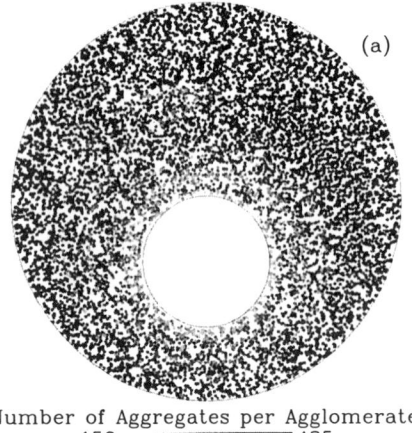

Number of Aggregates per Agglomerate
150 ━━━━━━━ 485

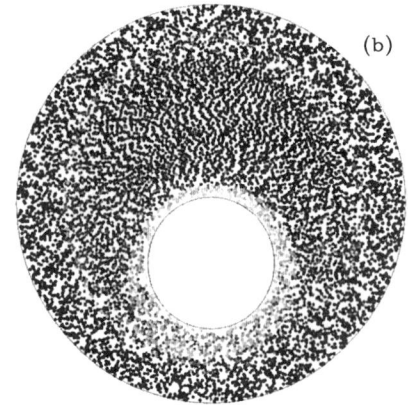

Number of Aggregates per Agglomerate
150 ━━━━━━━ 485

FIG. 36. The spatial variation of agglomerate sizes in simulations of erosion in the journal bearing flow. Initially there are 10000 agglomerates consisting of 400–500 aggregates. The grey scale represents the number of aggregates in the agglomerates. (a) Poorly mixed flow after four periods. (b) Well-mixed flow after one period (Hansen, *et al.* 1998).

small clusters near the inner cylinder where the shear rates are high. However, even the well-mixed system shows spatial variations of cluster sizes, indicating that the rate of mixing is not fast enough to eliminate the fluctuations produced by the spatially dependent fragmentation rate. Such spatial variations would undoubtedly cause deviations from predictions of mean field theories for the cluster size distribution.

*Illustration: **Size reduction by simultaneous erosion and rupture.***
Consider now continuous irreversible breakup by both rupture and erosion

mechanisms in a well-mixed system. As in the model of erosion discussed earlier, we assume that clusters consist of particles of mass ε, and that $\varepsilon \ll s(t)$, so that the discrete size distribution may be approximated as a continuum. The fragmentation equation for simultaneous rupture and erosion for the case of homogeneous rupture kernels is (Hansen and Ottino, 1996a)

$$\frac{\partial c(x,t)}{\partial t} = K\left\{\frac{\partial}{\partial x}[x^\lambda c(x,t)] + \frac{\delta(x-\varepsilon)}{\varepsilon}\int_\varepsilon^\infty y^\lambda c(y,t)\,dy\right\} \quad (66)$$

$$- x^\sigma c(x,t) + \int_x^\infty y^{\sigma-1} b_r(x/y) c(y,t)\,dy,$$

where λ and σ are the homogeneity indices for erosion and rupture, respectively. The first term on the right-hand side accounts for erosion, and the next two terms account for rupture.

A characteristic parameter of the process is the ratio of the overall rate of erosion relative to the overall rate of rupture given by

$$H \sim Ks(t)^{\lambda-\sigma-1}. \quad (67)$$

This ratio determines the applicability of the scaling solution: If $\lambda < \sigma + 1$, the scaling solution breaks down as $s(t)$ approaches zero.

Consider next the case of binary rupture when $\lambda = 0$, $s = 1$, and $b_r(x/y) = 2$. Assuming that fragments once eroded do not fragment further, an approximate form of the evolution equation for clusters larger than size ε is

$$\frac{\partial^2 c}{\partial x \partial t} = K\frac{\partial^2 c}{\partial x^2} - x\frac{\partial c}{\partial x} - 3c(x,t). \quad (68)$$

The mass of the eroded fragments can then be obtained from the mass balance. The solution breaks down as the average size (s) approaches the aggregate size (ε). Following Ziff and McGrady (1985), particle size distribution is obtained as

$$c(x,t) = (1 + t/s)^2 \exp[-(t+s)x - K(t^2/2 + ts)] \quad (69)$$

for the initial condition $c(x, 0) = e^{-sx}$. Since the kinetic equation (68) is linear, the solutions for other initial conditions are easily obtained by superposition.

HEURISTICS

- Erosion:

Occurs even at low fragmentation numbers, and over long time scales
Rate of erosion is directly proportional to the fragmentation number

and the surface area of the parent agglomerate

Bimodal cluster size distribution

Model computations indicate that for short times, polydispersity is inversely proportional to average size

Size distribution is not self-similar

- Rupture:

Occurs rapidly once the fragmentation number exceeds the critical value

Size distributions of fragments produced in a single rupture event are self-similar

For fractal agglomerates, the critical fragmentation number is inversely proportional to its volume $(R/a)^3$

Homogeneous breakage kernels with positive homogeneity index ($\lambda>0$) produce self-similar size distributions at long times

---

B. AGGREGATION

Aggregation of particles may occur, in general, due to Brownian motion, buoyancy-induced motion (creaming), and relative motion between particles due to an applied flow. Flow-induced aggregation dominates in polymer processing applications because of the high viscosities of polymer melts. Controlled studies—the conterpart of the fragmentation studies described in the previous section—may be carried out in simple flows, such as in the shear field produced in a cone and plate device (Chimmili, 1996). The number of such studies appears to be small.

At the simplest level, the rate of flow-induced aggregation of compact spherical particles is described by Smoluchowski's theory [Eq. (32)]. Such expressions may then be incorporated into population balance equations to determine the evolution of the agglomerate size distribution with time. However with increase in agglomerate size, complex (fractal) structures may be generated that preclude analysis by simple methods as above.

Let us illustrate first how different (idealized) aggregation processes may result in different structures. There is extensive literature on *diffusion-limited aggregation* (DLA) (for a comprehensive review, see Meakin, 1988). Three methods of simulation are common: (standard) diffusion-limited aggregation (DLA), reaction-limited aggregation (RLA), and linear trajectory aggregation (LTA). DLA structures are generated by placing a seed particle in the middle of a lattice. Other particles are placed in the lattice and follow a random walk trajectory. These moving particles bind onto the growing seed upon contact to form a cluster. In general, clusters produced by DLA simulations are fractal, and the fractal dimension of the clusters is approximately 2.5 (Fig. 37). The difference between RLA and DLA is that a sticking probability governs the binding in RLA. Thus, particles penetrate further into the growing agglomerate, and denser agglomerates are formed. The dimension of RLA clusters is greater than 2.5 and increases with a decreasing sticking probability. LTA is similar to DLA, with particles following a random linear trajectory instead of a random walk. LTA, which models agglomerates formed by mechanical motion such as a mixing processes, yields compact structures with a fractal dimension close to 3.

Unlike the simulations which only consider particle–cluster interactions discussed earlier, hierarchical cluster–cluster aggregation (HCCA) allows for the formation of clusters from two clusters of the same size. Clusters formed by this method are not as dense as clusters formed by particle-cluster simulations, because a cluster cannot penetrate into another cluster as far as a single particle can (Fig. 37). The fractal dimension of HCCA clusters varies from 2.0 to 2.3 depending on the model used to generate the structure: DLA, RLA, or LTA. For additional details, the reader may consult Meakin (1988).

The direct relevance of these diffusion-driven aggregation models to *flow*-driven aggregation is somewhat questionable, though they serve to highlight the influence of the aggregation process on the structures produced. These models have been used to synthesize structures to investigate how they break (Horwatt *et al.* 1992a,b).

Flow-induced aggregation may also result in the formation of agglomerates with complex (fractal) structure (Jiang and Logan, 1996). An understanding of the structures formed in the aggregation process is important—the kinetics of the aggregation is significantly affected by the type of structures formed, as we will see in the following sections. The kinetics in turn affect the evolution of the agglomerate size distribution, which is the quantity of primary interest from a practical viewpoint.

1. Global Scales: Agglomerate Size Distributions

Analytical approaches to obtain the agglomerate size distribution are possible for well-mixed systems and when the rate of aggregation of clusters is defined by simple functions. In general, irreversible aggregation in well-mixed systems is described by Smoluchowski's coagulation equation, which

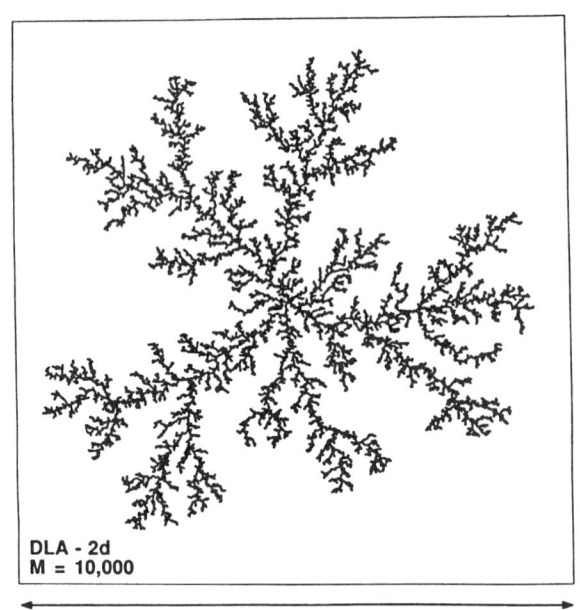

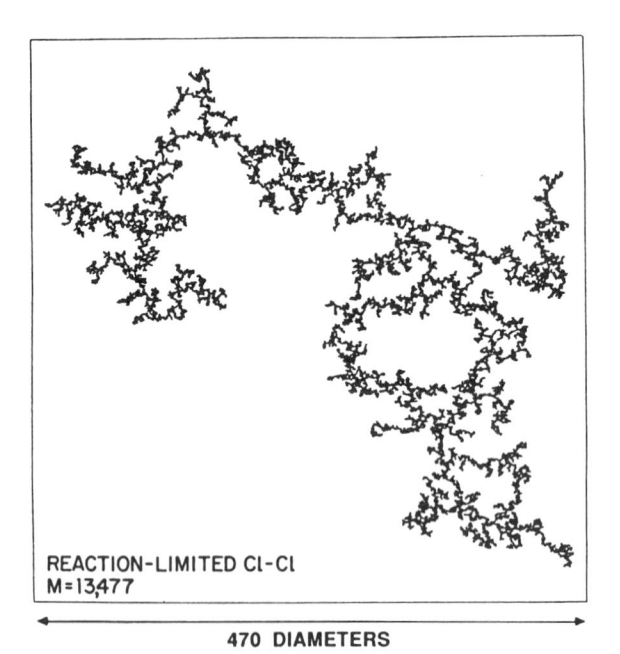

FIG. 37. Typical clusters obtained by diffusion-limited aggregation (DLA). *Top*: Two-dimensional diffusion-limited aggregation. *Bottom*: Reaction-limited hierarchical cluster–cluster aggregation (HCCA) (Meakin, 1988 with permission, from the *Annual Review of Physical Chemistry*, Vol. 39, © by Annual Reviews www.Annual/Reviews.org).

for a continuous distribution of cluster sizes can be written as

$$\frac{\partial c(x, t)}{\partial t} = \frac{1}{2}\int_0^x K(x - y, y)c(x - y, t)c(y, t)\, dy \qquad (70)$$

$$- c(x, t)\int_0^\infty K(x, y)c(y, t)\, dy,$$

where $c(x,t)$ is the concentration of clusters of mass x at time t and $K(x,y)$ is the rate of aggregation of clusters of masses x and y (van Dongen and Ernst, 1988). The first term on the right-hand side accounts for the formation of a cluster of size x due to aggregation of two clusters of size $x - y$ and y, whereas the second term represents the loss of clusters of mass x. There are significant parallels between this equation and the fragmentation equation.

Scaling solutions, as in the case of fragmentation, are possible if the kernel $K(x, y)$ is homogeneous, which requires that

$$K(\xi x, \xi y) = \xi^\kappa K(x, y), \qquad (71)$$

where κ is the homogeneity index. This condition is not very restrictive; for example, the aggregation rate given by Smoluchowski's theory [Eq. (32)] is homogeneous with $\kappa = 1$. Time-invariant scaled solutions are obtained in this case if the concentration is scaled as

$$c(x, t) = x^{-2}\psi(\eta), \qquad (72)$$

where $\eta = x/s(t)$ and $s(t)$ is the average size, as defined earlier. The relationship between the scaled concentration for fragmentation (ϕ) and ψ is then

$$\psi(\eta) = \eta^2\phi(\eta). \qquad (73)$$

This scaling (or self-similarity) is verified by numerous computational and theoretical studies (for a review, see Meakin, 1992).

For many well-studied rate of aggregation kernels, the average cluster size grows algebraically, $s(t) \sim t^z$. In fact, when $K(x, y)$ is homogeneous and κ, as defined in Eq. (73), is less than unity, we get $z = 1/(1 - \kappa)$. When $\lambda > 1$, loss of mass to the formation of infinite size clusters—the opposite of shattering in fragmentation, termed gelling—occurs in finite time.

Smoluchowski's equation, like the fragmentation equation, can be written in terms of the scaling distribution. Furthermore, general forms may be determined for the tails of the scaling distribution—limits of small mass, $x/s(t) \ll 1$, and large mass, $x/s(t) \gg 1$. The details can be found in van Dongen and Ernst (1988).

If the scaling form given earlier holds, then the bare and scaled moments are related by

$$M_\alpha = s(t)^{\alpha-1}m_\alpha. \qquad (74)$$

Hence, the polydispersity [Eq. (60)] is independent of time, provided $\int_0^\infty \phi(\eta)d\eta$ is bounded. Moreover, if $\phi(\eta) = e^{-\eta}$, as is expected for a rate of aggregation independent of mass (Schumann, 1940), the polydispersity is equal to 2. Also note that

$$\frac{M_\alpha}{M_{\alpha-1}} \sim \frac{M_2}{M_1}, \tag{75}$$

and therefore number average and weight average sizes are proportional to each other when scaling used for the fragmentation equation is applicable and the integral of $\phi(\eta)$ is bounded.

Illustration: Short-time behavior in well mixed systems. Consider the initial evolution of the size distribution of an aggregation process for small deviations from monodisperse initial conditions. Assume, as well, that the system is well-mixed so that spatial inhomogeneities may be ignored. Of particular interest is the growth rate of the average cluster size and how the polydispersity scales with the average cluster size.

This can be done by developing equations for the moments—for example, multiplying Smoluchowski's equation by $x^\alpha dx$, integrating from 0 to infinity, and manipulating the limits of integration yields (Hansen and Ottino, 1996b):

$$\frac{dM_\alpha}{dt} = \frac{1}{2}\int_0^\infty \int_0^\infty [(x+y)^\alpha - (x^\alpha + y^\alpha)]K(x,y)c(x,t)c(y,t)\,dxdy. \tag{76}$$

Expanding the new kernel,

$$Q_\alpha(x,y) = [(x+y)^\alpha - (x^\alpha + y^\alpha)]K(x,y), \tag{77}$$

in a series about the average cluster size yields

$$Q_\alpha(x,y) \approx s^{\kappa+\alpha}\left\{Q_\alpha(1,1) + \frac{\partial Q_\alpha(1,1)}{\partial x}\left(\frac{x}{s}-1\right)\right. \tag{78}$$

$$\left. + \frac{\partial Q_\alpha(1,1)}{\partial y}\left(\frac{y}{s}-1\right) + \ldots\right\},$$

where $\partial Q_\alpha(1,1)/\partial x$ denotes $\partial Q_\alpha/\partial x$ evaluated at $x = 1$ and $y = 1$ (this assumes that most of the clusters are relatively close to the average cluster size, i.e., the polydispersity is close to unity). Considering the number average cluster size $s(t) = M_1/M_0$, we determine that

$$\frac{ds}{dt} = \frac{M_1}{2M_0^2}\int_0^\infty \int_0^\infty K(x,y)c(x,t)c(y,t)\,dxdy, \tag{79}$$

which can be rewritten in terms of the moments as

$$\frac{ds}{dt} \approx \frac{M_1}{2}\left\{K(1,1) + \frac{\partial^2 K(1,1)}{\partial x^2}\left[\frac{M_2 M_0}{M_1^2} - 1\right] + \ldots\right\}s^\kappa, \quad (80)$$

where the higher-order terms include combinations of larger moments that are small when the polydispersity is near unity. Hence, when the polydispersity is close to unity the average cluster size is described by

$$s(t) = \begin{cases} s_0\left(1 + (1-\kappa)\dfrac{t}{\tau}\right)^{1/(1-\kappa)} & \text{for } \kappa \neq 1 \\ s_0 \exp\left(\dfrac{t}{\tau}\right) & \text{for } \kappa = 1 \end{cases} \quad \text{with } \tau = \frac{2s_0^{1-\kappa}}{M_1 K(1,1)}. \quad (81)$$

Here s_0 denotes the initial average cluster size.

Similar arguments can be used to determine the behavior of the zeroth and second moments. For example, the polydispersity evolves according to

$$P \approx \frac{1}{\beta}\left[\left(\frac{s(t)}{s_0}\right)^\beta - 1\right] + 1, \text{ where } \beta = \left[\frac{4}{K(1,1)}\frac{\partial K(1,1)}{\partial x} - 1\right], \quad (82)$$

for polydispersities near unity.

It should be noted that the predictions for the number average cluster size and polydispersity agree with analytical results for $K(x, y) = 1, x + y$, and xy. Furthermore, the short-time form of number average size in Eq. (81) matches the form of $s(t)$ predicted by the scaling ansatz. Computational simulations (Hansen and Ottino, 1996b) also verify these predictions (Fig. 38).

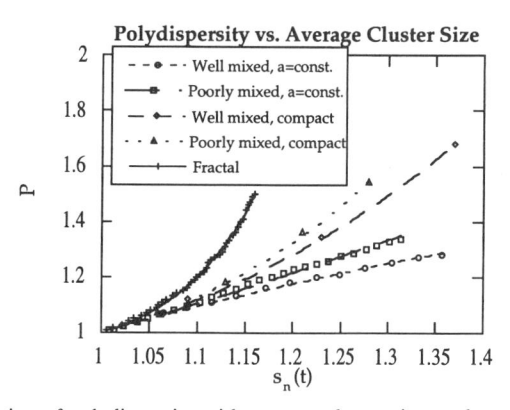

FIG. 38. Variation of polydispersity with average cluster size at short times in a journal bearing flow. The symbols are from simulations and the lines are fits from Eq. (82). The regular flow is the journal bearing flow with only the inner cylinder rotating (Hansen and Ottino, 1996b).

2. Size Distributions for Flow-Induced Aggregation

In this section, we consider flow-induced aggregation without diffusion, i.e., when the Péclet number, Pe ≡ VL/D, where V and L are the characteristic velocity and length and D is the Brownian diffusion coefficient, is much greater than unity. For simplicity, we neglect the hydrodynamic interactions of the clusters and highlight the effects of advection on the evolution of the cluster size distribution and the formation of fractal structures.

Torres, Russel, and Schowalter (1991a,b) have examined structure formation due to aggregation driven by shear and elongational flows. Two cases were computationally examined: *particle–cluster* aggregation, and *cluster–cluster* aggregation. In the case of particle–cluster aggregation, a particle is placed at a random position relative to the cluster and is advected by the flow. Rapid coagulation is assumed so that the particle aggregates with the cluster upon collision. The cluster–cluster aggregation procedure is similar but with the particle replaced by a cluster of the same size as the site cluster. Clusters formed in particle–cluster simulations are denser than suggested by experiments (Torres *et al.*, 1991a). On the other hand, the cluster–cluster simulations agree with experiments in shear flows carried out by the same investigators, suggesting that the formation of clusters is dominated by cluster–cluster aggregation. The fractal dimension of the aggregates formed in both shear and elongational flows by the cluster-cluster simulations is about 1.8 (Torres *et al.*, 1991b).

Only a few studies have considered aggregation in complex flows, in particular chaotic flows. Unlike studies of aggregation by kinetic equations, these simulations allow for spatial variations. A computational study (Muzzio and Ottino, 1988), focusing on compact clusters in a 2D chaotic flow (blinking vortex flow), shows that islands of regularity may cause spatial variations in the rate of aggregation, and that aggregation in "well-mixed" chaotic systems is similar mathematically to Brownian aggregation and can be described by Smoluchowski's equation. The effect of mixing on the fractal nature of clusters is considered explicitly by Danielson *et al.* (1991). They determined that the fractal dimension can be controlled by varying the degree of mixing in chaotic flows. The variation of fractal dimension with mixing is due to the nature of interactions of monomers and larger clusters in different mixing schemes. If the system is not well mixed (large islands), the large clusters, which are trapped in islands, do not interact with each other and the process resembles the particle–cluster aggregation of Torres *et al.* (1991b). However, if the system is well mixed (no islands), then larger clusters interact with each other and aggregation resembles cluster–cluster aggregation. Thus, the fractal dimension of a cluster is expected to decrease with better mixing.

We focus on aggregation in model, regular and chaotic, flows. Two aggregation scenarios are considered: In (i) the clusters retain a compact geometry—forming disks and spheres—whereas in (ii) fractal structures are formed. The primary focus of (i) is *kinetics and self-similarity* of size distributions, while the main focus of (ii) is the *fractal structure* of the clusters and its dependence with the flow.

Illustration: Aggregation in chaotic flows with constant capture radius.
Here we consider aggregation in a physically realizable chaotic flow, the journal bearing flow or the vortex mixing flow described earlier. The computations mimic fast coagulation; particles seeded in the flow are convected passively and aggregate upon contact. In this example the clusters retain a spherical structure and the capture radius is independent of the cluster size.

The evolution of the average cluster size is shown in Fig. 39a. In a well-mixed system the growth of the average cluster size is linear with time (Hansen and Ottino, 1996b). In a poorly mixed system, a system with small chaotic regions, the average cluster size grows less fast $(1 + t/\tau)^{0.6}$, with a small variation in the exponent with the capture radius of the clusters. The growth of the average cluster in a regular flow, i.e., no chaos present, is included for completeness. Intuitively, one may expect the aggregation in the poorly mixed system to be a hybrid of aggregation in the well-mixed flow and in the regular flow, but this is not the case. Surprisingly, the growth rate of the average cluster in the regular flow, approaching linear growth, is faster than in the poorly mixed flow.

As previously discussed, we expect the scaling to hold if the polydispersity, P, remains constant with respect to time. For the well-mixed system the polydispersity reaches about 2 when the average cluster size is approximately 10 particles, and statistically fluctuates about 2 until the mean field approximation and the scaling break down, when the number of clusters remaining in the system is about 100 or so. The polydispersity of the size distribution in the poorly mixed system never reaches a steady value. The ratio $M_3 M_1 / M_2^2$, which is constant if the scaling holds and mass is conserved, is also unsteady in the poorly mixed flow, indicating that the cluster size distribution is not self-similar in poorly mixed flows. The polydispersities of the clusters in the poorly-mixed and well-mixed flows are compared in Fig. 39b. The polydispersity of clusters in a regular flow, also included in Fig. 39b, initially increases, then slowly approaches unity as the number of clusters approaches 1. The higher polydispersity for clusters formed in the poorly mixed flow is indicative of the wider range of cluster sizes present in the system.

The self-similar nature of the cluster size distributions for the well-mixed flow is shown in Fig. 39c. Because of the discrete nature of cluster size in

Average Cluster Size

(a)

Polydispersity vs. Average Cluster Size

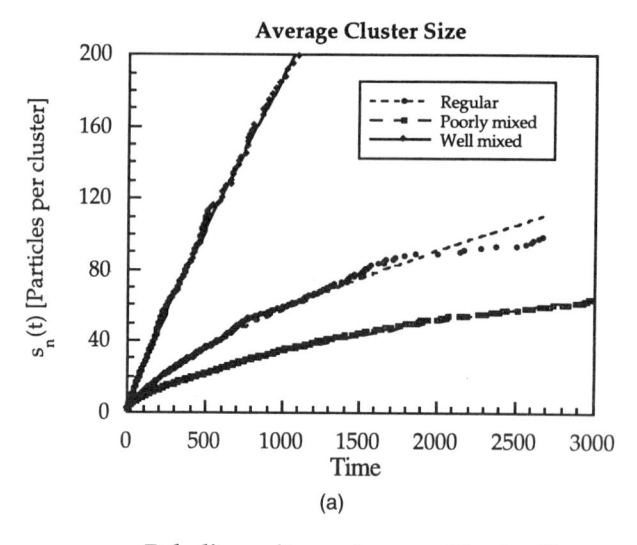

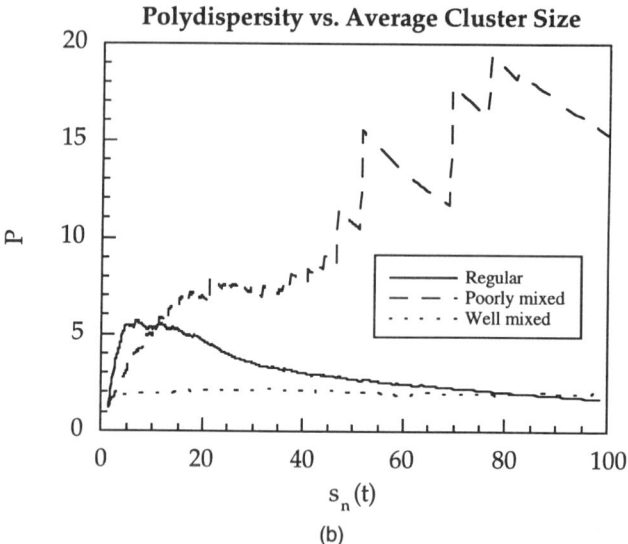

(b)

FIG. 39. (a) The growth of average cluster size for clusters with a constant capture radius in various 2D flows. (b) Variation of polydispersity with average cluster size. (c) Scaled distribution of the cluster sizes at different times. The regular flow is the journal bearing flow with only the inner cylinder rotating. One time unit is equivalent to the total displacement of the boundaries equal to the circumference of the outer cylinder (Hansen and Ottino, 1996b).

Scaled Size Distribution

[graph with y-axis $[s_n(t)]^2 c(x,t)/M_1$ from 0.00 to 1.20, x-axis $x/s_n(t)$ from 0 to 12, legend: • $s(t) = 8.78$, ■ $s(t) = 11.37$, ---- $\exp(-x/s(t))$]

(c)

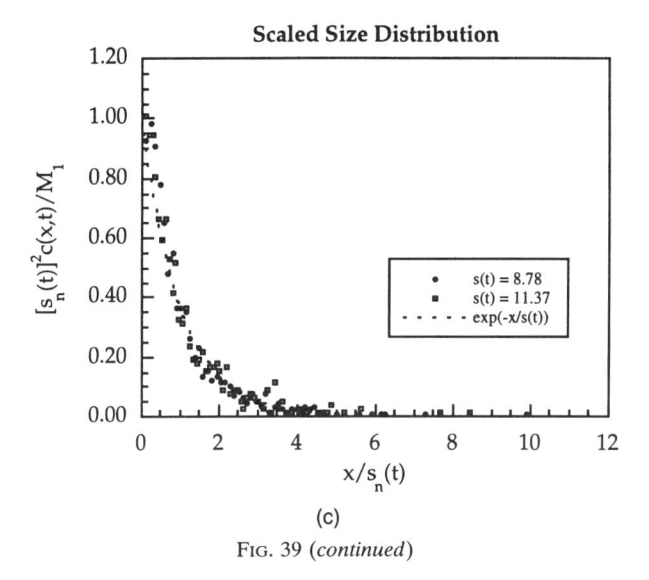

FIG. 39 (*continued*)

the simulations, this figure shows the cumulative mass fraction, as it is less prone to statistical fluctuations (a scaled cluster size distribution is also included for clarity). It should be noted that the cumulative size distribution scales as

$$M(x, t) = \int_0^x y c(y, t) \, dy = \int_0^{x/s(t)} \eta \phi(\eta) \, d\eta. \tag{83}$$

Thus, we plot $M(x,t)/M_1$ vs $x/s(t)$. As noted earlier, the cluster size distribution and the first moment of the size distribution are averaged over the entire journal bearing. As indicated by the behavior of P in Fig. 39b, the cluster size distribution becomes self-similar when the average size is about 10 particles per cluster.

Illustration: Aggregation of area-conserving clusters in two dimensional chaotic flows. Particles, convected passively in a two-dimensional chaotic flow, aggregate on contact to form clusters. The capture radius of the clusters increases with the size of the cluster. Since these simulations are in two dimensions, the area of the aggregating clusters is conserved.

The rate of aggregation of clusters with the same capture radius in two dimensions is proportional to the area of the cluster (Muzzio and Ottino, 1988). If this dependence of the rate of aggregation on cluster size holds for clusters with different capture radii, then the aggregation in these simulations is on the verge of gelation (van Dongen and Ernst, 1988). Simulations show that for aggregation of area-conserving clusters the formation of one large cluster predominates, but an infinite cluster is not formed since the system is finite. This large cluster is formed before the scaling distribution is reached.

In a system with a small chaotic region, the rate of aggregation in the regular region is significantly slower than in the chaotic region. Thus, the polydispersity of the clusters remains large until the regular region is "broken" by the increasing capture radius of the cluster. However, the regular region retains its identity if the capture radius of the clusters does not become large enough that clusters in the regular region aggregate with clusters in the chaotic region.

As shown in Fig. 40, the average cluster size in a well-mixed or chaotic system evolves as $(1 + t/\tau)^6$. There are small variations in the exponent with the area fraction of the clusters. The growth rate of the average cluster size in a poorly mixed system follow the same form; however, the exponent is quite diffrent. In a poorly mixed system, the exponent is dependent on the area fraction of clusters. For one flow studied, when the area fraction of clusters is 0.02 the exponent is 0.81. When the area fraction of clusters is increased to 0.1, the rate of aggregation increases dramatically—the exponent is 25. In fact, it is difficult to determine if this high rate of aggregation is algebraic or exponential. This implies that above some critical area fraction of clusters the large clusters are more likely to aggregate, whereas small clusters are more likely to aggregate below some critical area fraction of clusters. In the poorly mixed system, it appears that the growth rate changes as aggregation proceeds because clusters in the chaotic region become large enough to aggregate with clusters in the regular region.

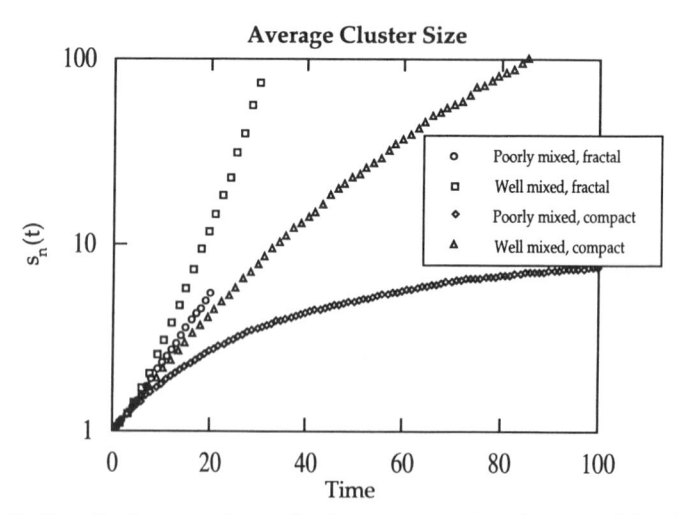

FIG. 40. Growth of average cluster size for area conserving clusters and fractal clusters in the journal bearing flow (Hansen and Ottino, 1996b).

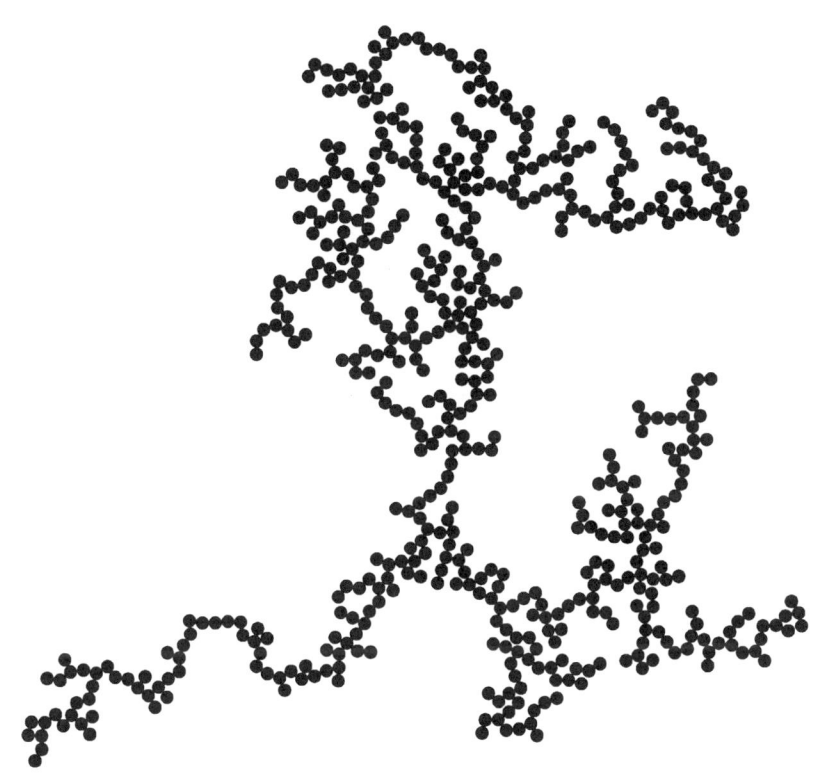

FIG. 41. Typical 2D fractal structure obtained by aggregation of particles in the journal bearing flow. Fractal dimension of the cluster is 1.54 (Hansen and Ottino, 1996b).

Illustration: Aggregation of fractal structures in chaotic flows. In a further study of aggregation in two-dimensional chaotic flows, the passively convected clusters retain their geometry after aggregation, i.e., fractal structures are formed. A typical fractal cluster resulting from these simulations is shown in Fig. 41.

The kinetics of fractal structures differ significantly from the kinetics of compact structures. As with the area-conserving compact structures, no self-similar distribution is observed. However, fractal clusters grow faster than compact structures. This is highlighted in Fig. 40, which illustrates that the average cluster size of fractal structures grows nearly exponentially, while the average cluster size of compact structures grows algebraically. The growth of fractal structures in the poorly mixed system is accurately described by $s(t) = s_0 \exp(t/\tau)$; however, aggregation in the well-mixed system may only be approximated by the exponential form. Interestingly, the overall rate of aggregation of fractal structures is first order, while the overall rate of aggregation of clusters with a constant capture radius is second order in the well-mixed system. The rate of aggregation of area-conserving clusters fits between first order and second order.

Note that only one system, the one corresponding to constant capture radius clusters in chaotic flows, behaves as expected via mean field predictions. In general, the average cluster size grows fastest in the well-mixed system. However, in some cases the average cluster size in the regular flow grows faster than in the poorly mixed system.

Furthermore, since the average cluster size of fractal clusters grows exponentially, we do not expect the scaling to hold. The argument is that κ would be unity if the scaling ansatz were applicable here, but that the scaling is not valid for $\kappa \geq 1$ and therefore not applicable to this case. Indeed, the polydispersity does not approach a constant value, indicating that the scaling does not apply. Again, significant differences between the polydispersities of fractal structures and compact structures are shown in Fig. 38. Initially, P increases rapidly; as is shown in Fig. 38, the polydispersity of fractal clusters grows faster than the polydispersity of their compact counterparts. Furthermore, in these simulations, P of fractal structures approaches unity sooner.

The long-time behavior of the polydispersity suggests that only one system, the well-mixed system with constant capture radius, may be described by the scaling ansatz. The scaling breaks down in poorly mixed systems because of spatial fluctuations. In other systems, aggregation of area-conserving or fractal clusters, the scaling breaks down because aggregation is dominated by formation of one or several large clusters.

The fractal nature of the structures is also of interest. Because of the wide range of flow in the journal bearing, a distribution of fractal clusters is produced. When the area fraction of clusters is 0.02, the median fractal dimension of the clusters is dependent on the flow, similar to the study by Danielson et al. (1991). The median fractal dimension of clusters formed in the well-mixed system is 1.47, whereas the median fractal dimension of clusters formed in the poorly mixed case is 1.55. Furthermore, the range of fractal dimensions is higher in the well-mixed case.

The results are different when the area fraction of clusters increases. The distribution of the fractal dimension of the clusters for a system with an area fraction of clusters of 0.10 is shown in Fig. 42. The median fractal dimension of the clusters is independent of the flow and is approximately

---

FIG. 42. Distribution of the fractal dimensions of the clusters generated by aggregation in the journal bearing flow for different flow types (Hansen and Ottino, 1996b).

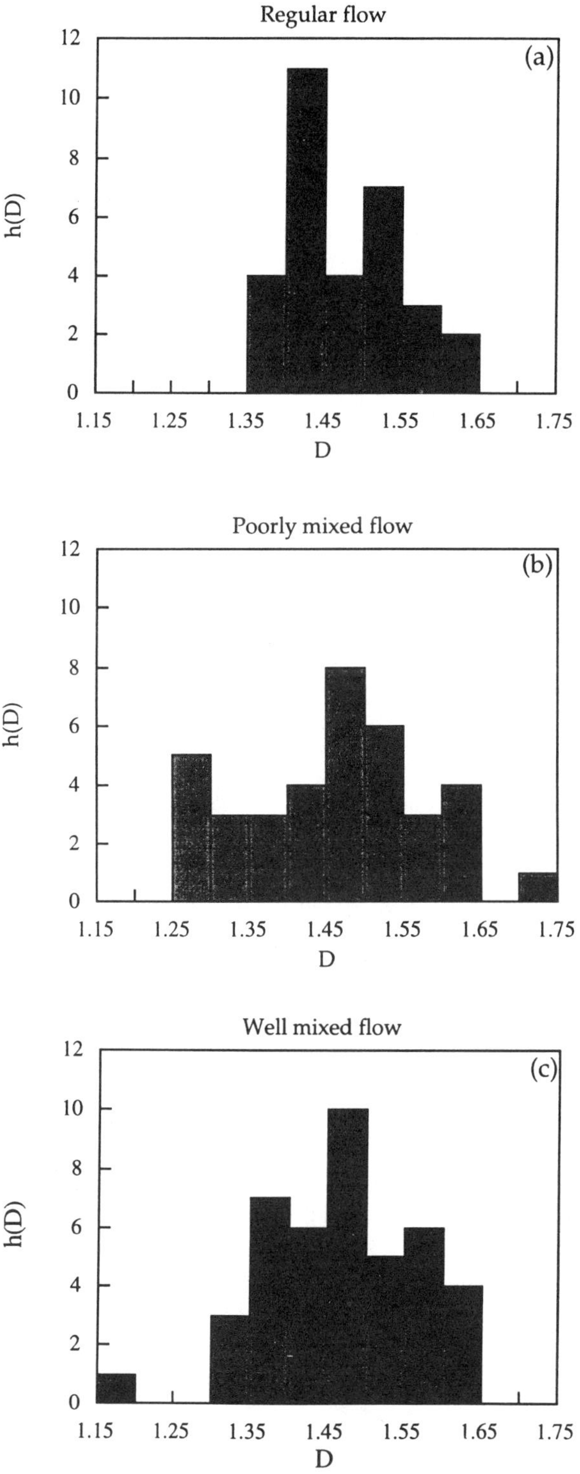

Regular flow
(a)
h(D)
D
Poorly mixed flow
(b)
h(D)
D
Well mixed flow
(c)
h(D)
D

1.47. Since the fractal dimension of the clusters is closer to the dimension of the clusters in the well-mixed system with a lower area fraction of clusters, this suggests that as the area fraction of clusters increases, the island of regularity gets broken up by the increasing capture radius of the clusters. Thus, aggregation in the poorly mixed system behaves similarly to that in the well-mixed system when aggregation occurs between the two disjoint regions of the flow.

HEURISTICS

- Rate of aggregation of compact clusters in well-mixed systems follows Smoluchowski's theory.
- Fractal dimension of clusters formed by flow-induced aggregation is independent of local flow type.
- Fractal dimension depends on mixing in chaotic flows: good mixing (no islands) gives lower fractal dimensions.
- Model computations indicate that the *average cluster size:*

Grows algebraically for compact structures; linearly for constant capture radius clusters in a well-mixed system.

Grows exponentially for fractal structures.

Has the fastest rate of growth in well-mixed systems.

- Model computations indicate that the *polydispersity:*

For short times is predicted by Eq. (82)

For long times is

  (i) Constant ~ 2 for constant capture radius clusters in a well-mixed system.
  (ii) Nonconstant for poorly mixed systems, and for area-conserving compact and fractal clusters.

IV. Concluding Remarks

The modeling of mixing processes has undergone exciting progress in the past few years. Computations have reached maturity and exploitation of concepts and results in the context of realistic devices is now a reality (Avalosse and Crochet, 1997a,b). But much, if not all, of the advances have been restricted to single-phase fluids. One would expect that similar advances will take place in dispersion of solids and liquids in viscous flows. One may also speculate as to whether modeling or theory will drive experiments or the other way around. One might argue that as things are now, theory is ahead and that more experiments are badly needed. But the theory picture, described at length in the previous sections, is clearly incomplete and one may argue that more realism is needed.

As pointed out earlier most of the examples involve "one-sided" interactions: the microstructures are acted on by the chaotic flow, but they do not modify the flow structure itself. Not much has been done regarding these interactions in the context of chaos and mixing, and it is probably useful to point out here whatever little is known. An example of coupling involves dilute viscoelastic fluids (polymer solutions) in chaotic flows. A general finding is that the *rate* of mixing seems always to be slowed down by elasticity. A second general example along the lines of microstructures and flow is the presence of a dispersed phase that alters the flow patterns, and computational studies have shown how regular islands can be broken as a consequence. Little is known, in general, in this case as well.

The caveats notwithstanding practical applications based on already existing theory seem possible. Zumbrunen *et al.* have studied experimentally the possibility of developing fine-scale structures of dispersed phased by exploiting chaotic three-dimensional flows (Liu and Zumbrunnen, 1996; Zumbrunnen, Miles, and Liu, 1996), and Liu and Zumbrunnen (1997) note that significant improvements in impact properties of a polystyrene matrix can be achieved by adding small volume fraction (~9%) of low-density polyethylene. As opposed to conventional dispersion methods, where the minor phase may consist of randomly distributed small droplets, controlled chaotic mixing produces highly connected stretched and folded structures that can be preserved upon solidification.

The interaction between the dispersed-phase elements at high volume fractions has an impact on breakup and aggregation, which is not well understood. For example, Elemans *et al.* (1997) found that when closely spaced stationary threads break by the growth of capillary instabilities, the disturbances on adjacent threads are half a wavelength out of phase (Fig. 43), and the rate of growth of the instability is smaller. Such interaction effects may have practical applications, for example, in the formation of monodisperse emulsions (Mason and Bibette, 1996).

Although the fundamentals of breakup of drops are reasonably well-understood, the physics of the fragmentation of clusters is not as well grounded. Characterization of the structure and the measurement of the strength of agglomerates need to be addressed in greater detail. However, as in the case of liquid–liquid dispersions, existing concepts provide inspiration for possible new technologies, as suggested, for example, by Danescu and Zumbrunnen (1997) for the case of conductive composites involving small volume fraction of a conductive component. Both extremes, no-

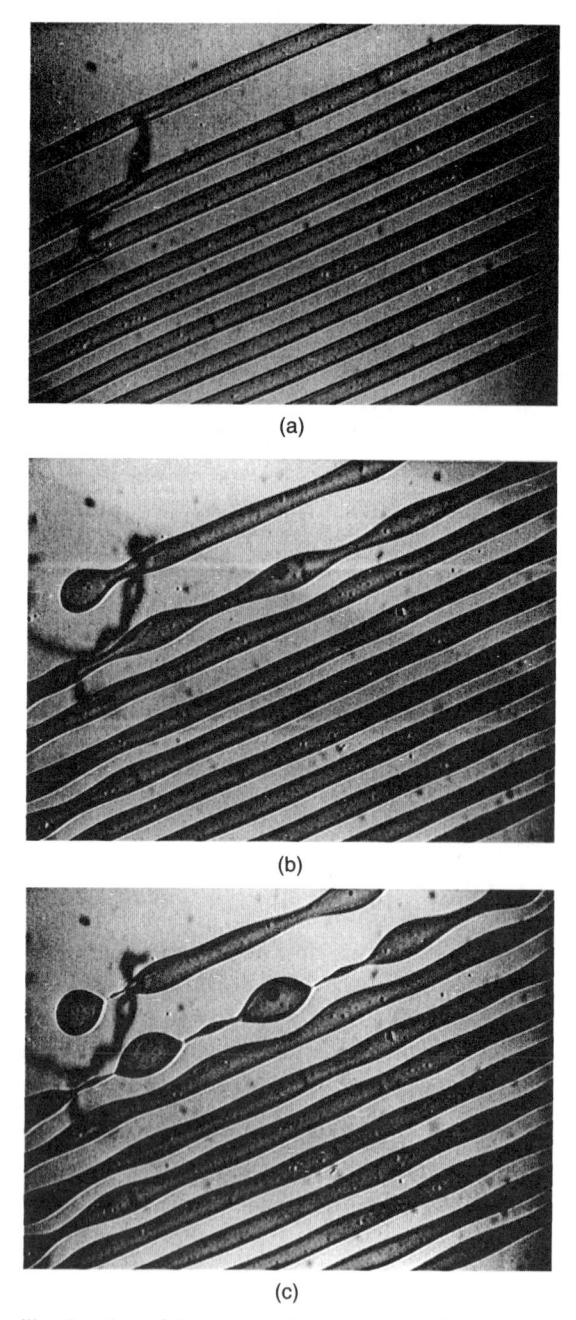

FIG. 43. Capillary breakup of closely spaced molten nylon-6 threads in molten polystyrene. Photograhs at different times are shown (frames "a" through "f" correspond to 0, 210, 270, 360, 390, and 510 s). The initial thread diameter is 70 μm (Elemans *et al.*, 1997).

(d)

(e)

(f)

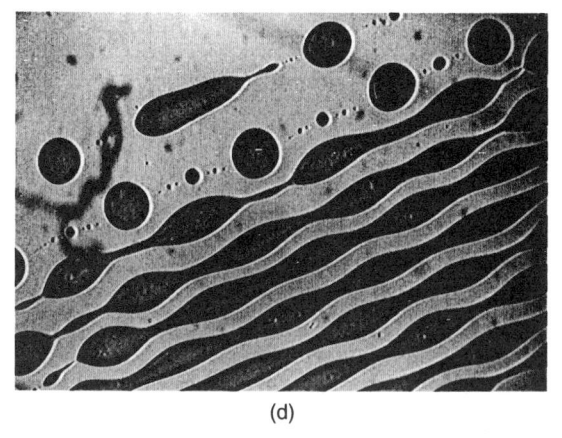

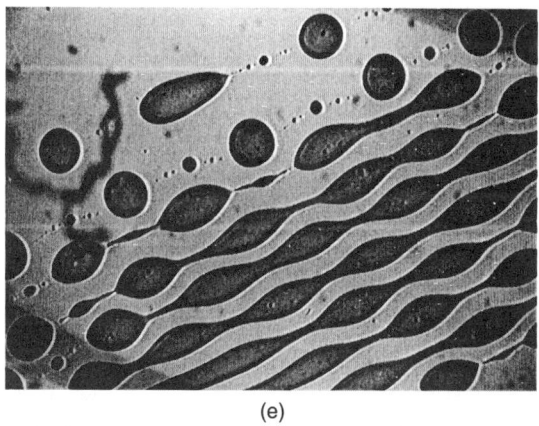

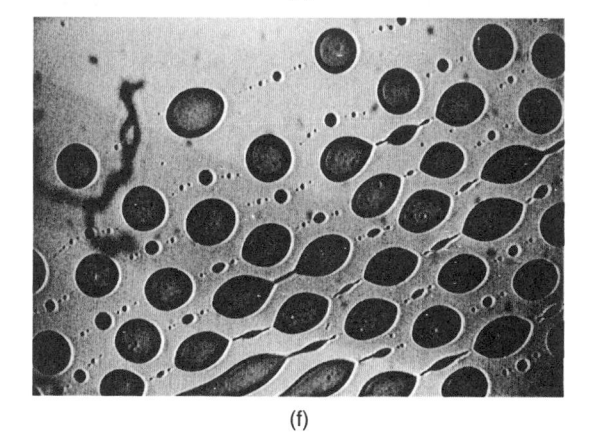

Fig. 43 (*continued*)

mixing and "perfect" mixing, lead to poor overall conductivities. No-mixing corresponds to the case of large isolated agglomerates in a polymer matrix; "perfect" mixing, to randomly placed aggregates having little contact with each other. At low volume fractions there are no sample-spanning clusters percolating through the sample, and the conductivity is poor. The best conductivities result when the particles are still recognizable as being aligned in long, thin interconnected structures, before further mixing randomizes the particles and connectivity is lost. There has been, for some time, industrial empirical evidence that "too much mixing" may not be good, and in this case, existing theory provides a rationalization of the observed events.

Integrating the fundamental processes at drop/cluster length scales to realistic flow fields is another area that is still developing, and the gap between the work presented here and more practical extrusion applications (Rauwendaal, 1994; White, 1990) needs to be bridged. Advances in this field would undoubtedly have an impact on the analysis and design of mixing equipment, as well as on optimization of processing conditions for mixers. All these issues require in-depth study before applications emerge.

It should be stressed again that the application of the concepts presented here is relevant to polymer and rubber processing and composite applications—which now form a fertile ground for the application of these ideas—as well to a variety of consumer products industries where product value is intimately tied to the creation of unique structures. Examples in this area may range from food products such as ice cream and margarine to laundry and personal care products such as dishwasher liquids, creams, and lotions. The creation of an organized body of knowledge geared toward the consumer product area presents considerable opportunities (Villadsen, 1997).

ACKNOWLEDGMENT

The research presented here has been supported over the past decade or so by grants to JMO awarded by the National Science Foundation, the Department of Energy—Basic Energy Sciences, and 3M.

REFERENCES

Acrivos, A., The breakup of small drops and bubbles in shear flows. *4th International Conference on Physicochemical Hydrodynamics, Ann. N. Y. Acad. Sci.,* **404,** 1–11 (1983).

Adler, P. M., and Mills, P. M., Motion and rupture of a porous sphere in a linear flow field. *J. Rheol.* **23,** 25–37 (1979).
Aref, H., Stirring by chaotic advection. *J. Fluid Mech.* **143,** 1–21 (1984).
Austin, L. G., A review introduction to the mathematical description of grinding as a rate process. *Powder Technol.* **5,** 1–17 (1971).
Avalosse, Th., Simulation numerique du melange laminaire par elements finis. Ph.D. Thesis, Université Catholique de Louvain, Belgium (1993).
Avalosse, Th., and Crochet, M. J., Finite element simulation of mixing: 1. Two-dimensional flow in periodic geometry. *AIChE J.* **43,** 577–587 (1997a).
Avalosse, Th., and Crochet, M. J., Finite element simulation of mixing: 2. Three-dimensional flow through a Kenics mixer. *AIChE J.* **43,** 588–597 (1997b).
Bagster, D. F., and Tomi, D., The stresses within a sphere in simple flow fields. *Chem. Eng. Sci.* **29,** 1773–1783 (1974).
Barthès-Biesel, D., and Acrivos, A., Deformation and burst of a liquid droplet freely suspended in a linear shear field. *J. Fluid Mech.* **61,** 1–21 (1973).
Batchelor, G. K., and Green J. T., The hydrodynamic interaction of two small freely-moving spheres in a linear flow field. *J. Fluid Mech.* **56,** 375–400 (1972).
Becher, P., and McCann, M., The process of emulsification: A computer model. *Langmuir* **7,** 1325–1331 (1991).
Bentley, B. J., and Leal, L. G., A computer-controlled four-roll mill for investigations of particle and drop dynamics in two-dimensional linear shear flows. *J. Fluid Mech.* **167,** 219–240 (1986).
Bidkar, U. R., and Khakhar, D. V., Collision rates in chaotic flows: Dilute suspensions. *Phys. Rev. A* **42,** 5964–5969 (1990).
Bigg, D., and Middleman, S., Laminar mixing of a pair of fluids in a rectangular cavity. *Ind. Eng. Chem. Fundam.* **13,** 184–190 (1974).
Brenner, H., Dissipation of energy due to solid particles suspended in a viscous liquid. *Phys. Fluids* **1,** 338–346 (1958).
Brunk, B. K., Koch, D. L., and Lion, L. W., Hydrodynamic pair diffusion in isotropic random velocity fields with application to turbulent coagulation. *Phys. Fluids* **9,** 2670–2691 (1997).
Chaiken, J., Chevray, R., Tabor, M., and Tan, Q. M., Experimental study of Lagrangian turbulence in *Stokes flow, Proc. Roy. Soc. Lond.* **A408,** 165–174 (1986).
Chakravarthy, V. S., and Ottino, J. M., Mixing of two viscous fluids in a rectangular cavity. *Chem. Eng. Sci.* **51,** 3613–3622 (1996).
Chella, R., and Ottino, J. M., Fluid mechanics of mixing in a single-screw extruder. *I&EC Fundamentals* **24,** 2, 170–180 (1985).
Chella, R., and Viñals, J., Mixing of a two phase fluid by cavity flow. *Phys. Rev. E* **53,** 3832–3840 (1996).
Cheng, Z., and Redner, S., Kinetics of fragmentation. *J. Phys. A: Math. Gen.* **23,** 1233–1258 (1990).
Chesters, A. K., The modeling of coalescence processes in fluid–liquid dispersions: A review of current understanding. *Trans. Inst. Engrs.* **69,** 259–270 (1991).
Chimmili, S., Shear induced agglomeration in particulate suspensions. M. Sci. thesis, West Virginia University (1996).
Danescu, R. I., and Zumbrunnen, D. A., Creation of conducting networks among particles in polymer melts by chaotic mixing. *J. Thermoplast Composites,* **11,** 299–320 (1998).
Danielson, T. J., Muzzio, F. J., and Ottino, J. M., Aggregation and structure formation in chaotic and regular flows. *Phys. Rev. Lett.* **66,** 3128–3131 (1991).
de Bruijn, R. A., Tipstreaming of drops in simple shear flows. *Chem. Eng. Sci.* **48,** 2, 277–284 (1993).

DeRoussel, P., Mixing and dispersion of viscous liquids. Ph.D. Thesis in progress, Northwestern University (1998).
Eggers, J., Nonlinear dynamics and breakup of free surface flows. *Rev. Modern Phys.* **69**, 865–930 (1997).
Elemans, P. H. M., van Wunnik, J. M., and van Dam, R. A., Development of morphology in blends of immiscible polymers. *AIChE J.* **43**, 1649–1651 (1997).
Erwin, L., Theory of mixing sections in single screw extruders. *Polym. Eng. Sci.* **18**, 7, 572–576 (1978).
Feke, D. L., and Manas-Zloczower, I., Rupture of inhomogeneous spherical clusters by simple flows. *Chem. Eng. Sci.* **46**, 2153–2156 (1991).
Filippov, A. F., On the distribution of the sizes of particles which undergo splitting. *Theory of Probab. and Its Appl.* (Engl. Transl.) **4**, 275–294 (1961).
Franjione, J. G., and Ottino, J. M., Feasibility of numerical tracking of material lines and surfaces in chaotic flows. *Phys. Fluids* **30**, 3641–3643 (1987).
Giesekus, H., Strömungen mit konstantem Geschwindigkeitsgradienten und die Bewegung von darin suspendierten Teilchen. *Rheol. Acta* **2**, 112–122 (1962).
Grace, H. P., Dispersion phenomena in high viscosity immiscible fluid systems and application of static mixers as dispersion devices in such systems. *3rd Eng. Found. Conf. Mixing, Andover, N. H.*; Republished in *Chem. Eng. Commun.* **14**, 225–227 (1982).
Hansen, S., Aggregation and fragmentation in chaotic flows of viscous fluids. Ph.D. Thesis, Northwestern University (1997).
Hansen, S., and Ottino, J. M., Agglomerate erosion: A nonscaling solution to the fragmentation equation. *Phys. Rev. E* **53**, 4209–4212 (1996a).
Hansen, S., and Ottino, J. M., Aggregation and cluster size evolution in nonhomogeneous flows. *J. Colloid Interface Sci.* **179**, 89–103 (1996b).
Hanson, S., Khakhar, D. V., and Ottino, J. N., Dispersion of solids in nonhomogeneous viscous flows, *Chem. Eng. Sci.* **53**, 1803–1817 (1998).
Hartley, P. A., and Parfitt, G. D., An improved split cell apparatus for the measurement of tensile strength of powders. *J. Phys. E, Sci. Instrum.* **17**, 347–349 (1984).
Hobbs, D. M., and Muzzio, F. J., The Kenics static mixer: a three-dimensional chaotic flow. *Chem. Eng. J.* **67**(3), 133–166 (1997).
Horwatt, S. W., Feke, D. L., and Manas-Zloczower, I., The influence of structural heterogeneities on the cohesivity and breakup of agglomerates in simple shear flow. *Powder Technol.* **72**, 113–119 (1992a).
Horwatt, S. W., Manas-Zloczower, I., and Feke, D. L., Dispersion behavior of heterogeneous agglomerates at supercritical stresses. *Chem. Eng. Sci.* **47**, 1849–1855 (1992b).
Huneault, M. A., Shi, Z. H., and Utracki, L. A., Development of polymer blend morphology during compounding in a twin-screw extruder. Part IV: A new computational model with coalescence. *Polym. Eng. Sci.* **35**(1), 115–127 (1995).
Jana, S. C., Metcalfe, G., and Ottino, J. M., Experimental and computational studies of mixing in complex Stokes flow—the vortex mixing flow and the multicellular cavity flow. *J. Fluid Mech.* **269**, 199–246 (1994).
Jana, S. C., Tjahjadi, M., and Ottino, J. M., Chaotic mixing of viscous fluids by periodic changes in geometry—baffled cavity flow. *AIChE J.* **40**, 1769–1781 (1994b).
Janssen, J. M. H., Dynamics of liquid–liquid mixing. Ph.D. Thesis, Eindhoven University of Technology (1993).
Janssen, J. M. H., and Meijer, H. E. H., Droplet breakup mechanisms: stepwise equilibrium versus transient dispersion. *J. Rheol.* **37**(4), 597–608 (1993).
Janssen, J. M. H., and Meijer, H. E. H., Dynamics of liquid–liquid mixing: A 2-zone mixing model. *Polym. Eng. Sci.* **35**(22), 1766–1780 (1995).

Jiang, Q., and Logan, B. E., Fractal dimensions of aggregates from shear devices. *J. AWWA,* February, pp. 100–113 (1996).
Kao, S. V., and Mason, S. G., Dispersion of particles by shear. *Nature* **253,** 619–621 (1975).
Kaper, T. J., and Wiggins, S., An analytical study of transport in Stokes flows exhibiting large scale chaos in the eccentric journal bearing. *J. Fluid Mech.* **253,** 211–243 (1993).
Kendall, K., Agglomerate strength. *Powder Metallurgy* **31,** 28–31 (1988).
Khakhar, D. V., and Ottino, J. M., Fluid mixing (stretching) by time periodic sequences for weak flows. *Phys. Fluids* **29**(11), 3503–3505 (1986a).
Khakhar, D. V., and Ottino, J. M., A note on the linear vector model of Olbricht, Rallison, and Leal as applied to the breakup of slender axisymmetric drops. *J. Non-Newtonian Fluid Mech.* **21,** 127–131 (1986b).
Khakhar, D. V., and Ottino, J. M., Deformation and breakup of slender drops in linear flows. *J. Fluid Mech.* **166,** 265–285 (1986c).
Khakhar, D. V., and Ottino, J. M., Breakup of liquid threads in linear flows. *Int. J. Multiphase Flow* **13**(1), 71–86 (1987).
Khakhar, D. V., Rising, H., and Ottino, J. M., An analysis of chaotic mixing in two chaotic flows. *J. Fluid Mech.* **172,** 419–451 (1986).
Khakhar, D. V., Franjione, J. G. and Ottino, J. M., A case study of chaotic mixing in deterministic flows: the partitioned pipe mixer. *Chem. Eng. Sci.* **42,** 2909–2926 (1987).
Kolmogorov, N. A., Über das logarithmisch normale Verteilungsgesetz der Dimensionen der Teilchen bei Zerstuckelung. *Doklady Akad. Nauk. SSSR* **31,** 99–101 (1941); translated into English by Levin in NASA-TT F-12,287 (1969).
Kuhn, W., Spontane Aufteilung von Flussigkeitszylindern in klein Kugeln. *Kolloid Z.* **132,** 84–99 (1953).
Kumar, S., and Homsy, G. M., Chaotic advection in creeping flow of viscoelastic fluids between slowly modulated eccentric cylinders. *Phys. Fluids* **8,** 1774–1787 (1996).
Kusch, H. A., and Ottino, J. M., Experiments on mixing in continuous chaotic flows. *J. Fluid Mech.* **236,** 319–348 (1992).
Lee, Y. J., Feke, D. L., and Manas-Zloczower, I., Dispersion of titanium dioxide agglomerates in viscous media. *Chem. Eng. Sci.* **48,** 3363–3372 (1993).
Leong, C. W., and Ottino, J. M., Experiments on mixing due to chaotic advection in a cavity *J. Fluid Mech.* **209,** 463–499 (1989).
Leong, C. W., Chaotic mixing of viscous fluids in time-periodic cavity flows. Ph.D. Thesis, Univ. Massachusetts, Amherst (1990).
Levich, V. G., "Physicochemical Hydrodynamics." Prentice-Hall, Englewood Cliffs, NJ, 1962.
Liu, Y. H., and Zumbrunnen, D. A., Emergence of fibrillar composites due to chaotic mixing of molten polymers. *Polymer Composites* **17,** 187–197 (1996).
Liu, Y. H., and Zumbrunnen, D. A., Toughness enhancement in polymer blends due to the in-situ formation by chaotic mixing of fine-scale extended structures, *J. Mater. Sci.* **34,** 1921–1931 (1999).
Manas-Zloczower, I., Dispersive mixing of solid additives, *in* "Mixing and Compounding of Polymers—Theory and Practice." (I. Manas-Zloczower and Z. Tadmor, Ed.) Hanser Publishers, Munich, 1994, pp. 55–83.
Manas-Zloczower, I., and Feke, D. L., Analysis of agglomerate separation in linear flow fields. *Intern. Polym. Process. II* 3/4, 185–190 (1988).
Manas-Zloczower, I., Nir, A., and Tadmor, Z., Dispersive mixing in internal mixers—a theoretical model based on agglomerate rupture. *Rubber Chem. Tech.* **55,** 1250–1285 (1982).
Manas-Zloczower, I., Nir, A., and Tadmor, Z., Dispersive mixing in rubber and plastics. *Rubber Chem. Tech.* **57,** 583–620 (1984).

Mason, T. G., and Bibette, J., Emulsification in viscoelastic media. *Phys. Rev. Lett.* **77,** 3481–3484 (1996).

McGrady, E. D., and Ziff, R. M., Shattering transition in fragmentation. *Phys. Rev. Lett.* **58,** 892–895 (1987).

McGrady, E. D., and Ziff, R. M., Analytical solutions to fragmentation equations with flow. *AIChE J.* **34,** 2073–2076 (1988).

Meakin, P., Models for colloidal aggregation. *Ann. Rev. Phys. Chem.* **39,** 237–269 (1988).

Meakin, P., Aggregation kinetics. *Physica Scripta* **46,** 295–331 (1992).

Meijer, H. E. H., and Janssen, J. M. H., Mixing of immiscible liquids, *in* "Mixing and Compounding of Polymers—Theory and Practice." (I. Manas-Zloczower and Z. Tadmor, Ed.). Hanser Publishers, Munich, 1994, pp. 85–147.

Meijer, H. E. H., Lemstra, P. J., and Elemans, P. M. H., Structured polymer blends. *Makromol. Chem., Makromol. Symp.* **16,** 113–135 (1988).

Mikami, T., Cox, R. G., and Mason, S. G., Breakup of extending liquid threads. *Int. J. Multiphase Flow* **2,** 113–138 (1975).

Morse, P. M., and Feshbach, H., "Methods of Theoretical Physics." McGraw-Hill, New York, 1953.

Muzzio, F. J., and Ottino, J. M., Coagulation in chaotic flows. *Phys. Rev. A* **38,** 2516–2524 (1988).

Muzzio, F. J., Swanson, P. D., and Ottino, J. M., The statistics of stretching and stirring in chaotic flows. *Phys. Fluids A* **5,** 822–834 (1991a).

Muzzio, F. J., Tjahjadi, M., and Ottino, J. M., Self-similar drop size distributions produced by breakup in chaotic flows. *Phys. Rev. Lett.* **67,** 54–57 (1991b).

Neil, A. U., and Bridgwater, J., Attrition of particulate solids under shear. *Powder Technol.* **80,** 207–209 (1994).

Niederkorn, T. C., and Ottino, J. M., Mixing of viscoelastic fluids in time-periodic flows. *J. Fluid Mech.* **256,** 243–268 (1993).

Niederkorn, T. C., and Ottino, J. M., Mixing of shear thinning fluids in time-periodic flows. *AIChE J.* **40,** 1782–1793 (1994).

Olbricht, W. L., Rallison, J. M., and Leal, L. G., Strong flow criteria based on microstructure deformation. *J. Non-Newtonian Fluid Mech.* **10,** 291–318 (1982).

Ottino, J. M., "The Kinematics of Mixing: Stretching, Chaos, and Transport." Cambridge Univ. Press, Cambridge, U.K., 1989.

Ottino, J. M., Mixing, chaotic advection and turbulence. *Ann. Rev. Fluid Mech.* **22,** 207–253 (1990).

Parfitt, G. D., The mixing of cohesive powders, *in* "Mixing in the Process Industries" (N. Harnby, M. F. Edwards, and A. W. Nienow, Eds.). Elsevier, North-Holland, 1992, pp. 321–348.

Patlazhan, S. A., and Lindt, J. T., Kinetics of structure development in liquid–liquid dispersions under simple shear flow. Theory. *J. Rheol.* **40,** 1095–1113 (1996).

Powell, R. L., and Mason, S. G., Dispersion by laminar flow. *AIChE J.* **28,** 286–293 (1982).

Rallison, J. M., A numerical study of the deformation and burst of a viscous drop in general shear flows. *J. Fluid Mech.* **109,** 465–482 (1981).

Rallison, J. M., The deformation of small viscous drops and bubbles in shear flows. *Ann. Revs. Fluid Mech.* **16,** 45–66 (1984).

Rauwendaal, C., "Polymer Extrusion," 3rd ed. Hanser Publishers, Munich, 1994.

Redner, S., Fragmentation, *in* "Statistical Models for the Fracture of Disordered Media," (H. J. Herrman and S. Roux, Eds.). Elsevier, North-Holland, 1990, pp. 321–348.

Robinson, D. J., and Earnshaw, J. C., Experimental study of aggregation in two dimensions. I. Structural aspects. *Phys. Rev. A* **46,** 2045–2054 (1992).

Rumpf, H., The strength of granules and agglomerates in "Agglomeration" (W. A. Knepper, Ed.). Wiley Interscience, New York, 1962.
Rumscheidt, F. D., and Mason, S. G., Particle motions in sheared suspensions. XII. Deformation and burst of fluid drops in shear and hyperbolic flow. *J. Colloid Sci.* **16,** 238–261 (1961).
Rwei, S. P., Manas-Zloczower, I., and Feke, D. L., Observation of carbon black agglomerate dispersion in simple shear flows. *Polym. Eng. Sci.* **30,** 701–706 (1990).
Rwei, S. P., Manas-Zloczower, I., and Feke, D. L., Characterization of agglomerate dispersion by erosion in simple shear flows. *Polym. Eng. Sci.* **31,** 558–562 (1991).
Rwei, S. P., Manas-Zloczower, I., and Feke, D. L., Analysis of the dispersion of carbon black in polymeric melts and its effect on compound properties. *Polym. Eng. Sci.* **32,** 130–135 (1992).
Schepens, F. A. O., Chaotic mixing in the extended periodic cavity flow. Masters Thesis, Eindhoven University of Technology, 1996.
Schowalter, W. R., Stability and coagulation of colloids in shear fields. *Ann. Rev. Fluid Mech.* **16,** 245–261 (1984).
Schuman, T. E. W., Theoretical aspects of the size distribution of fog particles. *Quart. J. Meteorological. Soc.* **66,** 195–208 (1940).
Smoluchowski, M., Versuch einer mathematischen Theorie der Koagulationskinetik kolloider Losungen. *Z. Phys. Chem.* **92,** 129–168 (1917).
Sonntag, R. C., and Russel, W. B., Structure and breakup of flocs subject to fluid stresses. I. Shear experiments. *J. Colloid Interface Sci.* **113,** 399–413 (1986).
Sonntag, R. C., and Russel, W. B., Structure and breakup of flocs subject to fluid stresses. II. Theory, *J. Colloid Interface Sci.* **115,** 378–389 (1987a).
Sonntag, R. C., and Russel, W. B., Structure and breakup of flocs subject to fluid stresses. III. Converging flow. *J. Colloid Interface Sci.* **115,** 390–395 (1987b).
Spielman, L. A., Viscous interactions in Brownian coagulation. *J. Colloid Interface Sci.* **33,** 562–571 (1970).
Srinivasan, M. P., and Stroeve, P., Subdrop ejection from double emulsion drops in shear flow. *J. Membrane Sci.* **26,** 231–236 (1986).
Stone, H. A., Dynamics of drop deformation and breakup in viscous fluids. *Ann. Revs. Fluid Mech.* **26,** 65–102 (1994).
Stone, H. A., and Leal, L. G., Relaxation and breakup of an initially extended drop in an otherwise quiescent fluid. *J. Fluid Mech.* **198,** 399–427 (1989).
Stone, H. A., Bentley, B. J., and Leal, L. G., An experimental study of transient effects in the breakup of viscous drops. *J. Fluid Mech.* **173,** 131–158 (1986).
Stroeve, P., and Varanasi, P. P., An experimental study of double emulsion drop breakup in uniform shear flow. *J. Coll. Int. Sci.* **99,** 360–373 (1984).
Sundaraj, U., Dori, Y., and Macosko, C. W., Sheet formation in immiscible polymer blends: model experiments on an initial blend morphology. *Polymer* **36,** 1957–1968 (1995).
Swanson, P. D., and Ottino, J. M., A comparative computational and experimental study of chaotic mixing of viscous fluids, *J. Fluid Mech.* **213,** 227–249 (1990).
Tanner, R. I., A test particle approach to flow classification for viscoelastic fluids. *AIChE J.* **22,** 910–918 (1976).
Taylor, G. I., The viscosity of a fluid containing a small drops of another fluid. *Proc. R. Soc.* **A138,** 41–48 (1932).
Tjahjadi, M., and Ottino, J. M., Stretching and breakup of droplets in chaotic flows. *J. Fluid Mech.* **232,** 191–219 (1991).
Tjahjadi, M., Stone, H. A., and Ottino, J. M., Satellite and subsatellite formation in capillary breakup. *J. Fluid Mech.* **243,** 297–317 (1992).

Tomotika, S., On the instability of a cylindrical thread of a viscous liquid surrounded by another viscous fluid. *Proc. R. Soc.* **A150,** 322–337 (1935).

Tomotika, S., Breaking up of a drop of viscous liquid immersed in another viscous fluid which is extending at a uniform rate. *Proc. R. Soc.* **A153,** 302–318 (1936).

Torres, F. E., Russel, W. B., and Schowalter, W. R., Floc structure and growth kinetics for rapid shear coagulation of polystyrene colloids. *J. Colloid Interface Sci.* **142,** 554–574 (1991a).

Torres, F. E., Russel, W. B., and Schowalter, W. R., Simulations of coagulation in viscous flows. *J. Colloid Interface Sci.* **145,** 51–73 (1991b).

Ulbrecht, J. J., Stroeve, P., and Pradobh, P., Behavior of double emulsions in shear flows. *Rheol. Acta* **21,** 593–597 (1982).

van Dongen, P. G. J., and Ernst, M. H., Scaling solutions of Smoluchowski's coagulation equation. *J. Stat. Phys.* **50,** 295–329 (1988).

Varanasi, P. P., Ryan, M. E., and Stroeve, P., Experimental study on the breakup of model viscoelastic drops in uniform shear flow. *I&EC Research* **33,** 1858–1866 (1994).

Verwey, E. J., and Overbeek, J. T. G., "Theory of the Stability of Lyophobic Colloids." Elsevier, 1948.

Villadsen, J., Putting structure into chemical engineering. *Chem. Eng. Sci.* **52,** 2857–2864 (1997).

Wang, H., Zinchenko, A., and Davis, R. H., The collision rate of small drops in linear flow fields. *J. Fluid Mech.* **265,** 161–188 (1994).

White, J. L., "Twin Screw Extrusion." Hanser, Munich, 1990.

Williams, M. M. R., An exact solution of the fragmentation equation. *Aerosol Sci. and Tech.* **12,** 538–546 (1990).

Zeichner, G. R., and Schowalter, W. R., Use of trajectory analysis to study stability of colloidal dispersions in flow fields. *AIChE J.* **23,** 243–254 (1977).

Zhang, D. F., and Zumbrunnen, D. A., Chaotic mixing of two similar fluids in the presence of a third dissimilar fluid. *AIChE J.* **42,** 3301–3309 (1996a).

Zhang, D. F., and Zumbrunnen, D. A., Influences of fluidic interfaces on the formation of fine scale structures by chaotic mixing. *J. Fluids Eng.* **118,** 40–47 (1996b).

Ziff, R. M., New solutions to the fragmentation equation. *J. Phys. A: Math. Gen.* **24,** 2821–2828 (1991).

Ziff, R. M., and McGrady, E. D., The kinetics of cluster fragmentation and depolymerisation. *J. Phys. A: Math. Gen.* **18,** 3027–3037 (1985).

Ziff, R. M., and McGrady, E. D., Kinetics of polymer degradation. *Macromolecules* **19,** 2513–2519 (1986).

Zumbrunnen, D. A., Miles, K. C., and Liu, Y. H., Auto-processing of very fine-scale materials by chaotic mixing of melts. *Composites* **27A,** 37–47 (1996).

APPLICATION OF PERIODIC OPERATION TO SULFUR DIOXIDE OXIDATION

Peter L. Silveston

Department of Chemical Engineering, University of Waterloo, Waterloo, Ontario, Canada N2L 3G1

Li Chengyue

Department of Chemical Engineering, Beijing University of Chemical Technology, Beijing 100029, China

Yuan Wei-Kang

Department of Chemical Engineering, East China University of Chemical Technology, Shanghai 201107, China

I.	Introduction	206
II.	Air Blowing of the Final Stage of a Multistage SO_2 Converter	208
	A. Experimental Studies	208
	B. Mechanistic Model	215
	C. Application to the Final Stage of SO_2 Converter with Composition Forcing	216
	D. Application to an Isothermal Backmixed Reactor	217
III.	SO_2 Converters Based on Periodic Reversal of the Flow Direction	223
	A. Industrial Applications	225
	B. Experimental Results	230
	C. Modeling and Simulation	234
	D. Overview	248
IV.	Conversion of SO_2 in Trickle-Bed Catalytic Scrubbers Using Periodic Flow Interruption	248
	A. Experimental Studies	249
	B. Modeling and Simulation	256
	C. Application to Stack-Gas Scrubbing	261
	D. Physical Explanation	269
V.	The Future of Periodic Operations for SO_2 Oxidation	272
	Nomenclature	273
	References	278

Three different versions of periodic operation—composition modulation or catalyst flushing, flow direction switching, and flow interruption—have been tested on SO_2 oxidation over the past two decades. When applied to recovering sulfuric acid from SO_2 in plant emissions, improvements over steady-state operation are found for each version. However, among the different versions, only periodic switching of flow direction has been commercialized. This review examines plant data for flow direction switching as well as laboratory and modelling results for all versions of periodic operation. Comparing these versions one against the other for the same reaction provides a compact overview of possible industrial applications of periodic operations. Each version exploits a different process consideration and process improvements result from different physical and kinetic mechanisms.

I. Introduction

Sulfur dioxide oxidation is the essential catalytic reaction in the manufacture of one of the world's most important bulk chemicals: sulfuric acid. It is also a key step in processes now under consideration for removing sulfur from industrial stack gases. It is not surprising, then, that this reaction has been chosen to see if the emerging technique of periodic operation of chemical reactors would be useful commercially. What is surprising, however, is that three different types of periodic operation have been considered for this application: (1) periodic air blowing of the final stage of a multistage catalytic converter, (2) periodic reversal of flow direction in a single-stage converter, and (3) periodic interruption of liquid flow in a trickle-bed reactor. Research on these applications has attracted worldwide participation as Table I shows. There is a large literature on periodic flow reversal; only a few of the references are given. Others will be found in a review by Matros and Bunimovich (1996).

Sulfuric acid is produced commercially from a dust-free ore smelter offgas or from a SO_2-rich gas stream obtained by burning sulfur. A multistage, near-adiabatic reactor employing trays of a granular, potassium-promoted vanadia catalyst is employed. The oxidation is highly exothermic, so heat must be removed. Various methods are used: heat exchange with a waste heat boiler or with the fresh gas feed, or cold gas injection. Cooling is usually carried out between stages, always prior to the last stage and frequently before the next-to-last stage. Conversion of SO_2 is equilibrium limited so, despite cooling prior to the final stage, SO_2 is not completely

TABLE I
GEOGRAPHIC DISTRIBUTION OF THE RESEARCH EFFORT ON THE APPLICATION OF PERIODIC OPERATION TO SO₂ OXIDATION

Type of periodic operation	Country	References
Periodic air blowing of the final stage of a multistage catalytic converter	Canada, Russia, USA	Briggs et al. (1977, 1978); Silveston and Hudgins (1981); Silveston et al. (1994; Strots et al. (1992)
Periodic reversal of flow direction in a single-stage converter	Russia, Bulgaria, China, Japan, USA	Boreskov et al. (1977); Boreskov and Matros (1984); Matros (1985, 1989); Sapundzhiev et al. (1990); Bunimovich et al. (1990); Isozaki et al. (1990); Snyder and Subramaniam (1993); Bunimovich et al. (1995); Xiao and Yuan (1996); Wu et al. (1996); Matros and Bunimovich (1996)
Periodic interruption of liquid flow in a trickle-bed reactor	Canada, Argentina, Russia, USA, Korea, Germany	Haure et al. (1989); Haure et al. (1990); Stegasov et al. (1992); Gangwal et al. (1992); Metzinger et al. (1994); Lee et al. (1995)

oxidized to SO_3. SO_3 leaving the multistage reactor is washed with quantitative capture of SO_3. Improvements to the current technology would be lower operating temperatures or other means of increasing SO_2 conversion in the reactor and lower-cost heat exchange.

Two of the three versions of periodic operation considered in this review were conceived to improve sulfuric acid production technology. Periodic air blowing is a method of reducing SO_2 emissions from sulfuric acid plants. The primary application of periodic flow reversal is to exhaust streams from metal smelters in which SO_2 concentrations are typically below 4.5 vol%. SO_2 can be oxidized with this periodic operation for large volumes of a low-temperature gas without requiring additional fuel or the use of expensive shell and tube heat exchangers. Periodic flow reversal also provides a declining temperature profile axially in the reactor that increases conversion when just a single reversible and highly exothermic reaction occurs. The third version, periodic interruption of liquid flow in a trickle bed, is aimed at removing parts-per-million amounts of SO_2 from stack gases while providing a moderately concentrated sulfuric acid product. It is one of several processes being considered for supplementing or even replacing lime addition to to coal-burning furnaces. Lime addition cannot meet the targets of 98% SO_2 removal that are now being sought.

Why are we comparing different versions of periodic operation for the same reaction? First of all, a comparison provides a compact review of this type of operation without dealing in generalities or having to discuss details of different chemical systems. Process improvements achieved for SO_2 oxidation by each version typify what can be expected in applications to other reaction systems. Finally, this comparison demonstrates that the mechanisms that lead to process improvement are quite different for each version of periodic operation.

II. Air Blowing of the Final Stage of a Multistage SO_2 Converter

A. EXPERIMENTAL STUDIES

In 1973 Unni *et al.* (1973) published one of the first experimental studies of the periodic operation of a catalytic reactor. The authors chose SO_2 oxidation as their test reaction. A commercial catalyst containing 9.1 wt% V_2O_5, 10.1 wt% K_2O, and 0.45 wt% Fe_2O_3, supported on a diatomaceous earth, was employed, and experiments were run at 405°C and 1 bar so that conditions were similar to those found at the inlet to a commercial SO_2 converter. Catalyst size was reduced to 30/40 U.S. mesh, however, to avoid transport interference. Variables in the Unni study were cycle period, amplitude, and the time average feed composition. Both amplitude and the time average feed composition were expressed in terms of the $SO_2:O_2$ volume ratio. The remarkable finding was that SO_2 oxidation is excited at cycle periods measured in hours, as may be seen in Fig. 1. The largest rate enhancements were obtained at a time-average $SO_2:O_2$ ratio = 0.6 with an amplitude of 0.3 measured in terms of the $SO_2:O_2$ ratio. Higher and lower time average ratios (0.9 and 0.3) resulted in a significantly poorer performance, but this seems due to the lower amplitudes available at these averages. Symmetrical cycles were used by Unni *et al.,* but because ratios rather than concentrations were used, amplitudes for SO_2 and O_2 are not the same. Regardless of the time average $SO_2:O_2$ ratio, the largest enhancements were achieved at cycle periods between 4 and 6 h.

Strategies in which one reactant concentrations is held constant were examined in several runs at $\tau = 240$ min and are shown in the figure. These strategies appear to be comparable to variation of both reactants at a time average $SO_2:O_2$ ratio = 0.9, but variation of O_2 and N_2 holding SO_2 constant at a mean $SO_2:O_2$ ratio = 0.6 gave a value of Ψ less than half the value obtained when both reactants are cycled. Cycle amplitude with symmetrical forcing is important and it appears that a threshold amplitude between 0.1 and 0.2 as a $SO_2:O_2$ ratio exists. Too large an amplitude, which occurs

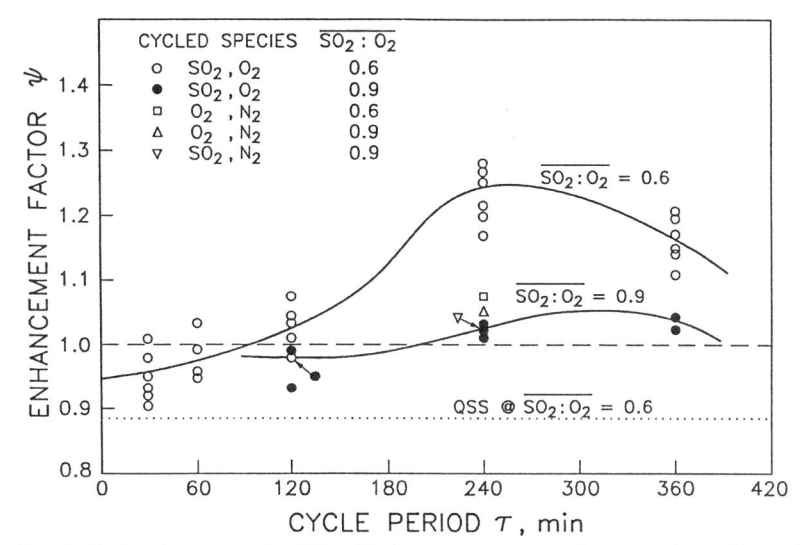

FIG. 1. Rate enhancement for SO$_2$ oxidation over a commercial promoted vanadia catalyst as a function of forcing period, time average feed composition as SO$_2$:O$_2$ ratio, and forcing manipulation (1 bar and 405°C, $s = 0.5$). (Figure adapted from Unni *et al.*, 1973, with permission of the authors.)

when one reactant is not present in the feed for a half cycle, diminishes the rate enhancement.

Briggs *et al.* (1977, 1978) attempted to apply the composition forcing used by Unni to decreasing emissions from SO$_2$ converters. They found, however, that with high levels of SO$_3$ in the feed no rate enhancement was achieved. Periodic blowing of the final stage in the converter to reduce the level of SO$_3$ on the catalyst offered a possible means to achieve higher rates, because SO$_2$ oxidation is inhibited by SO$_3$ at high SO$_2$ conversions under steady-state operation, so this approach was pursued.

The experiments undertaken by Briggs *et al.* (1977) and interpreted by Silveston and Hudgins (1981) explored the application of composition forcing to multistaged SO$_2$ converters. The same commercial catalyst employed by Unni *et al.* was chosen, but used in an integral reactor containing 30 g of 12/20 U.S. mesh particles. The initial experiments examined composition forcing at high conversion of a feed stream containing 12.4 vol% SO$_2$. Feed for the cycled stage was prepared in a separate converter containing a Pt/Al$_2$O$_3$ catalyst at 470°C. This converter, simulating the first three stages and operating at steady state, converted about 90% of the SO$_2$ in the feed to SO$_3$. Figure 2 shows the results obtained using symmetrical cycling with $\tau = 26$ min on the final converter stage. The 12.4 vol% SO$_2$ feed to the

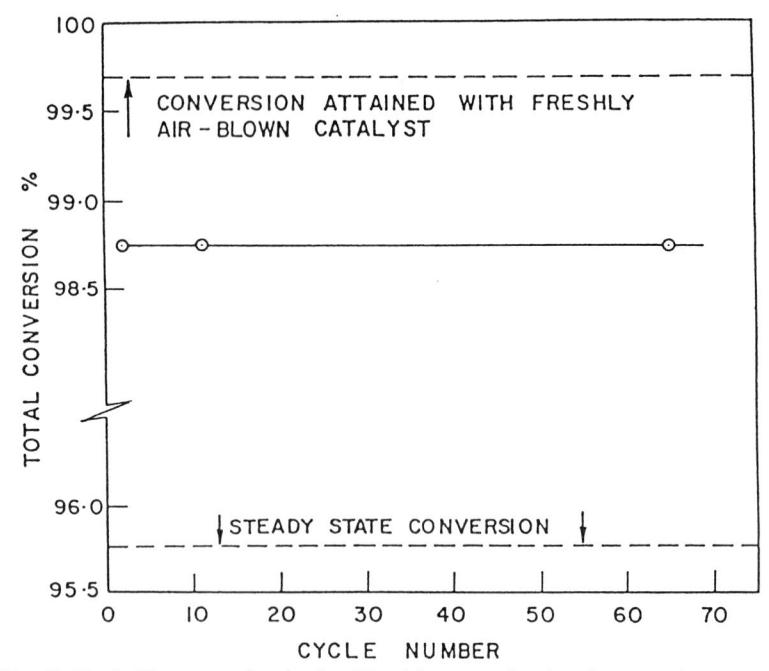

FIG. 2. Total SO_2 conversion in the SO_2-rich stream leaving the second reactor as a function of cycle number and type of reactor operation for a time average feed containing 12.4 vol% SO_2, $T = 406°C$. With composition forcing, measurement follows switch to the SO_3/SO_2 feed by 12 min. Reference lines (---) show limiting conversions for steady-state operation and under periodic forcing. (Figure adapted from Briggs et al., 1977, with permission of the authors.)

steady-state preconverter is representative of acid plant feeds obtained by burning sulfur. When the catalyst bed operated at steady state, it raised the SO_2 conversion to 95.8%. The conversion under modulation, 12 min after introducing the $SO_3:SO_2$ mixture to the reactor, was 98.8%. The time average conversion in the half cycle was higher. Step change observations indicate that an even higher time average conversion could have been attained by shortening the half cycle, but the 12 min was dictated by the necessity of saturating the 97 wt% acid absorber with SO_2 before measurements could be made. Conversion in Fig. 2 is plotted against cycle number. The figure demonstrates that this high half-cycle conversion is maintained over some 30 h of operation. Only traces of SO_2 were detected leaving the vanadia catalyst bed 2 min after switching back to air. Consequently, no significant amounts of SO_2 are desorbed from the catalyst in the stripping step. Therefore, the figure represents the performance under composition modulation even though just half-cycle data are shown. It is notable that the higher conversion achieved by modulation implies a 53% increase in catalyst activity.

The upper dashed line in Fig. 2 shows the initial conversion on switching in the step change experiment discussed earlier. It gives the limiting performance for final stage periodic air flushing at 406°C with feed typical of a sulfur burning acid plant. This conversion, 99.7%, exceeds slightly the equilibrium conversion of 99.4% calculated from the NBS tables of thermodynamic data at the temperature of the experiment.

Replacing air by either N_2 or O_2 did not affect the conversion obtained in the experiment just described. Consequently, the higher conversion does not result from catalyst reoxidation. It must be due solely to stripping SO_3 from the catalyst.

In order to explore composition modulation of the final stage of a converter further, Briggs et al. (1978) added a second integral reactor, also holding about 30 g of the vanadia catalyst. With the preconverter in place, this system was operated on a typical feed from sulfur burning, with a $SO_2:O_2:N_2$ composition in vol% of 10.8:15.2:74, and from a smelter effluent with a composition of 8.0:6.2:85.8. The cycled beds of vanadia catalyst were held in a fluidized sand bath at 401°C for the former feed and at 405°C for the latter one. The space velocity for both the air and the SO_3/SO_2 mixture was about 24 min^{-1} (STP). Table II summarizes the experimental results for the cycle periods tested.

SO_2 emitted from the modulated bed goes through a minimum after switching to the SO_3/SO_2 mixture. Lowest values are obtained 2 min after the composition change for the sulfur burning feed and they are about 8% of the steady-state emission, whereas for the smelter effluent feed, the lowest emission is about 13% of the steady-state value. Evidently, a cycle period of 4 to 5 min would be optimum for the conditions used, yielding a performance some 10% better than that shown at $\tau = 10$ in Table II.

The explanation of the delayed minimum in the SO_2 concentration is that data are for the combined streams emerging from the two parallel reactors, one of which is converting the $SO_2/SO_3/O_2$ mixture from the preconverter while the other is undergoing air blowing. The initially high SO_2 originates from the blowing step taking place in one of the beds. This drops with time in the half cycle and the SO_2 concentration approaches zero. The rising SO_2 concentration in the latter part of the half cycle comes from the other bed of the converter, where SO_2 reacts to form SO_3. Adsorption of SO_2 on the air-blown catalyst surface where a portion is oxidized to SO_3 keeps the SO_2 effluent from the active bed low. The trioxide is adsorbed (more strongly than SO_2) so this bed eventually saturates and the SO_2 concentration starts to climb.

A second group of experiments measured temperature variation in the

TABLE II
Composition Cycling Results

	Sulfur burning feed	Smelter off-gas feed
Feed: Mole ratio SO_2/O_2	0.709	1.30
Composition (mole frac.)		
SO_2	0.108	0.080
O_2	0.152	0.062
N_2	0.740	0.858
Bath temperature, final stage	401°C	405 ± 1°C
Mass of catalyst, final stage	25 g	23 g
Conversion (fractional)		
In preconverter	0.917	0.883
Final stage, steady stage	0.952	0.913
Final stage, cycling		
(mean value)		
for $\tau_{1/2}$ = 10 min	0.994	0.984
for $\tau_{1/2}$ = 20 min	0.992	0.981
for $\tau_{1/2}$ = 24 min	—	0.976
Equilibrium, steady state	0.995[a]	0.987[a]
$(SO_2)^{eye}/(SO_2)^{steady}$ (mean molar ratio)		
from final stage		
for $\tau_{1/2}$ = 10 min	0.122	0.184
for $\tau_{1/2}$ = 20 min	0.164	0.229
SO_2 concentration in gas leaving scrubber (SO_3-free) at steady state	6060 ppm	7790 ppm

[a] Calculated from NBS Table of Thermodynamic Data.

bed and the relative concentration of SO_3 leaving the bed during the two halves of a cycle (Briggs et al., 1977). These measurements are shown in Fig. 3. The drop in temperature as SO_3 is stripped from the catalyst suggests either desorption from a melt phase or decomposition of a complex in that phase. Desorption and decomposition are endothermic. When the SO_3/SO_2 reactant mixture is reintroduced (Fig. 3b), the breakthrough behavior indicates SO_3 absorption. This is exothermic and it accounts for at least part of the temperature rise seen in the figure. Lack of symmetry in the feed and air flushing steps is notable because the duration of each step is the same. SO_3 desorption is incomplete after 13 min of air exposure, whereas the SO_3 breakthrough in the other cycle step takes less than 5 min. Temperature drop on SO_3 desorption is about 8°C, while the rise in the feed step is about 16°C. Oxidation of SO_2 when it is present in the feed step, however, explains this difference. The temperature rise after the minimum and the fall after the maximum temperatures in the cycle are consequences of placing the reactor in a fluidized sand bath held at 408°C. In an adiabatic reactor the temperature changes would be even larger.

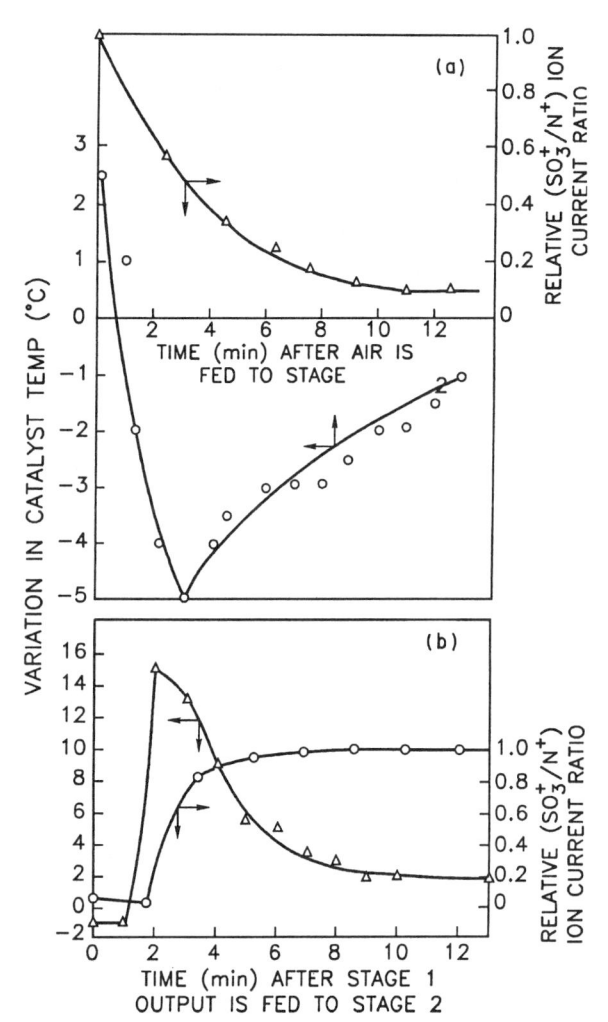

FIG. 3. Time variation of the catalyst bed temperature and the relative SO$_3$ signal in the stream leaving the cycled bed for composition forcing of the final stage of a SO$_2$ converter with an air stream and effluent from the previous stage: (a) half cycle with air feed, (b) half cycle with SO$_3$/SO$_2$ feed. Feed to the system contains 12.4 vol% SO$_2$, conversion in the first stage = 90% and τ = 26 min, s = 0.5. (Figure adapted from Briggs *et al.*, 1977, with permission, © 1977 Elsevier Science Publishers.)

Figure 3 suggests that an unsymmetrical cycle with the duration of air flushing twice the duration of reactant feed would provide even greater reduction of SO$_2$ emissions. The drawback of unsymmetrical cycling is that more than two catalyst beds in parallel are needed. An alternative would be to use a symmetrical cycle, but increase the airflow rate. This, however, would increase the loading on scrubbers in the acid plant. It is possible to avoid overloading the scrubbers by using the flushing air to burn sulfur. This scheme is shown in Fig. 4. It does not significantly affect the heat balance in the plant, but since the feed to the first bed of the converter will now contain a small amount of SO_3 there might even be a performance benefit, because low levels of SO_3 raise catalyst activity and the first stage is rarely equilibrium limited. Silveston and Hudgins (1981) used this scheme in their evaluation of the Briggs work. They concluded that the stiffest emission standards could be met at a cost below that for a double contact–double absorption plant for either a grass-roots situation or for a retrofit one when acid is produced by sulfur burning. The poorer performance of periodic air flushing with a smelter off-gas and the inability to recycle air probably mean that feed composition forcing is not attractive for producing acid from smelter off-gas. This exploratory work so far has failed to attract industrial interest despite its simplicity and apparent low cost.

SO_2 oxidation is one of the few reactions studied under periodic operation for which a mechanistic model is available. Moreover, a reactor model incorporating the mechanism and its kinetic model has been tested against forcing data. This modeling success means a model is available for exploring applications of periodic air flushing. An example of such an application where the objective is to examine different forcing strategies will be discussed in what follows.

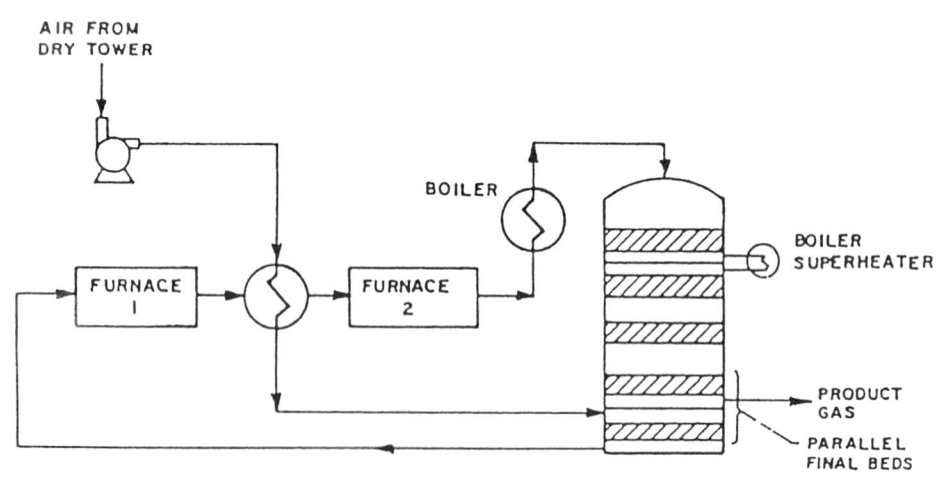

FIG. 4. Flow diagram of an SO_2 converter with periodic air flushing of the final catalyst stage and recycle of the air stream to the sulfur burner (in the diagram, air stripping of one of the final beds is shown). (Figure taken from Briggs *et al.*, 1978, with permission of the authors.)

B. MECHANISTIC MODEL

It has been known for decades that commercial potassium oxide–promoted vanadia catalysts function in a melt phase under reaction conditions and that SO_2 oxidation proceeds through a redox mechanism. Potassium oxide is converted to the pyrosulfate in the presence of SO_3 and this compound acts as a fluxing agent for vanadia. In recent years, a detailed mechanism has emerged (Balzhinimaev et al., 1985, 1989). Potassium-promoted vanadia catalysts oxidize SO_2 through a redox cycle in which three different binuclear vanadium complexes participate. These are (1) an oxo complex with oxygen bridging two V atoms, (2) a sulfite complex in which SO_3 forms the bridge, and (3) a peroxo complex involving O_2 in the bridge. These complexes exist as oligomers in a melt with $K_2S_2O_7$ as the fluxing agent. The mechanism involves transformation of these complexes into one another through oxidation, SO_2 absorption, and the oxidation of SO_2 to SO_3. Fast side reactions involving $S_2O_7^{2-}$ remove vanadium from the melt phase by precipitation. A slow reduction step leads to V^{4+}, which is catalytically inactive. The mechanism is summarized in Table III. The bulk of the SO_3 produced comes from transfer between the V^{5+} complexes and does not involve a change in the coordination of vanadium. Dissolution of SO_3 in the melt is rapid as indicated in (1) of the table and can be treated as an equilibrium step. SO_3 is released from the melt by decomposition of the pyrosulfate. Oxygen enters the catalytic network through step (4), and its absorption in the melt may be rate controlling.

TABLE III
VANADIA COMPLEX TRANSFORMATION IN THE BALZHINIMAEV MECHANISM[a]

$$SO_3(g) + (SO_4^{2-})^m \xrightarrow{rapid} (S_2O_7^{2-})^m \quad (1)^b$$

$$(V_2^{5+}O_2^{2-})^m + SO_2^{(m)} \underset{k_{-1}}{\overset{k_1}{\rightleftharpoons}} (V_2^{5+}O^{2-})^m + SO_3^m \quad (2)$$

$$(V_2^{5+}O^{2-})^m + SO_2^m \underset{k_{-2}}{\overset{k_2}{\rightleftharpoons}} (V_2^{5+}SO_3^{2-})^m \quad (3)$$

$$(V_2^{5+}SO_3^{2-})^m + O_2^m \underset{k_{-3}}{\overset{k_3}{\rightleftharpoons}} (V_2^{5+}O_2^{2-})^m + SO_3^m \quad (4)$$

$$(V_2^{5+}SO_3^{2-})^m \underset{k_{-4}}{\overset{k_4}{\rightleftharpoons}} (V_2^{4+})^m + SO_3^m \quad (5)$$

$$V_k + (S_2O_7^{2-}) \xrightarrow[K_{in}]{rapid} V_k^{in} \text{ (inactive form)} + (S_2O_7^{2-})^m \quad (6)$$

[a] m = molten or glass-like catalyst phase (species exists in this phase)
[b] All SO_3^m exists in the melt as $S_2O_7^{2-}$; thus, $C_{SO_3}^m = C_{S_2O_7^{2-}}^m$.

216   P. SILVESTON ET AL.

The kinetic model of Ivanov and Balzhinimaev (1987) and Balzhinimaev et al. (1989) assumes that the steps in the mechanism of Table III are elementary and that the active and inactive forms of the complexes are in equilibrium. Table IV presents the model. Parameters for the model are to be found in Table VI.

Application of the Balzhinimaev model requires assumptions about the reactor and its operation so that the necessary heat and material balances can be constructed and the initial and boundary conditions formulated. Intraparticle dynamics are usually neglected by introducing a mean effectiveness factor; however, transport between the particle and the gas phase is considered. This means that two heat balances are required. A material balance is needed for each reactive species (SO_2, O_2) and the product (SO_3), but only in the gas phase. Kinetic expressions for the Balzhinimaev model are given in Table IV.

C. Application to the Final Stage of SO_2 Converter with Composition Forcing

Silveston et al. (1994) use a one-dimensional plug flow model to represent the packed bed in the final stage. Because the intent of their work was to model the experiments of Briggs et al. discussed earlier, they allowed for heat loss or gain in the bench scale reactor used by Briggs through wall

TABLE IV
Kinetic Expression for the Balzhinimaev Model

$$\sum_k y_k + \sum_k y_k^{in} = 1 \qquad (7)^a$$

$$y_k^{in} = K_{in} C_{SO_3}^m y_k \qquad (8)$$

$$\Sigma y_k^{in} = \frac{K_{in} C_{SO_3}^m}{1 + K_{in} C_{SO_3}^m} \qquad (9)$$

$$r_1 = \frac{1}{1 + K_{in} C_{SO_3}^m} \left(k_1 P x_{SO_2} y_1 - \frac{k_{-1}}{K_H^3} C_{SO_3}^m y_2 \right) \qquad (10)$$

$$r_2 = \frac{1}{1 + K_{in} C_{SO_3}^m} (k_2 P x_{SO_2} y_2 - y_3) \qquad (11)$$

$$r_3 = \frac{1}{1 + K_{in} C_{SO_3}^m} \left(k_3 P x_{SO_2} y_3 - \frac{k_3}{K_H^3} C_{SO_3}^m y_1 \right) \qquad (12)$$

$$r_4 = \frac{1}{1 + K_{in} C_{SO_3}^m} \left(k_4 y_3 - \frac{k_{-4}}{K_H^3} C_{SO_3}^m y_4 \right) \qquad (13)$$

[a] Indices on y: 1, $(V_2^{5+}O_2^{2-})^m$; 2, $(V_2^{5+}O^{2-})^m$; 3, $(V_2^{5+}SO_3^{2-})^m$; 4, $(V_2^{4+})^m$.

transport terms employing overall bed-to-wall coefficients. They observed from comparison of the magnitude of terms in the balances that gas phase storage terms could be safely neglected. Table V gives their model. Note that these researchers incorporated reactions in the particle in their material balances rather than treating them as boundary conditions. Remaining boundary conditions and the initial conditions are given in the table. Each change of feed initiates a new solution so that two sets of initial conditions are needed: (1) $t = 0$, (2) $t = n\tau/2$, $n\tau$, $3n\tau/2$, etc.

Parameters representing the catalyst bed were taken from Briggs *et al.* (1977, 1978); transport coefficients for heat and mass were calculated from well-established correlations, whereas the kinetic parameters came from measurements on Russian catalysts (Ivanov and Balzhinimaev, 1987) of about the same composition as the commercial catalyst used by Briggs. These parameters are collected in Table VI. The model used by Silveston *et al.* contained, therefore, no adjustable parameters.

Figure 5 gives the simulation results with the model given for the conditions used by Briggs *et al.* to obtain Fig. 3. Data points are shown in Fig. 5b, but not in 5a. Mass spectrometer readings were not calibrated, and only normalized data are shown in Fig. 3a. The simulation estimates the shape of the midbed temperature and the SO_3 vol% variations successfully. It also reproduces the initial bed temperature lag for the first minute after introduction of the SO_3/SO_2 reactant mixture (Fig. 5b), as well as the absence of a lag when air is introduced to the catalyst bed displacing the reactant mixture (Fig. 5a). The model also gives the slow adjustment of the bed temperature after the maximum and minimum temperatures, although the rates of cooling and heating are not correct. The most serious deficiency of the model is that it overestimates the temperature rise and drop by 15 and 8°C, respectively.

Integration of Eq. (6) for SO_2 in Table V estimates the conversion achieved. Simulation of periodic symmetrical switching between a reactant mixture and air gave an estimate of 99.4% at 12 min after the switch to the SO_3/SO_2 reactant mixture in reasonable agreement with the overall conversion of 98.8% measured by Briggs *et al.* (1977). With respect to model sensitivity, it was found that bed midpoint temperature was sensitive to the wall and gas to particle heat transfer coefficients. An extensive study of sensitivity, however, was not undertaken.

D. Application to an Isothermal Back-Mixed Reactor

Strots *et al.* (1992) undertook a study of composition forcing employing the Balzhinimaev model given in Table IV for the simplest reactor situation:

TABLE V
Model of a Nonadiabatic Packed Bed Reactor for SO₂ Oxidation Incorporating the Balzhinimaev Mechanism[a]

Heat balance on the catalyst:

$$(1 - \varepsilon_B)c_p^s \rho_{cat} \frac{\partial T_s}{\partial t} = \varepsilon_m(1 - \varepsilon_B)H_3^{dis}\frac{\partial C_3^m}{\partial t} + \varepsilon_m(1 - \varepsilon_B)s^v \sum r_j H_j$$
$$+ \varepsilon_m(1 - e_B)s^v H_{in}\frac{\partial \sum y_k^{in}}{\partial t} + h_p S_v(T - T_g) + h_w S_w(T_w - T_s) \quad (14)$$

Heat balance on gas phase:

$$0 = -u\rho_g c_p^g \frac{\partial T}{\partial z} + h_p S_v(T_g - T) + h_w^g S_w(T_w - T) \quad (15)$$

Mass balance on SO₃ in the melt phase:

$$\varepsilon_m(1 - \varepsilon_B)\frac{\partial C_3^m}{\partial t} = k_m S_m \left(\frac{K_H^3 P x_3^P}{1 + K_H^3 P x_3^P/C^o} - C_3^m\right)$$
$$+ \varepsilon_m(1 - \varepsilon_B)s^v(r_1 + r_2 + r_4) - \varepsilon_m(1 - \varepsilon_B)s^v\frac{\partial \sum y_k^{in}}{\partial t} \quad (16)$$

Mass balance on SO₃ in the particle void space:

$$0 = -k_m S_m \left(\frac{K_H^3 P x_3^P}{1 + K_H^3 P x_3^P/C^o} - C_3^m\right) + k_p^3 S_v(C_3^g - C_3^p) \quad (17)$$

Mass balances on SO₂ and O₂ in the particle void space:

$$0 = \varepsilon_m(1 - \varepsilon_B)s^v \sum \nu_{ij}r_j + k_p^i S_v(C_i^g - C_i^p),$$
$$i = 1, 2 \quad (18)$$

Gas phase mass balances on reactants:

$$0 = -u\frac{\partial C_i^g}{\partial z} + k_p^i S_v(C_i^p - C_i^g),$$
$$i = 1, 2, 3 \quad (19)$$

Number balances on vanadia complexes:

$$\frac{\partial y_k}{\partial t} = \Sigma \mu_{kj} r_j, \quad (20)$$
$$k = 1\text{-}4$$

and

$$\Sigma y_k + \Sigma y_k^{in} = 1. \quad (21)$$

TABLE V (Continued)

Boundary conditions:

$z = 0$

$(n-1)\tau < t < n\tau/2$: $T = T^\circ$ (22)

$n = 1, 2, \ldots$ $C_i = 0$ $i = 1-3$

$n\tau/2 < t < n\tau$ $T = T^\circ$

$n = 1, 2, \ldots$ $C_i = C_i^\circ$ $i = 1, 2$ (23)

                 $C_3 = 0$

Initial conditions:

                 $C_1 = 0$ $i = 1, 2, 3$

$t = 0; 0 \leq z \leq L$: $C_3^m = 0$

                 $T_s = T = T^\circ$

                 $y_k, y_1^{in}, y_k^{in} = 0, k = 2, 3, 4$

                 $y_1 = 1$

$z \geq 0$:

$t = n\tau/2, n\tau$ $C_i^g(t^+) = C_i^g(t^-), C_i^p(t^+), = C_i^p(t^-), C_3^m(t^+) = C_3^m(t^-),$

$n = 1, 2, 3$

$y_k(t^+) = y_k(t^-),$ $y_k^{in}(t^+) = y_k^{in}(t^-)$ (24)

$T(t^+) = T(t^-),$ $T_s(t^+) = T_s(t^-)$

[a] i = species index: 1 = SO_2, 2 = O_2, 3 = SO_3; j = reaction index: see Table III; k = complex index: see Table III.

a back-mixed reactor operating both isobarically and isothermally. These investigators also assumed equilibrium between the gas phase and the catalyst phase so heat and mass transport were neglected. With these assumptions, the equations in Table V simplify substantially to give

$$(\varepsilon_B(1 + \varepsilon_p(1 - \varepsilon_B)) + \varepsilon_m H_i RT)(dx_i/dt) \quad (25)$$
$$= ((x_i)_o - x_i)/t_s + (\varepsilon_m C_V RT/P)(S\nu_{ik}r_k),$$
$$i = SO_2, O_2, SO_3$$

$$dy_j/dt = \Sigma \mu_{jk} r_k \quad (26)$$
$$j = V_2^{5+}O_2^{2-}, V_2^{5+}O_2^-, V_2^{5+}SO_3^{2-}, V_2^{4+},$$

TABLE VI
MODEL PARAMETERS

Specific heats

$$c_p^s = 0.21 \text{ kcal/kg °C}, \quad c_p^g = 1.08 \text{ kcal/kg °C}$$

Heat effects

$$H_{in} = 26 \text{ kcal/mol}, \quad H_3^{dis} = 12 \text{ kcal/mol}$$

Solubility

$$K_H^3 = 0.0372 \text{ mol/cm}^3 \text{ atm}$$

Complex, sulfate, pyrosulfate concentrations

$$s^v = 0.0024 \text{ mol/cm}^3, \quad C^o = 0.02 \text{ mol/cm}^3$$

Bed and catalyst properties

$\rho_{cat} = 264.7 \text{ kg/m}^3$ $\quad$ $\rho_g = 0.25 \text{ kg/m}^3$

$S_v = 3600 \text{ m}^{-1}$ $\quad$ $\varepsilon_B = \varepsilon_p = 0.4$

$\varepsilon_m = 0.20$ $\quad$ $u = 0.1 \text{ m/s}$

Transport properties

$h_p = 0.528 \text{ kcal/m}^2 \text{ s °C}$ $\quad$ $h_w = 2.2 \text{ kcal/m}^2 \text{ s °C}$

$h_{wg} = 0.5 \text{ kcal/m}^2 \text{ s °C}$ $\quad$ $k_p = 0.315 \text{ m/s}$

$k_m S_m = 6 \text{ s}^{-1}$

---

where the index k signifies the reaction rates shown in Table IV. SO_3 is also stored in complexes present in the melt so that an additional term has to be added to the left side of Eq. (25). This term is

$$((\varepsilon_m C_V K_L H_{SO3} RT)/(1 + H_{SO3} P x_{SO3}((1/C_o) + K_L))^2)(dx_{SO3}/dt).$$

Strots *et al.* used the back-mixed reactor model to identify operating methods capable of improving the rate of this reaction for a given feed and set of operating conditions. They focused most of their effort on a forcing strategy using switching between an inert and the reaction mixture. The initial conditions expressing this strategy are

$$n\tau \leq t \leq (n + 1/2)\tau: \quad (x_i)_o = 0$$
$$n = 0, 1, 2, 3, \ldots \quad (27)$$
$$(n + 1/2)\tau \leq t \leq (n + 1)\tau: \quad (x_i)_o = 2(x_i)_o$$

For their simulation of SO_2 oxidation, Strots *et al.* assumed $\varepsilon_B(1 + \varepsilon_p(1 - \varepsilon_B)) = 0.61$, $\varepsilon_m = 0.14$, $C_v = 0.002$, $C_o = 0.02$, $x_{SO3} = 0$ and $x_{SO2} =$

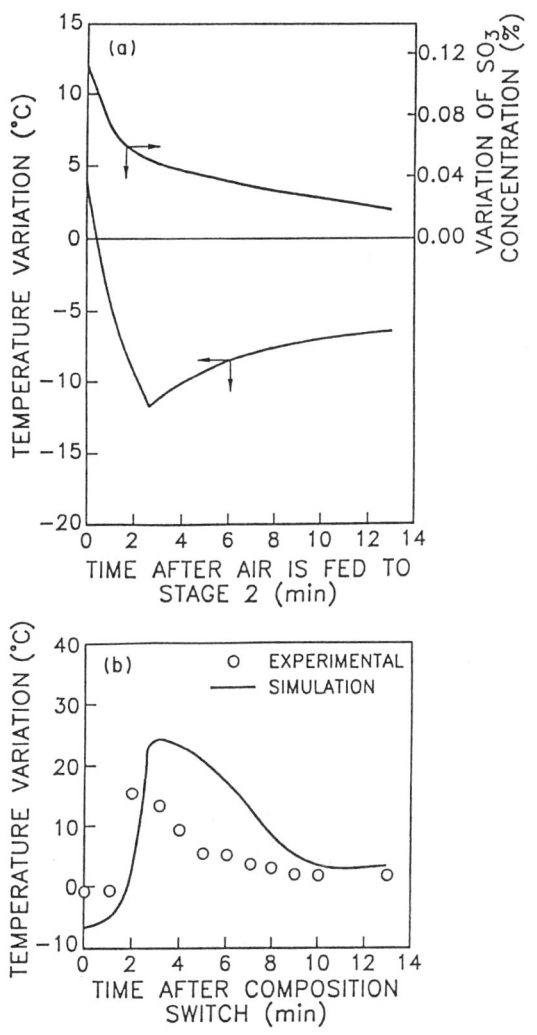

APPLICATION OF PERIODIC OPERATION TO SO₂ OXIDATION   221

FIG. 5. Simulation of the composition forcing experiment of Briggs et al., (1977) using the model of Table V with parameters from Table VI: (a) half cycle with air feed, (b) half cycle with SO₃/SO₂ feed. (Figure adapted from Silveston et al., 1990, with permission, © 1990 VSP, Utrecht, The Netherlands.)

$x_{O2} = 0.1$. They examined the effect of temperature (673 to 798 K), space-time ($0.01 \leq \tau_s \leq 4$ s), and cycling strategy on rate enhancement. Resonance was observed for $\tau_s = 0.1$ s with $\Psi > 1$ for temperatures up to 770 K at a cycle period of about 1000 s. Above 770 K, $r_{qss} > r$ so that resonance disappeared. The quasi-steady-state rate, r_{qss}, decreased rapidly with decreasing temperature below 770 K, and this seemed to be responsible for the resonance observed.

Multiple resonance was observed at $T = 723$ K and $\tau_s = 0.01$ s. The enhancement, Ψ, exceeds 1 at a cycle period between 0.2 and 0.3 s and also at greater than 1000 s, although quasi steady state is also attained at about that period. Resonance also occurs at two different cycle periods at $\tau_s = 0.02$ s, Ψ is much smaller. As the space-time increases above 0.25 s, the rate enhancement becomes very small and the multiple resonance phenomenon seems to disappear.

The influence of cycling strategy on enhancement is given in Fig. 6 for $T = 723$ K and $\tau_s = 0.01$ s. Curve (3) shows forcing by switching between an inert and a reactant feed, that is, the strategy given by the initial conditions expressed in Eq. (27). This was the strategy used by Briggs et al. (1977,1978) in the experiments discussed earlier. Ψ is somewhat over 1.1. Curves (4) and (5) in Fig. 6 show the effect on enhancement of adding SO_3 to the reactor feed. To remain comparable to other curves in the figure, the condition $x_{SO2} + x_{SO3} = 1$ and $(x_{SO3} + x_{O2})/2 = 1$ is imposed. For $0.09 \leq x_{SO3}$ (corresponding to conversion of 90% of the SO_2 in feed to a preconverter to SO_3) the enhancement is negligible. However, the enhancement is over 30% for $x_{SO3} \geq 0.096$ (corresponding to 96% conversion of the SO_2 to SO_3) at a cycle

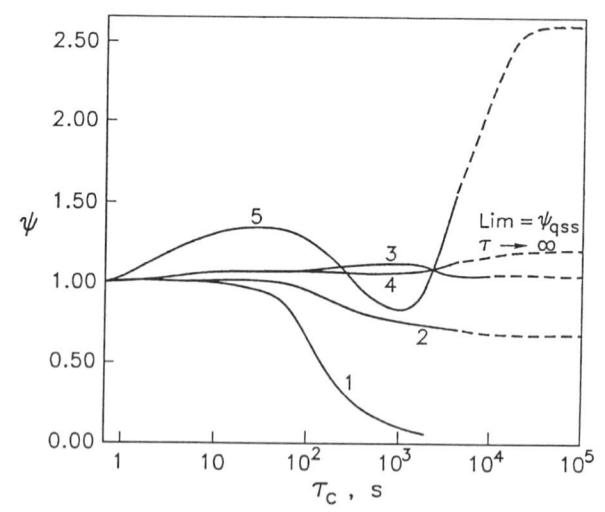

FIG. 6. Influence of the cycling strategy on the enhancement factor in a CSTR simulation at 723 K, $\tau_s = 0.1$ s. Curves: (1) SO_2 and O_2 concentration varied 180° out of phase, (2) only SO_2 concentration varied, (3) SO_2 and O_2 concentration varied in phase, (4) SO_3 present in the feed so that $x_{SO2} + x_{SO3} = 0.1$ and $(x_{O2} + x_{SO3})/2 = 0.1$ and $x_{SO3} = 0.09$, (5) as in (4) but with $x_{SO3} = 0.096$. (Figure taken from Strots et al., 1992, with permission, © 1992 Elsevier Science Publishers.)

period between 20 and 100 s, and multiple resonance seems to arise. This result recalls Briggs' experiments on forcing the final stage of an SO_2 converter where enhancements of the order of 50% were observed at cycle periods in the range predicted by the Strots simulation.

If the SO_2 and O_2 concentrations are switched 180° out of phase so that SO_2 is absent from the reactor feed during one half cycle and O_2 is absent in the other half cycle, Fig. 6 shows that Ψ is less than 1 regardless of the cycle period. Forcing just the SO_2 concentration at a constant O_2 concentration also fails to enhance the rate of SO_2 oxidation in a back-mixed reactor. Even though the experiments of Unni *et al.* (1973), discussed earlier, were performed under isothermal conditions and differentially so that they could have been simulated by Strots' model, the strategy used by Unni was different from those investigated. Nevertheless, one of the experiments undertaken by Unni switched between a reactant mixture and a feed that did not contain SO_2. This experiment exhibited $\Psi < 1$. Strots' model predicts this observation.

Strots *et al.* (1992) studied the influence of model kinetics on enhancement under composition forcing and observed that Ψ significantly exceeds 1 when SO_3 desorbs faster than it decomposes because of the cycles involving vanadium complexes. This suggested raising the rate of oxidation by increasing the space velocity during the flushing half cycle to strip out more SO_3. Strots introduces a dissymmetry factor, D, to measure this velocity change, where D is the ratio of space velocities in the half cycles. Changing velocities was also mentioned discussing Briggs' study of forcing the final stage of a SO_2 converter by periodically flushing the bed in that stage with air. Strots noted that raising the space velocity by shortening the flushing portion of the cycle keeps the same mean feed composition. Strots and co-workers found that modulating the catalyst with a reactant mixture at a low space velocity and an inert stream at a high space velocity significantly increases the conversion attainable and may make it exceed the equilibrium limit that would be encountered in steady-state operation.

III. SO_2 Converters Based on Periodic Reversal of the Flow Direction

Development of this type of periodic operation has proceeded primarily at the Russian Academy of Science's Institute of Catalysis in Novosibirsk (Boreskov and Matros, 1984; Matros, 1985, 1989). Application to acid production has been discussed many times by Russian authors (see Matros, 1989).

Flow reversal can be exploited in various arrangements of one or more catalyst beds. The simplest scheme (Boreskov and Matros, 1984) uses a single catalyst bed and requires four quick-opening valves. For the best performance these valves should be leak-tight because of differing differential pressures across the valves in the on and off position. When flow occurs, as shown in Fig. 7, with cold feed entering the top of the bed, a front of decreasing temperature moves from the top of the bed to the bottom. After half a period, $\tau/2$, valves labeled "1" in the figure close and valves labeled "2" open. This changes the flow direction from downward to upwards and the temperature front now moves from the bottom to the top of the bed. This scheme is appropriate for reactant streams with a small adiabatic temperature rise (ΔT_{ad}). Problems are that the valves and associated piping may be subjected to large temperature variations during a cycle if ΔT_{ad} exceeds 80 to 100°C. In this case maximum bed temperature also may be high and catalyst damage could result. Of course, placing lift valves, which open only on flow, in the line can isolate the quick-opening,

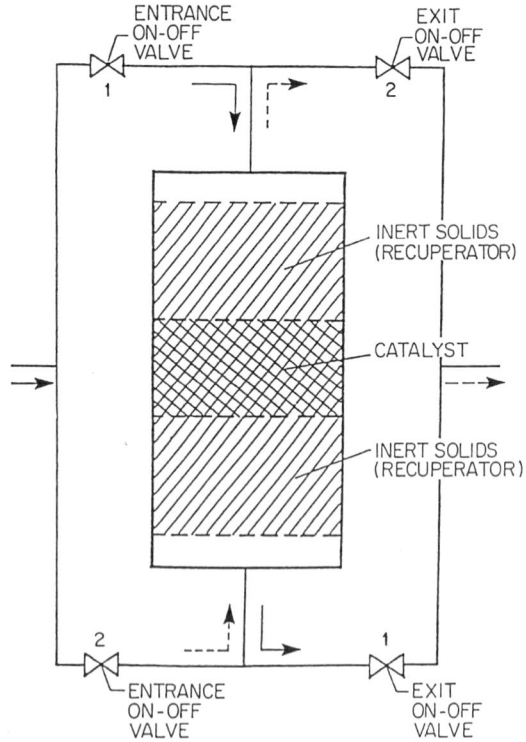

FIG. 7. Single-bed version of flow reversal with catalyst and inert packing zones. (Figure adapted from Matros, 1989, with permission, © 1989 Elsevier Science Publishers.)

leak-tight valves and substantially reduce the temperature variation they experience.

When $\Delta T_{ad} < 30°C$, heat recuperation is important. The outlet switching temperature will be low so the reaction contribution of the inlet and outlet regions is small. In this situation, inert packing with high heat capacity and low porosity can be used in place of catalyst. The variant is shown in Fig. 7.

Flow reversal performance is controlled weakly by the period τ. Flow reversal is an autothermal operation and as such exhibits parametric sensitivity. Greater stability can be ensured by the conventional expedient of providing cooling in the catalyst bed. It can also be done through bypassing part of the reactor effluent gas around the recuperator section.

A flow direction switch occurs at $\tau/2$ for a cycle period τ set by the plant operator, or it may be initiated by the reactor or recuperator outlet temperature. Startup of the flow-reversal system requires a heating device to bring the catalyst bed or at least the frontal portion up to ignition temperature. This temperature is about 350°C for SO_2 oxidation using conventional catalysts.

Catalyst circulation with gas flow always in the same direction is an alternative to periodic flow direction reversal. Circulating recuperator beds, usually with structured solids, have been used industrially for low-temperature applications for decades. If a catalytic material replaces the inert packing, most of the advantages of flow reversal can be retained. Although fast fluidization can be used for catalyst circulation, a rotating bed is the usual means of providing circulation. Rotating the catalyst avoids valve operation in a hostile environment, but brings with it new problems of moving solids at temperatures as high as 400°C and providing seals to minimize by passing.

A. INDUSTRIAL APPLICATIONS

As of 1990, seven industrial-scale plants were operating in Russia (Bunimovich et al., 1990). The first plant started up in 1982. The design of these plants generally follows the schematic given in Fig. 7. Some employ an intermediate heat exchanger, but details have not been published. Designs are marketed around the world by several organizations.

Performance data have been published on most of the SO_2 oxidation flow reversal plants operating in Russia (Matros, 1989; Bunimovich et al., 1990). These are summarized in Table VII. More than 14 years of operating data has been collected on the Krasnouralsk plant. The SO_2 concentration in the feed has varied considerably. Typically, during a working shift the variation is between 1 and 3%, but short-term variations have been much

TABLE VII
INDUSTRIAL-SCALE ACID PLANTS IN GREATER RUSSIA USING FLOW REVERSAL (BUNIMOVICH et al., 1990)

No.	Enterprise, year of startup	Gas capacity, 10^3 m³/h	SO_2 concentration (%)	Type of scheme	Cycle duration (min)	Conversion (%)	T_{max} in catalyst beds (°C)	Pressure drop (mm water)
1	Krasnouralsk Copper Smelter, 1982	40	1–2.5	Single-bed reactor	30–60	94–96	480–530	600–860
2	Mednogorsk Copper Smelter and Acid Plant, 1984	40	3–7	Reactor with intermediate heat removal	60–90	92–94	560–590	400–500
3	Krasnouralsk Copper Smelter, 1987	40	3–5	Reactor with intermediate heat removal	93–95	93–95	530–550	400–600
4	Ust-Kamenogorks Lead–Zinc Smelter, 1985	65	4–8	Reactor with damping bed	60–120	96–97.5	540–590	700–800
5	Alta Verdi Smelter, 1985	30	~4	DC/DA scheme with single bed, flow reversal used in second stage with SO_2 conc. 0–1%	80–100	95–96	520–560	~600
6	Pechenga-Nickel Smelter, 1987	90	1.5–3.5	Reactor with intermediate heat removal		95–96	480–520	600–800
7	Balkhas Copper Smelter, 1989	70	9–11	DC/DA scheme with single bed, flow reversal used in second stage with SO_2 conc. 0–1%		>99.5	460–480	

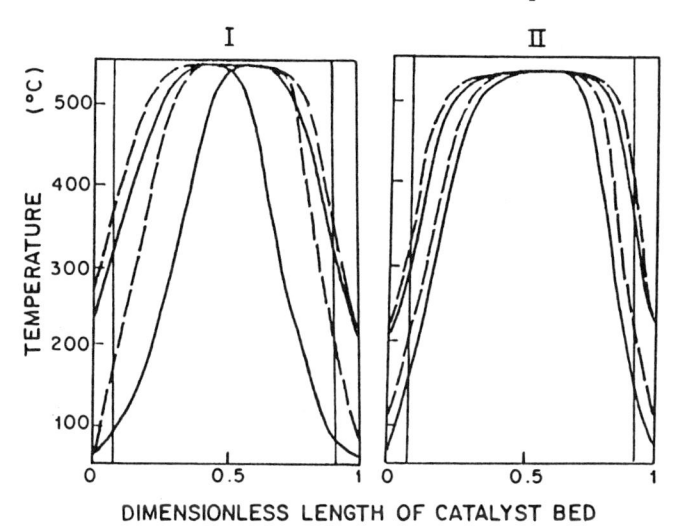

FIG. 8. Observed temperature profiles in a single, flow-reversal SO₂ oxidation reactor processing (I) 40,000 m³/h, (II) 20,000 m³/h of a smelter effluent containing 2.3 vol% SO₂. (Figure adapted from Matros, 1989, with permission, © 1989 Elsevier Science Publishers.)

larger. Hourly means ranged from 0.2 to 4.5%. Gas loading varies, too. Occasionally the reactor operates on half the loading given in the table (Matros, 1989; Bunimovich et al., 1990). Figure 8 shows measured temperature profiles at extremes of gas loading for a feed gas containing 2–3% SO₂. Cycle period is about 25 min for the measurements in the figure. Feed enters the single-bed, reverse-flow reactor at temperatures between 50 and 70°C. Contiguous beds of inert packing are used at the reactor inlet and outlet as shown in Fig. 7. Flow direction switching is initiated by thermocouples at the top and bottom of the bed. At a low volumetric loading, the figure shows that about half the catalyst operates at virtually constant temperature. The largest temperature variations are confined to inert packing at the entrance and outlet. Increasing the volumetric loading twofold sharply narrows the constant temperature region (ca. 10% of the catalyst) and also raises the maximum temperature. Conversion is reduced.

Several pilot plants have been built to test periodic flow direction reversal. Pilot-scale reactors with bed diameters from 1.6 to 2.8 m were operated with flow reversal for several years. The units, described by Bunimovich et al. (1984, 1990) and Matros and Bunimovich (1996), handled 600 to 3000 m³/h and operated with cycle periods of 15 to 20 min. Table VIII shows the performance of these plants for different feeds and potassium oxide promoted vanadia catalysts. The SVD catalyst was granular; the IK-1-4 was in the form of 5 (i.d.) × 10-mm cylinders, while the SVS catalyst was

TABLE VIII
Pilot Plant Performance on Different SO$_2$ Feed Mixtures with Comparison to Model Simulation (Bunimovich et al., 1990)

No.	Range of input SO$_2$ conc. (%)	Approx. O$_2$ conc. (%)	Catalyst type	Bed height (m)	Linear velocity (m)	Cycle duration (min)	Conversion (%)	Max. temp. (°C)	ΔP (mm water)	Assumed SO$_2$ (%)	Calc. conversion (%)	Calc. T_{max} (°C)	Calc. ΔP (mm water)
1	0.70–0.85	10	SVS	2.7	0.41	20	96–97	450	300	0.8	98.0	484	320
			SVD		0.36	20	95–96	450	260	0.95	97.7	494	250
2	0.9–1.0	10	SVS	2.7									
			SVD										
3	1.2–1.3	10	SVS	2.7	0.20	20	95–96	460	80–100	1.2	97.0	493	85
			SVD										
4	1.6–1.8	10	IK-1-4	2.0	0.23	30	94–95	500–510	200	1.7	95.6	527	160
5	2.1–2.6	11	IK-1-4	2.0	0.23	30	91–94	540–550	200	2.3	92.3	560	194
6	3.0–3.5	12.5	IK-1-4	2.0	0.25	25	90–93	590	280	3.2	90.0	596	230
7	3.1–3.3	12.5	SVS	2.7									
			SVD	1.7	0.20	50	93–94	540–550	80–100	3.2	91.0	580	95
8	3.1–3.3	12.5	SVD$_{exp}$	4.6	0.20	50	92–95	520	10–20	3.2	92.5	560	6
9	4.0–4.5	12.5	SVD$_{exp}$	1.7	0.13	30–40	90–92	580	10–15	4.2	90	588	12
10	5.9–6.1	15	SVD$_{exp}$	1.7	0.06	120	90–93	580	10–15	6	92.6	582	2
11	5.9–6.3	14.7	SVD	0.85	0.07	120	85–87	630–640	—	6	91.6	600	17
12	8.3–8.6	12.6	SVD$_{exp}$	1.75	0.06	120	85	610	—	8.4	86.2	612	2

25 (i.d.) × 50-mm hollow cylinders with a 2.6-mm wall thickness. Inlet temperatures were 30 to 50°C.

Table VIII demonstrates the inverse relationship of conversion to SO_2 concentration in the feed that is a consequence of applying flow reversal to SO_2 oxidation using a single reactor. As the SO_2 concentration in the table moves from 0.8 to over 8 vol%, the conversion drops from 96–97% down to 85%. At the same time, the maximum bed temperature changes from 450 to 610°C. For an equilibrium-limited, exothermic reaction, this behavior is explained by variation of the equilibrium conversion with temperature.

Xiao and Yuan (1996) also published pilot-plant measurements. These investigators employed a bypass stream from a full-scale acid plant. Their reactor consisted of three stages, each 800 mm in diameter and packed to a depth of 500 mm with a commercial Type S101 catalyst. Some cooling was provided after the first and second stages to minimize peak temperatures in the bed. Experiments were conducted with a feed gas containing about 8 vol% SO_2. Briggs et al. (1977, 1978) also used a feed of this composition in their experiments on periodic air blowing of the final stage of an SO_2 converter. Operating at a superficial velocity of 0.07 m/s, Xiao and Yuan report 77 to 80% conversion and a maximum bed temperature of about 640°C. Temperature profiles for upflow and down flow just before switching of the flow direction are shown in Fig. 9 for a cycle period of 120 min after the system has become cyclically invariant. The profiles are nearly symmetrical. The effect of interstage cooling is clearly seen. The curious "M" shape of the profiles appears to be due to completion of the reaction in the first stage as the temperature profile in the second stage varies by just 20°C. The temperature rise in the third stage is largely a "ghost" of the previous half cycle. Increasing the flow rate by about 40% had little effect on conversion and peak bed temperatures, but had a large effect on the contribution of cooling, thereby increasing the temperature in the second bed.

Isozaki et al. (1990) describe a double contacting-double absorption process utilizing flow reversal in both contacting stages. Figure 10 is a schematic flow diagram of the Hitachi Zosen design. The Isozaki paper gives performance details for a pilot plant operated in Saganoseki, Japan, in 1988. The first stage plant was designed to process 3000 to 3600 Nm^3/h of an acid plant feed containing 5–9 vol% SO_2 and 8–14 vol% O_2 available at 50 to 65°C. It achieved an 85 to 90% SO_2 conversion using a 50- to 60-min flow reversal cycle. The second stage operated on a feed with a SO_2 content of about 0.5 to 1.5 vol%. Intermediate cooling was not required as the figure shows. Initial tests on this plant at 3000 Nm^3/h showed conversions of 94 to 96% depending on the SO_2 concentration in the feed. Cycle periods

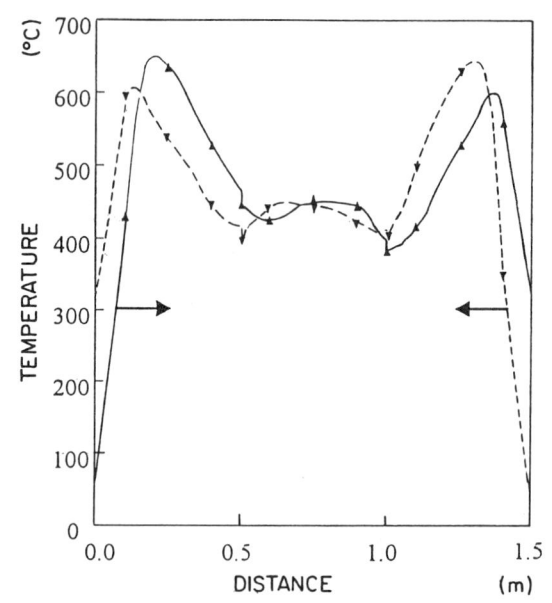

FIG. 9. Invariant cycling state temperature profiles for a three-stage, 800 mm diameter pilot plant SO$_2$ converter with interstage cooling. Profiles were measured just before switching of the flow direction and are for SO$_2$ at 7.6 vol% and a superficial velocity of 0.07 m/s. Catalyst was a commercial potassium-promoted vanadia, Type S101. (Figure adapted from Xiao and Yuan, 1996, with permission of the authors.)

varied between 20 and 36 min, increasing with higher SO$_2$ content in the feed. Operating both stages together would bring SO$_2$ conversion to just over 99.5%.

Application of flow reversal is particularly attractive for the second stage of a double absorption system, according to Matros (1985, 1986, 1989). A conventional first stage results in a feed gas for the second stage containing 0.5 to 1.0 vol% SO$_2$. At this level, maximum bed temperature will be under 500°C; temperature drops toward the outlet, so conversion of about 96% of the SO$_2$ entering the stage is possible. Over two stages, conversion reaches 99.6%. This agrees well with the Japanese results just mentioned. Two full-scale plants using flow reversal in the second stage have been operating in Russia since 1989 (Table VII).

B. Experimental Results

Results for bench-scale flow-reversal units operated in Russia (Boreskov *et al.,* 1982: Matros, 1989) and China (Wu et al., 1996; Xiao and Yuan,

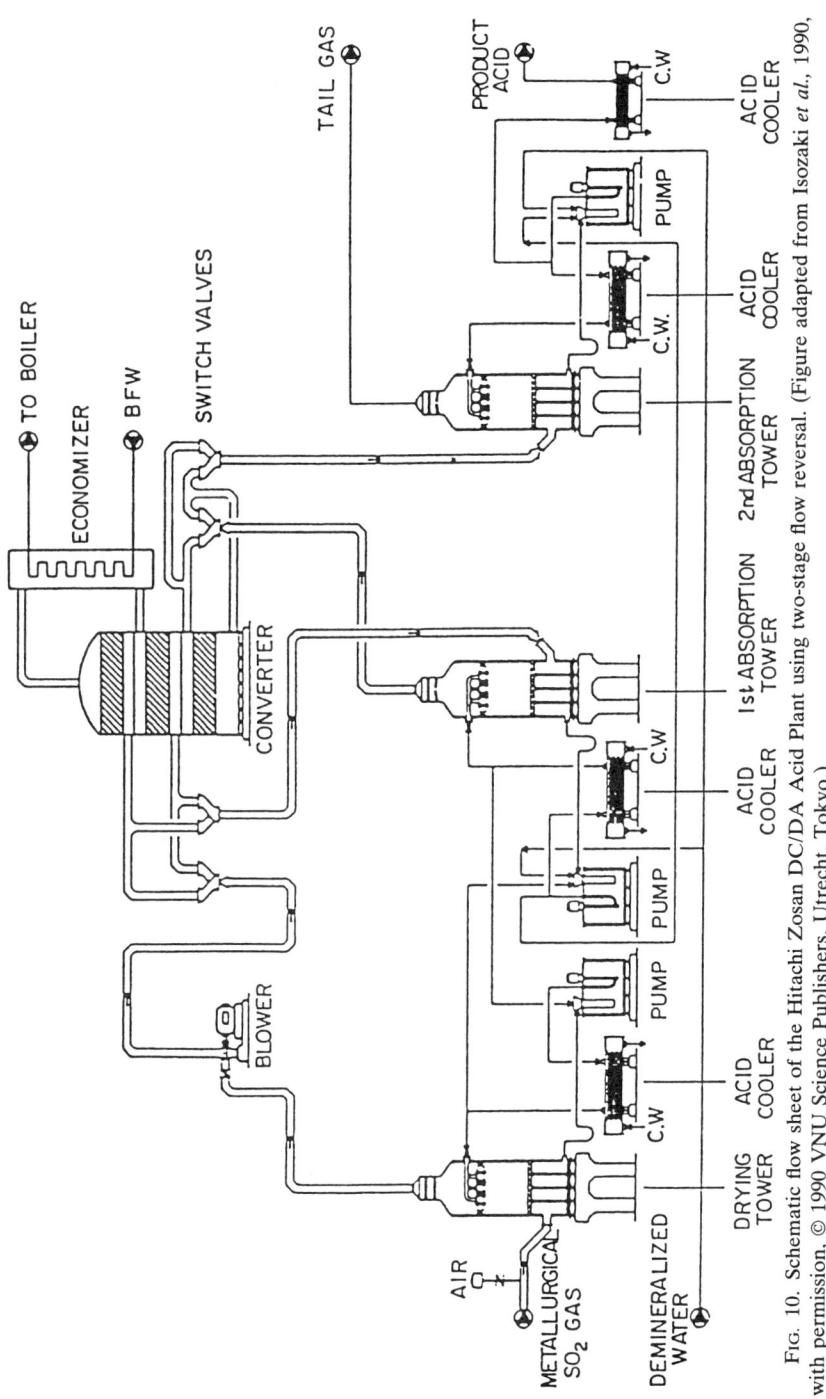

FIG. 10. Schematic flow sheet of the Hitachi Zosan DC/DA Acid Plant using two-stage flow reversal. (Figure adapted from Isozaki et al., 1990, with permission, © 1990 VNU Science Publishers, Utrecht, Tokyo.)

1996) have been published. The experimental system used by Xiao and Yuan (1996) appears in Fig. 11. The reactor was 32 mm (i.d.) × 1100 mm and was packed with a Type S101 potassium-promoted vanadia catalyst to a depth of 900 mm. Inert packing, 100 mm, was placed above and below the catalyst. Wu *et al.* (1996) used the same commercial catalyst but employed a 50 mm (i.d.) × 4000 mm reactor, about half full of catalyst. By contrast, Russian researchers used a 175 mm (i.d.) × 1300 mm reactor. Development of the temperature profile after a switch in the direction of flow is shown in Fig. 12 for stationary, reproducible operation of a 2 m deep converter bed (Wu et al., 1996). Results for different SO_2 concentrations, superficial

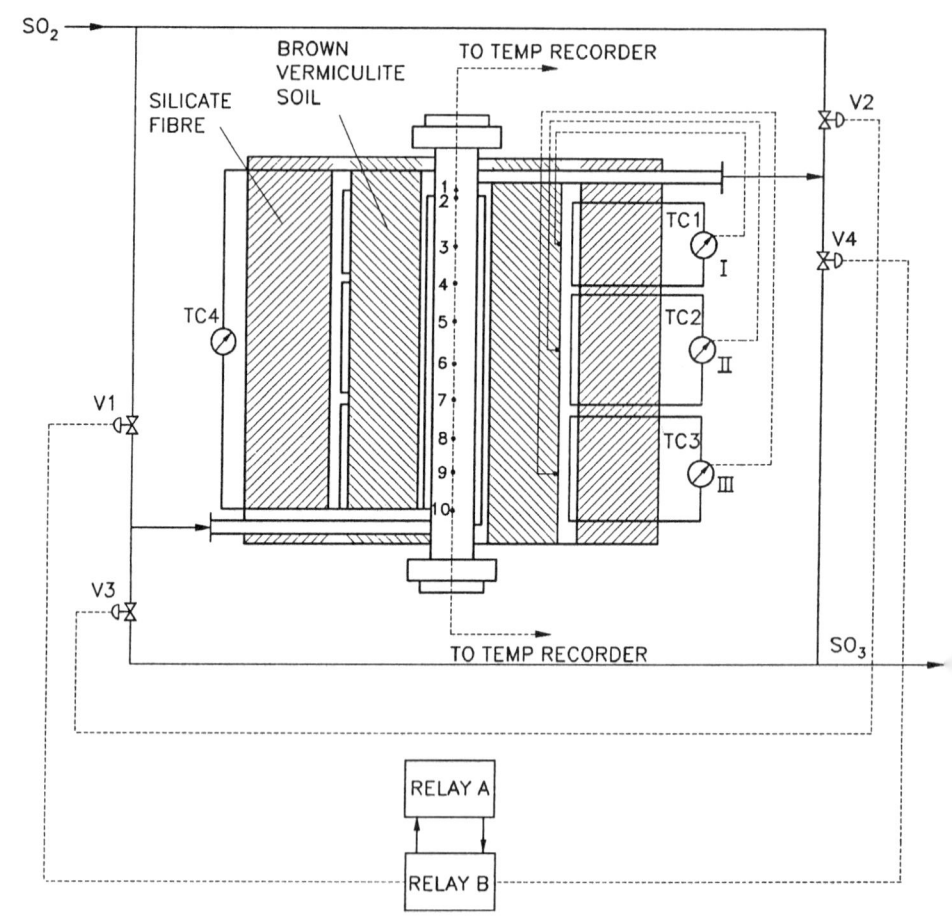

FIG. 11. Schematic of a research flow-reversal reactor showing switching valves and instrumentation. Heavy insulation was used to try to obtain adiabatic operation. (Figure adapted from Xiao and Yuan, 1996, with permission of the authors.)

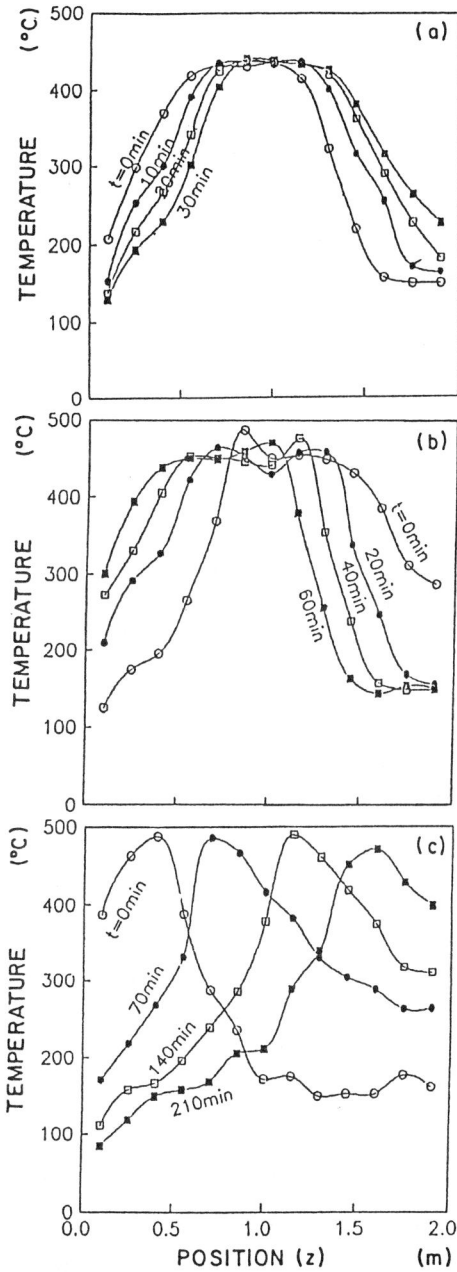

FIG. 12. Time change of temperature profiles in a catalyst bed under an invariant cycling state for an SO_2-containing feed entering the reactor at 25°C. Results for the Chinese S101 catalyst. (Figure adapted from Wu et al., 1996, with permission of the authors.)

velocities, and cycle periods are given. In Fig. 12a, the curves for 0 and 30 min represent the start and end of a half cycle. A comparison of the (a) and (b) sections of the figure shows that increasing the throughput of reactant, just as for higher SO_2 levels in the feed and higher flow rates, raises the maximum temperature and extends the width of the high temperature zone. Figure 12a shows that about half of the bed is held at constant temperature. This temperature exceeds the adiabatic temperature rise for the feed by about 300°C. Movement of the temperature front through the bed can be seen in (c) of the figure. Increasing the cycle period above 420 min would have extinguished the reaction because the cold front, started at the time of switching by cold feed entering the bed, reaches the end of the bed.

C. Modeling and Simulation

If kinetic processes on catalytic surfaces in SO_2 oxidation are assumed to be at steady state, temperature and concentration fields in a radially symmetrical, adiabatic catalyst bed are described by the equations collected in Table IX for the reactor space $0 \leq z \leq L$ and time $t \geq 0$ (Matros, 1989; Matros and Bunimovich, 1996). The subscript p indicates a condition at the surface of a catalyst particle of radius R_p. A spherical particle has been assumed in Eq. (31). Rates are expressed in units of particle volume. Multiplying ρ_B converts them to mass units. Assuming plug flow through the reactor and neglecting radial temperature gradients reduces dimensionality. Matros (1989) contends that allowing for radial gradients in large-diameter, well-packed beds without wall anomalies has a negligible influence on front movement. Radial heat losses will be negligible if the reactor is insulated even when narrow beds are used.

Equations in Table IX are written per unit of bed volume; $(\lambda_l)_g$ is a time averaged, mean axial bed conductivity. D_l is a longitudinal diffusivity and λ_l allows for particle to particle conductivity. Not all the terms in the model as given in the table are important. For example, Wu et al. (1995, 1996) and Xiao and Yuan (1996) neglect the accumulation and dispersion terms in Eq. (30) and the accumulation and conduction terms in Eq. (28).

Because the term $r(C_A, T)$ is exponentially dependent on T and can be nonlinear as well, a numerical solution or piecewise linearization must be used. To simplify the numerical manipulations, equations in Table IX are normalized by $\xi = z/L$, $\tau' = ut/L$, and $x = 1 - C_i/(C_i)_o$, where i is normally SO_2. y also is a normalized quantity. The Péclet numbers for mass and heat are written $Pe_M = 2R_p u/D_l$ and $Pe_H = 2R_p c_p u/\lambda_l$ for a spherical particle. They are also written in terms of bed length as Bodenstein numbers. It is

TABLE IX
Model for a Packed-Bed Catalytic Reactor with Periodic Reversal of Flow Direction

Gas phase:

$$(\bar{C}_p)_g \frac{\partial T}{\partial t} = +(\bar{\lambda}_l)_g \frac{\partial^2 T}{\partial z^2} - (\bar{C}_p)_g u \frac{\partial T}{\partial z} + h_p S_{ext}(T_p - T) \quad (28)$$

Solid phase:

$$(C_p)_p \frac{\partial T_p}{\partial t} = +\bar{\lambda}_l \frac{\partial^2 T_p}{\partial z^2} - h_p S_{ext}(T_p - T) + \frac{\Delta H_{ad}\bar{\eta}}{(1 - \varepsilon_B)} r(C_{A_{r=R_p}}, T_p) \quad (29)$$

Gas phase:

$$\varepsilon_B' \frac{\partial C_A}{\partial t} = D_l \frac{\partial^2 C_A}{\partial z^2} - u \frac{\partial C_A}{\partial z} - k_p S_{ext}(C_A - C_{A_{r=R_p}}) \quad (30)$$

Solid phase:

$$k_p(C_A - C_{A_{r=R_p}}) = \frac{R_p \bar{\eta}}{3} r(C_{A_{r=R_p}}, T_p) \quad (31)$$

Boundary conditions:

$$Z = 0 \quad (-\bar{\lambda}_l)_f \frac{\partial T}{\partial z} = (c_p)_g u + (T - T^o) \quad (32)$$

$$\frac{\partial T_p}{\partial z} = 0 \quad (33)$$

$$-D_A \frac{\partial C_A}{\partial z} = u^+(C_A - C_A^o) \quad (34)$$

$$z = L, \quad (-\bar{\lambda}_l)_g \frac{\partial T}{\partial z} = (C_p)_g u^-(T - T^o) \quad (35)$$

$$\frac{\partial T_p}{\partial z} = 0 \quad (36)$$

$$-D_A \frac{\partial C_A}{\partial z} = u^-(c_A - c_A^o) \quad (37)$$

$$t = t_o + \frac{n\tau}{z}, \quad n = 1, 2, 3\ldots, \quad T(z, t^-) = T(z, t^+) \quad (38)$$

$$T_p(z, t^-) = T_p(z, t^+) \quad (39)$$

$$C_A(z, t^-) = C_A(z, t^+) \quad (40)$$

where

$$u^+ = \max(u(t), 0), \quad u^- = \max(-u(t), 0) \quad (41)$$

With flow from the left, u^+ will be positive and $u^- = 0$. When the flow is from right to left, $-u(t)$ is positive so $u^- = u(t)$, and $u^+ = 0$; t^-, t^+ designate instants before and after a flow direction switch.

TABLE X
SIMPLIFIED, DIMENSIONLESS FORM OF THE FLOW REVERSAL MODEL

$$C_r \frac{\partial \theta}{\partial \tau'} = \frac{1}{\text{Phs}} \frac{\partial^2 \theta}{\partial \xi^2} + \beta_1 r(\theta, y) - \alpha(\theta - T) \tag{42}$$

$$0 = \frac{1}{\text{Phg}} \frac{\partial^2 T}{\partial \xi^2} - \frac{\partial T}{\partial z} + \alpha(\theta - T) \tag{43}$$

$$\gamma(y - x) = \beta_2 r(\theta, y) \tag{44}$$

$$\frac{\partial x}{\partial \xi} = \beta_2 r(\theta, y) \tag{45}$$

with boundary conditions:

$$\xi = 0, \tau' > 0 \begin{cases} \dfrac{\partial \theta}{\partial \xi} = 0 \\ -\dfrac{1}{\text{Phg}} \dfrac{\partial T}{\partial \xi} + T = T_{\text{in}} \\ x = 0 \end{cases} \tag{46}$$

$$(\xi = 1, \tau' > 0) \begin{cases} \dfrac{\partial \theta}{\partial \xi} = 0 \\ \dfrac{\partial T}{\partial \xi} = 0 \end{cases} \tag{47}$$

$$\tau = \frac{n}{2}\tau, \quad n = 1, 2, 3$$

$$T(\xi, \tau^-) = T(\xi, \tau^+)$$
$$\theta(\xi, \tau^-) = \theta(\xi, \tau^+) \tag{48}$$
$$x(\xi, \tau^-) = x(\xi, \tau^+)$$
$$y(\xi, \tau^-) = y(\xi, \tau^+),$$

this formulation that is used in Table X that gives the dimensionless form of the model (Xiao and Yuan, 1994).

Temperature and concentration differences between gas and catalyst can be neglected to give a pseudo-homogeneous model,

$$\frac{\partial X}{\partial \xi} + \kappa(1 - X)\exp\left(\frac{\theta}{B\theta + 1}\right) = 0 \tag{50}$$

$$\gamma \frac{\partial \theta}{\partial \tau'} = \frac{1}{(\text{Pe}_H)} \frac{2R_p}{L} \frac{\partial^2 \theta}{\partial \xi^2} - \frac{\partial \theta}{\partial \xi} + \Delta\theta_{\text{ad}}\kappa(1 - X)\exp\left(\frac{\theta}{B\theta + 1}\right). \tag{51}$$

the higher order term allows for conductivity in the bed only.

TABLE X (continued)
SIMPLIFIED, DIMENSIONLESS FORM OF THE FLOW REVERSAL MODEL

where τ^- and τ^+ are the dimensionless times just prior to and after a direction switch.

$$\theta(z, 0) = \text{initial temperature profile} \quad (49)$$

$$\beta_1 = (1 - \varepsilon_B)\frac{(-\Delta H)}{(\rho C_P)_g T_0}\frac{L}{u}$$

$$\beta_2 = (1 - \varepsilon_B)\frac{L}{u}\frac{1}{C_{in}}$$

$$\alpha = (1 - \varepsilon_B)\frac{S_v h_P}{(\rho C_P)}\frac{L}{u}$$

$$\gamma = (1 - \varepsilon_B)S_v k_P \frac{L}{u}$$

$$C_r = (1 - \varepsilon_B)\frac{(\rho C_P)_s}{(\rho C_P)_g}$$

$$\frac{1}{\text{Phs}} = (1 - \varepsilon_B)\frac{(\lambda_l)}{(\rho C_P)_s u L}$$

$$\frac{1}{\text{Phg}} = \varepsilon_B \frac{(\lambda_l)_g}{(\rho C_P)_g u L}$$

$$\frac{1}{\text{Pmg}} = \varepsilon_B \frac{D_e}{u L}$$

Often in industrial-scale reactors, the lengths of reactors and gas velocities lead to large values of $Pe(L/2R_p)$ that make the contribution of the dispersion terms negligible. In this situation, the order of the set of PDEs is reduced and an accurate description of front movement can be made through piecewise linearization of the rate term and use of the elegant method of characteristics (Rhee and Amundson, 1974). Boreskov et al. (1977) note that the pseudo-homogeneous model gives an adequate approximation for flow reversal. Eigenberger and Nieken (1988) also used this packed-bed model in their study of flow reversal.

The ultimate simplification is

$$\frac{(\lambda_l)_g}{(c_p)_g u L}\frac{d\theta}{d\xi} = \frac{1}{2}\Delta T_{ad}(X_1 - X_2) \quad (52)$$

$$v_{sv}\frac{dX_1}{d\xi} = \frac{1}{2}r(\theta, X_1) \quad (53)$$

$$v_{sv}\frac{dX_2}{d\xi} = \frac{1}{2}r(\theta, X_2). \quad (54)$$

The boundary conditions are

$$\xi = 0: \quad X_2 = X_f, \quad \theta = T° + \Delta T_{ad} X_f \qquad (55)$$
$$X_1 = 0$$
$$\xi = 1: \quad X_1 = X_2. \qquad (56)$$

In these equations, $\xi = 0$ is the bottom of the catalyst bed and X_1 is the conversion in the flow direction from bottom to top, while X_2 is the conversion in the opposite flow direction. Bunimovich et al. (1990) suggest using Eqs. (52) to (54) for an initial estimate of the temperature profiles in order to speed up conversion on integration of the full model equations in Table X. This step would only be taken if it were the stationary cyclic state profiles that are wanted.

A substantial literature, mainly in Russian, discusses the simulation of various industrial processes operating under flow reversal. Much of this work deals with SO_2 oxidation. For the rate term, Russian researchers (Boreskov et al., 1982) used the expression

$$r_{SO_2} = k \frac{P_{SO_2} - (P_{O_2})^{1/2}(1 - \beta)}{(1 + K_{ad}(P_{SO_3})^{1/2})}. \qquad (57)$$

The β term allows for the reverse reaction, and it is usually written

$$\beta = \frac{P_{SO_3}}{K P_{SO_2}(P_{O_2})^{1/2}}. \qquad (58)$$

The rate model contains four adjustable parameters, as the rate constant k and a term in the denominator, K_{ad}, are written using the Arrhenius expression and so require a preexponential term and an activation energy. The equilibrium constant can be calculated from thermodynamic data. The constants depend on the catalyst employed, but some, such as the activation energy, are about the same for many commercial catalysts. Equation (57) is a steady-state model: the low velocity of temperature fronts moving through catalyst beds often justifies its use for periodic flow reversal.

The calculated conversions presented in Table VIII used Eq. (57). They are quite remarkable. They reproduce experimental trends of lower conversion and higher peak bed temperature as the SO_2 content in the feed increases. Bunimovich et al. (1995) compared simulated and experimental conversion and peak bed temperature data for full-scale commercial plants and large-scale pilot plants using the model given in Table IX and the steady-state kinetic model [Eq. (57)]. Although the time-average plant performance was predicted closely, limiting cycle period predicted by the model are much longer than experimental limits. Periods used in plant operation were between 25 and 60% of this limit.

There are many other SO_2 simulation using the models described in Tables IX and X or their further simplifications (Boreskov and Matros, 1984; Matros, 1989; Bunimovich *et al.*, 1990; Sapundzhiev *et al.*, 1990; Snyder and Subramanian, 1993; Xiao and Yuan, 1994, 1996; Zhang *et al.*; 1995; Wu *et al.*, 1996). These simulations are capable of reproducing operating data as demonstrated in Table VIII and in the preceding discussion. They have been useful in understanding the application of periodic flow reversal to SO_2 oxidation, as we shall see.

A simplified version of the model in Table IX, neglecting accumulation of mass and heat as well as dispersion and conduction in the gas phase, predicts dynamic performance of a laboratory SO_2 converter operating under periodic reversal of flow direction quite well. This is shown by Fig. 13 taken from Wu *et al.* (1996). Data show the temperature profiles in a 2-m bed of the Chinese S101 catalyst once a stationary cycling state is attained. One set of curves shows the temperature distribution just after switching direction and the second shows the distribution after a further 60 min. Simulated and experimental profiles are close. The surprising result is that the experimental maximum temperatures equal or exceed the simu-

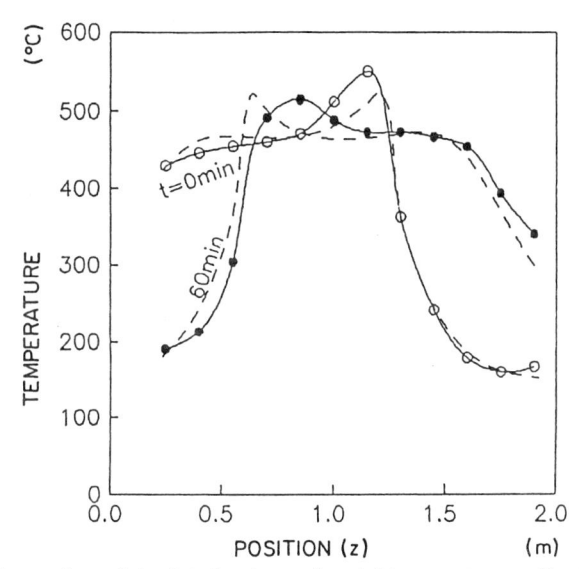

FIG. 13. Comparison of simulated and experimental temperature profiles in a 2-m, near-adiabatic, packed-bed SO_2 reactor using a Chinese S101 catalyst and operating under periodic reversal of flow direction with $\tau = 180$ min, SV = 477 h^{-1}, and inlet SO_2 = 3.89 vol% and T = 25°C. (Figure adapted from Wu *et al.*, 1996, with permission of the authors.)

lated values. The model is adiabatic, but the reactor cannot be operated without heat loss, so experimental values should fall below the simulated results. In an earlier paper employing the same model, Zhang *et al.* (1995) explored the effect of catalyst bed length, *L,* inlet SO_2 concentration, space velocity, and cycle period τ on SO_2 conversion and the peak bed temperature. With respect to concentration and space velocity, simulation results agreed with Russian work and experimental data shown in Table VIII: Increasing SO_2 vol% and SV decrease conversion, but increase the peak bed temperature. Bed length has the opposite effect: Conversion increases and peak temperature decreases as the depth of the catalyst bed is augmented. Increasing τ increases both conversion and peak temperature, but the change is small over a 3 × change of cycle period. Both Zhang *et al.* (1995) and Wu *et al.* use a version of the rate expression given by Eq. (57) in their simulations.

Xiao and Yuan (1994, 1996) used a model similar to that given in Table X for a comprehensive investigation of SO_2 oxidation in an adiabatic packed bed employing flow reversal. Figure 14 illustrates the effect of inlet SO_2 concentration on reactor operation during periodic switching of the flow direction. Figure 14a nicely shows the increase in maximum bed temperature and in width of a zone of high, near-constant temperature as the inlet vol% SO_2 rises. Figure 14b gives the effect of inlet concentration on maximum temperature, SO_2 conversion, and the velocity of the moving temperature fronts. The figure shows that there is a minimum inlet concentration of about 1.2 vol% for a stable reaction. This minimum depends on O_2 concentration, inlet temperature, catalyst, and space velocity. Increasing space velocity reduces the minimum. The reason for its existence, shrinking of the zone of high and near-constant temperature, can be seen in Fig. 14a. Above the minimum inlet concentration, the peak bed temperature increases rapidly with rising SO_2 concentration in the feed. The temperature extends from 400°C because this was the startup temperature. It is often convenient to express the concentration effect in terms of the adiabatic temperature rise, ΔT_{ad}. This approach is favored by Russian investigators (Matros, 1989). Returning to Fig. 14b and noting that the SO_2 concentration is proportional to ΔT_{ad}, then once the minimum ΔT_{ad} is reached, the reaction is stabilized. Increasing ΔT_{ad} only increases the maximum temperature in the catalyst bed. This increase is about the same as the increase in ΔT_{ad}. Increasing temperature causes a maximum in SO_2 conversion that is shown in the plot. This reflects the importance of the reverse reaction at high temperatures. Finally, Fig. 14b shows that the variation of velocity of the temperature front with the inlet SO_2 vol% is small. All these results are for $\tau = 30$ min and SV = 514 h^{-1}.

Xiao and Yuan (1996) also explored the effect of replacing the front

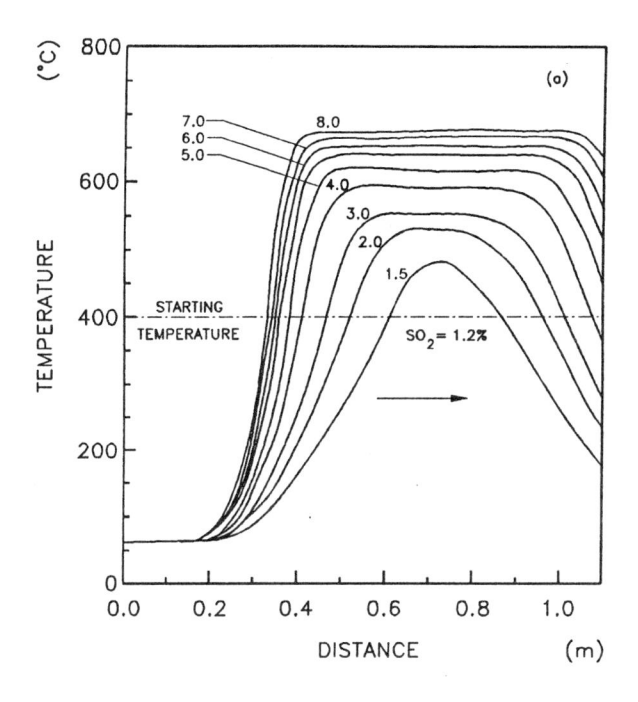

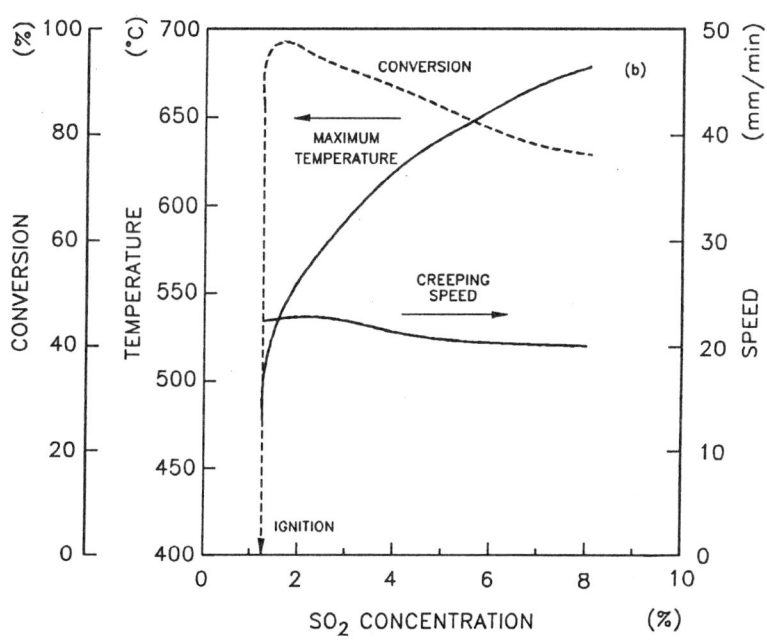

FIG. 14. Influence of inlet SO$_2$ concentration on behavior and performance in adiabatic, packed-bed SO$_2$ converters operating under periodic flow reversal. Simulation results for τ = 30 min, SV = 514 h^{-1}, T_o = 25°C: (a) effect of inlet SO$_2$ vol% on the temperature profile in the catalyst bed, (b) influence of inlet SO$_2$ on converter performance and the velocity of the temperature front. (Figure adapted from Xiao and Yuan, 1996, with permission of the authors.)

and back portions of the bed by an inert packing as depicted in Fig. 7. This packing functions as a heat recuperator. The simulation of Xiao and Yuan demonstrates that these zones are below the temperature required for the reaction to proceed for most of a cycle so that catalyst in these zones is nonfunctional. Furthermore, these zones experience temperature fluctuations reaching up to 300°C when there is a flow direction switch. The inert zones augment the temperature of the gas entering the catalyst bed and thereby increase temperatures in the bed. This reduces conversion. There is a strong time-in-cycle effect caused by inert packing as shown in Fig. 15. Higher temperature at the end of the catalyst section as the half cycle progresses decreases SO_2 conversion through the reverse reaction, causing the drop shown in the figure for 0.2 m of noncatalytic packing on each end of the catalyst bed. The simulation assumes tau; = 30 min and SV = 514 h^{-1}. The inlet SO_2 concentration is 2.62 vol%. Matros (1989) observed the effect of recuperator zones earlier. He also noted that the presence of these zones extends the maximum cycle period before blowout of the reaction occurs.

Xiao and Yuan (1994) also studied startup, and a large wrong-way behav-

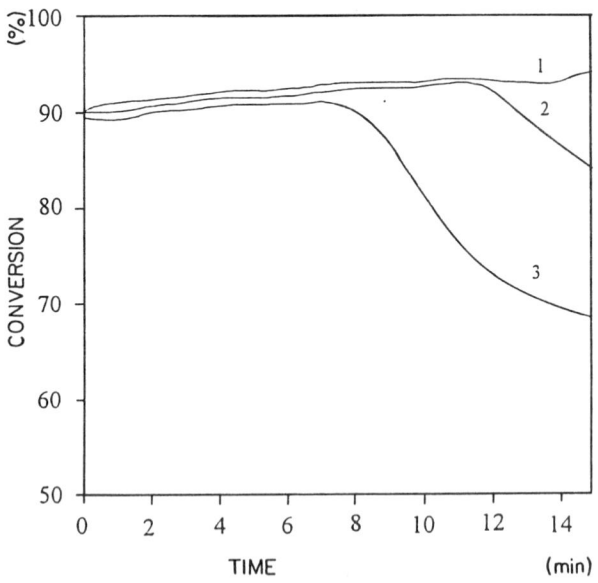

FIG. 15. Variation of SO_2 conversion with time within a half cycle based on simulation of the reactor in Fig. 14 and for the conditions given in that figure with inlet SO_2 = 2.62 vol%. Curve 1, full 1.4-m bed filled with catalyst; curve 2, 0.1-m regions filled with noncatalytic packing at both ends of the catalyst bed; curve 3, 0.2-m regions instead of 0.1-m ones. (Figure adapted from Xiao and Yuan, 1996, with permission of the authors.)

ior was seen when cold feed is introduced. A sharp temperature jump occurs near the entrance to the bed that could be large enough to damage catalyst. The magnitude of the temperature jump depends strongly on the SO_2 concentration as well as other operating conditions. Xiao and Yuan successfully simulated the dynamic operation of their laboratory reactor and comment that the match between model and data can be improved by allowing for radial heat loss in their model. This was accomplished by adding a heat transfer term to Eq. (28) in Table IX. The reactor model remains one-dimensional.

Bunimovich et al. (1984) point out that if the period of flow reversal, τ, is very small relative to the time required for the temperature front to creep through the bed and the high-temperature zone occupies most of the bed; a relaxed steady state is achieved in which the temperature profile is constant through most of the bed. This profile can be calculated and leads to a steady-state model for this extreme variant of flow reversal.

Both Matros (1989) and Eigenberger and Nieken (1988) mention potential operating problems identified using the model summarized in Table IX. For any operation there is a minimum feed temperature for ignition. At temperatures above this minimum, a stable temperature profile is achieved after several reversal cycles. Flow reversal can tolerate lower concentrations and even the absence of reactants for time periods $\Delta t < \tau$. Concentration increases, for exothermic reactions, lead to abrupt increases in the maximum bed temperature that may cause catalyst deactivation or even damage to the reactor vessel. The problem of high reactant concentrations has inspired a number of multibed flow-reversal schemes that are discussed by Matros (1989). When process upsets upstream of the flow-reversal reactor cause reactant concentration excursions; control can be handled by recycle to dilute the stream, by diverting reactant flow at points within the reactor, or by adding thermal ballast in the form of a boiler or a liquid-phase heat exchanger. Purwono et al. (1994) have examined the last suggestion experimentally and through simulation. A theoretical lower limit on cycle period exists (Van den Bussche et al., 1993) and there is an upper limit discussed when experimental results were considered. This period is predictable. There is also a maximum period at startup in order to ignite the reaction. Eigenberger and Nieken (1988) found that these maximum periods are not the same.

A concern in the application of periodic flow reversal to converting SO_2 in smelter emissions is the variability of these emissions. Both SO_2 concentration and gas volume may vary, often irregularly. Like startup, this is a matter that could profit from simulation studies using models discussed in this section.

Bunimovich et al. (1995) have applied the Balzhinimaev model for SO_2

oxidation over vanadia catalysts to a nonisothermal tubular reactor with periodic flow reversal. The model is probably too complicated for wide use, as it appears from the preceding discussion that steady-state models give adequate results for most simulation studies. These earlier models assume that the steady-state kinetics apply under the slowly changing conditions of periodic flow reversal, but they overpredict the time average SO_2 conversions by about 1% at low inlet SO_2 levels. They also fail to predict an experimentally observed low level in the SO_3 content of the effluent after switching followed by a slow rise until the concentration slightly exceeds the SO_2 concentration in the converter feed. After this maximum, SO_3 sinks slowly until it falls below the inlet SO_2 level. (Steady-state models do not show this, as may be seen in Fig. 15.) Allowing for the dynamics of the catalyst using the kinetic model given in Table IV, Bunimovich et al. (1995) demonstrate that predicted conversions fall below those given by the steady-state kinetic model thereby realizing better agreement with experimental data. They also show that dissolution of SO_3 from the melt is slow, and it is this that accounts for the initial low level of SO_3 in the effluent and its subsequent rise and overshoot. Improved dynamic prediction that the unsteady-state kinetic model of Balzhinimaev offers is important for the design of control systems for flow reversal, but is not necessary for exploring applications of periodic flow reversal to acid plant or smelter effluents where only the time-average performance is of interest.

Bunimovich et al. (1995) lumped the melt and solid phases of the catalyst but still distinguished between this lumped solid phase and the gas. Accumulation of mass and heat in the gas were neglected as were dispersion and conduction in the catalyst bed. This results in the model given in Table V with the radial heat transfer, conduction, and gas phase heat accumulation terms removed. The boundary conditions are different and become identical to those given in Table IX, expanded to provide for inversion of the melt concentrations when the flow direction switches. A dimensionless form of the model is given in Table XI. Parameters used in the model will be found in Bunimovich's paper.

It is instructive to examine predictions of the Bunimovich et al. model. The temperature profiles in Fig. 16a at times within a half cycle are about the same as the profiles for 6 vol% SO_2 predicted by Xiao and Yuan (1996) and shown in Fig. 14a. Note that curves 1 and 5 in Fig. 16 are those at the switching time and show the inversion when the flow direction changes. The remarkable prediction of the Table XI model is the sharp variation of the SO_3 concentration in the melt with position in the bed that develops about 1 min after the direction changes and persists for about 13 min. Curves (c) to (f) show there is a substantial change in the complexes present in the melt with position in the bed and that the complex concentrations

TABLE XI
Periodic Flow Reversal Model Employing the Balzhinimaev Unsteady State Kinetic Model for SO₂ Oxidation Over Vanadia Catalysts
(Bunimovich et al., 1995)

$$\gamma_\Theta \frac{\partial \Theta}{\partial \tau} = \gamma_D \Delta T_D \frac{\partial C_3^L}{\partial \tau} + \Lambda \sum_{i=1}^{4} \Delta T_i r_i - \alpha(\Theta - T) \tag{59}$$

$$\gamma_D \frac{\partial C_3^L}{\partial \tau} = \beta_3 \left[x_3^f - \frac{C_3^L}{H_3 P(1 - C_3^L/C_o)} \right] + \Lambda \sum_{i=1}^{4} \nu_{3_i} r_i \tag{60}$$

$$\gamma_\Theta \frac{\partial \theta_m}{\partial t} = \sum_{i=4}^{4} \mu_i r_i, \quad m = \overline{1,3} \tag{61}$$

$$\frac{\partial x_j^f}{\partial \xi} = \beta_j(x_j^f - x_j), \quad j = \overline{1,3} \tag{62}$$

$$\frac{\partial T}{\partial \xi} = \alpha(\Theta - T) \tag{63}$$

$$\Lambda \sum_{i=1}^{4} \nu_{ji} r_i = \beta \left(x_j^f - \frac{C_j}{H_j P} \right), j = 1, 2, \tag{64}$$

with boundary and initial conditions

At $\xi = 0$:

$$T = T_{in}, \quad x_j^f = (x_j^f)_{in}.$$

At $\tau = 0$:

$$\Theta = \Theta_0(\xi); x_3^L = (x_3^L)_0(\xi); \quad \theta_m = (\theta_{m,0})(\xi) \tag{65}$$

$$(m = \overline{1,3})$$

$$\tau = t_c, \frac{n\tau}{2}, n\tau \quad n = 1, 2, 3, \ldots$$

$$\theta_m(\xi) = \theta_m(1 - \xi), \quad x_j^f(\xi) = x_j^f(1 - \xi)$$
$$C_3^L(\xi) = C_3^L(1 - \xi), \Theta(\xi) = \Theta(1 - \xi) \tag{66}$$
$$T(\xi) = T(1 - \xi)$$

can be described in terms of sharply defined moving fronts. V^{4+} represents the concentration of catalytically inactive vanadium. Relatively high levels shown in Fig. 16f reflect low temperatures in the front of the bed and the high level of SO_3 in the outlet. Both conditions cause vanadium reduction. The SO_2 and O_2 curves in Figs. 16g and 16h show a concentration front moving through the bed with a narrow reaction zone. The SO_3 profile is unusual: Just after the switch SO_3 leaving the catalyst bed drops to zero.

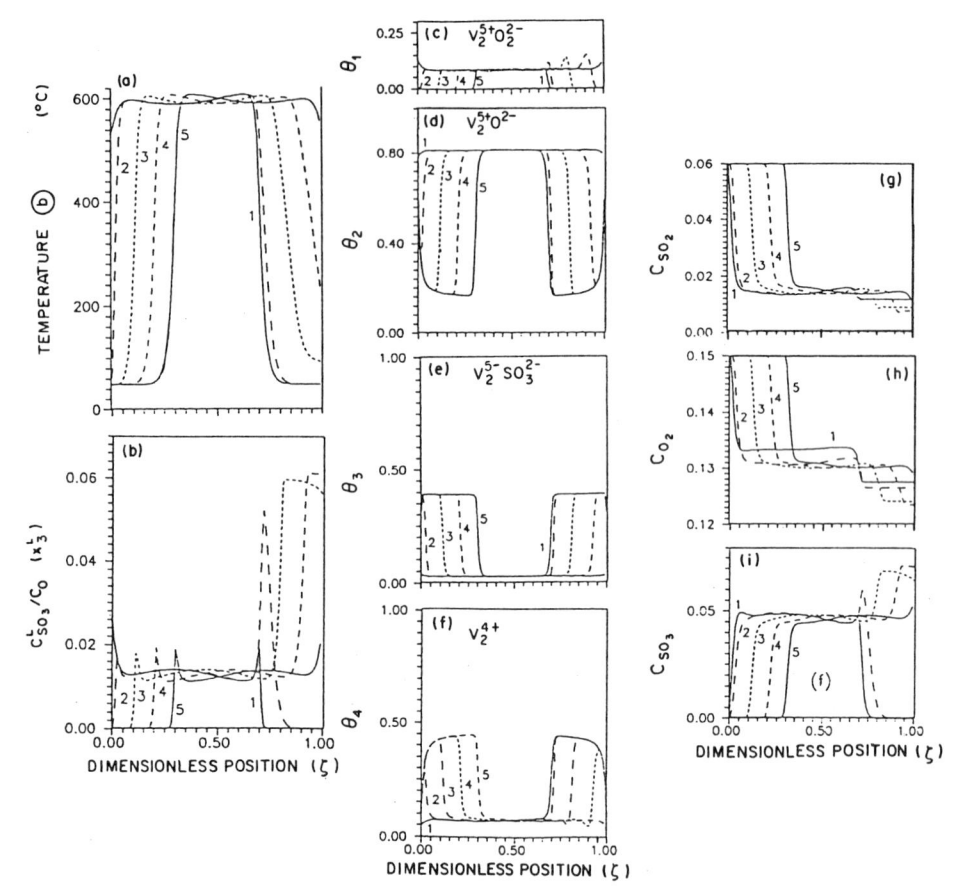

FIG. 16. Variation in a stationary cycling state of catalyst temperature, SO$_3$, and complex concentrations in the melt phase and the concentration of gas phase species with time in a half cycle in the forward flow portion of a reactor operating under periodic reversal of flow direction with τ = 40 min, SV = 900 h^{-1}, (C$_{SO2}$)$_o$ = 6 vol%, (C$_{O2}$)$_o$ = 15 vol%, T_o = 50°C. Curves: 1, just after switching flow direction; 2, 1 min; 3, 6.6 min; 4, 13.3 min, and 5, 20 min after a switch in flow direction. (Figure adapted from Bunimovich et al., 1995, with permission, © 1995 Elsevier Science Ltd.)

By 6 min into the half cycle, buildup of the SO$_3$ concentration in the melt (see Fig. 16b) and decreasing SO$_3$ solubility with rising temperature results in a significant rise in SO$_3$ leaving the catalyst bed. This continues for some 13 to 14 min into the half cycle and then drops.

Model dynamics were forced to steady state by setting derivatives for the melt complexes in Eq. (61) to zero (Bunimovich et al., 1995). This should make the model behave as though the steady-state kinetic model

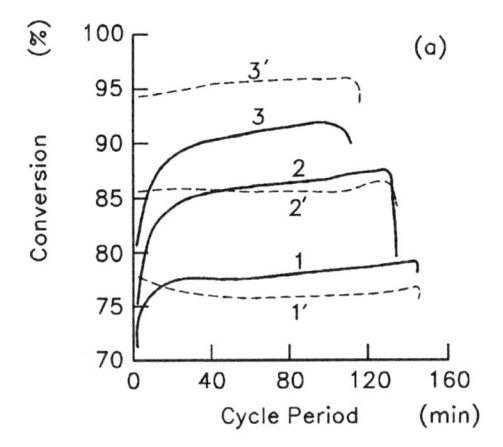

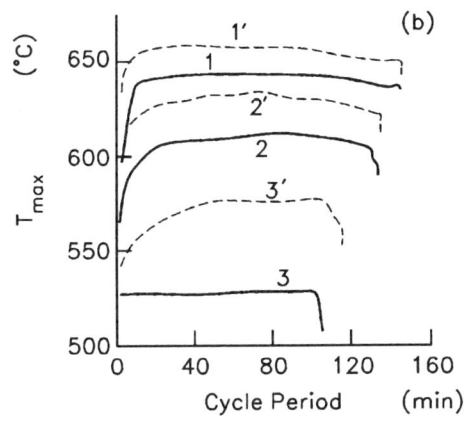

APPLICATION OF PERIODIC OPERATION TO SO₂ OXIDATION   247

FIG. 17. Comparison of the variation of the time-average SO₂ conversion and the maximum bed temperature predicted for stationary cycling condition by an unsteady-state and a steady-state kinetic model for a packed-bed SO₂ converter operating with periodic flow reversal under the input conditions given in Fig. 16: unsteady-state model—, steady state model-------. Curves: 1,1', $(C_{SO2})_o$ = 9 vol%; 2,2', $(C_{SO2})_o$ = 6 vol%; 3,3', $(C_{SO2})_o$ = 3 vol%. (Figure adapted from Bunimovich et al., 1995, with permission, © 1995 Elsevier Science Ltd.)

were in use, except for SO₃ solubility variations as temperature changes. Figure 17 compares SO₂ conversion and the peak bed temperatures for the dynamic kinetic model with those for its steady-state version. With relatively high inlet SO₂ levels, 9 vol%, conversion predictions differ by about 3% with the dynamic model providing higher conversions. Temperatures are only 10 to 15°C apart. With 6 vol% SO₂ in the feed, both unsteady-state and steady-state kinetic models predict the same conversions. Maximum temperature prediction differences increase to about 20°C. When $(C_{SO2})_o = 3$ vol%, conversions predicted by the steady-state kinetic model were about 5% greater than those predicted by the unsteady-state model. Differences in the predicted maximum temperatures grew to 40°C. Bunimovich *et al.* point out that the simulation using the steady-state kinetic model gave poorer predictions of the experimental data in Table VIII when the inlet SO_2 level dropped under 3 vol%.

D. Overview

Simulation as well as experimental measurements demonstrate features of periodic flow reversal that account for its successful application to SO_2 oxidation: (1) a region of uniform high temperature in the reactor center that is essentially time invariant, (2) a profile consisting of a maximum temperature followed by diminishing temperature in the flow direction, which results in higher conversions for equilibrium-limited, exothermic reactions, (3) overall conversion or outlet concentrations that show little variation with time, and (4) outlet temperatures close to inlet temperatures. Reactors employing periodic reversal of flow direction offer lower capital cost because heat recovery is performed by the catalyst bed or by a packed-bed recuperator instead of a shell and tube heat exchanger. Reactors using flow reversal appear to be able to handle small variations of operating conditions without resorting to process control because of the large thermal inertia they provide. For exothermic, equilibrium-limited reactions, such as SO_2 oxidation, the declining temperature profile toward the end of the reactor is probably the primary advantage offered.

IV. Conversion of SO_2 in Trickle-Bed Catalytic Scrubbers Using Periodic Flow Interruption

Periodic operation of three-phase reactors is a new area of study. Mass transfer of gaseous reactants through the liquid phase limits many trickle-bed processes. One way of increasing the rate of mass transfer and thereby rates in three-phase reactors is to segregate the gas and liquid phases using structured packing formed from a catalytically active, porous membrane. This arrangement provides access to the catalyst surface without transport through the liquid phase. Yang and Cussler (1987) discuss such a system.

Plate type packing to separate the phases is discussed by Carlsson *et al.* (1983) and by Hatziantoniu *et al.* (1986). De Vos *et al.* (1982, 1986) describe use of a monolithic porous catalyst with vertical and horizontal channels. The liquid phase flows downward through an array of parallel channels in the monolith, while gas moves in cross flow through a separate set of channels. Another approach treats the catalyst to make part of the surface hydrophobic or lyophobic (Berruti *et al.*, 1984). The gas phase has direct access to the surface on these unwetted portions of the surface, resulting in partial, spatial segregation of the phases.

Periodic flow interruption can be thought of as a type of partial segregation. It is a temporal segregation rather than a spatial one. The converse of periodic flow interruption is periodic flooding. Hot spots are observed in industrial-scale trickle beds and are thought to occur because of nonuniform irrigation of the catalyst by the liquid phase, often a consequence of low liquid loading. Periodic flooding fully irrigates the bed and is thus a means of avoiding hot spots. To carry this concept further, a low irrigation rate causes local temperatures to rise. If the liquid is near its boiling point, evaporation can take place. This removes the liquid phase from the catalyst surface and allows direct access for the gas phase. In situations where mass transfer through the liquid is limiting, direct access raises the temperature and thereby the rate of reaction substantially. Once the liquid phase evaporates completely, the rate of reaction diminishes and the location cools. It can then be reflooded by the liquid phase. This process can be replicated for an entire trickle bed by stopping and restarting liquid flow through the bed. Interrupting liquid flow in this way causes a temporary increase in the temperature of the catalyst bed, and thus higher rates and perhaps a selectivity improvement if higher temperatures favor better selectivity. It can also lead to catalyst overheating and runaway if the duration of the flow interruption is too long. Choice of cycle period and cycle split, however, control the duration and thus the temperature rise.

For their experimental investigation of flow interruption, Haure *et al.* (1989) chose the catalytic oxidation of SO_2 over a high-surface-area activated carbon catalyst. Several research groups have studied this catalytic reaction and kinetics are available. It proceeds rapidly at 25°C and is controlled, at least partially, by O_2 transport through the liquid phase.

A. EXPERIMENTAL STUDIES

A 40-mm i.d. Lucite column served as the reactor for the experiments of Haure *et al.* (1989). Reactor and feed system are shown schematically in Fig. 18. Attention was paid to the distribution of the liquid phase: liquid was introduced onto the top of an array of parallel tubes that distributed the phase uniformly over a packing of fine glass helices. The helices were in direct contact with the 14/30-mesh, BPL activated carbon, a coal based material, produced by the Calgon Carbon Corp. Bed depth was kept constant at 160 mm in all experiments. Thermocouples were located at 30, 80, and 140 mm below the top of the carbon bed, but the points just penetrated the Lucite walls and did not extend into the bed. Water, the liquid phase, was saturated with air prior to entering the reactor. An overflow tank and system of two-way solenoid valves, in parallel, and a needle valve provided flow interruption and precise flow control. The gas, made up of 1.3 vol% anhydrous SO_2 and 98.7 vol% dry air, flowed continuously through the reactor at a rate of 25.4 ml (STP)/s. This corresponds to a space velocity of 800 h^{-1}.

In Haure's work, the gas phase leaving the reactor was not analyzed.

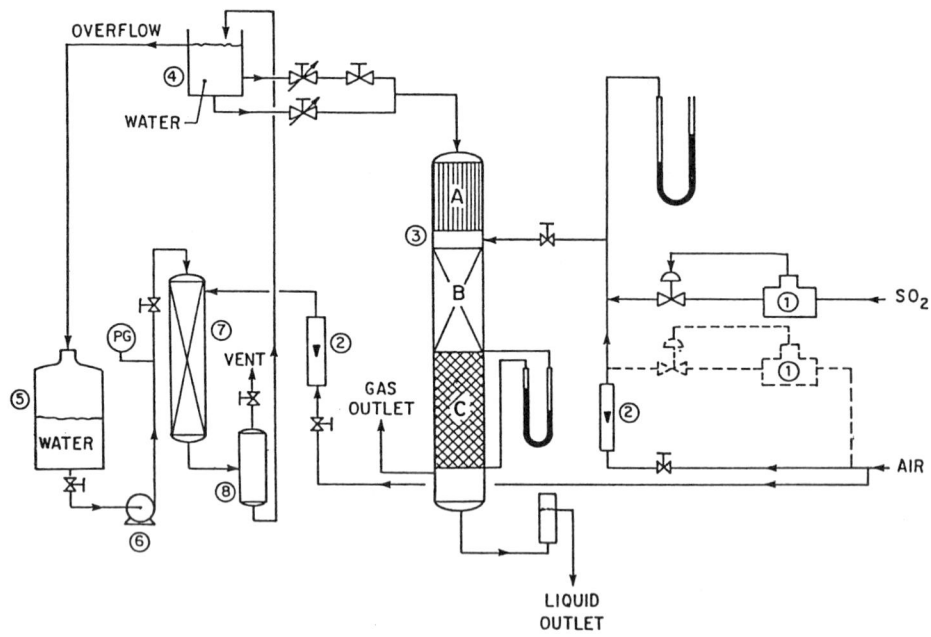

FIG. 18. Experimental trickle-bed system: A, tube bundle for liquid flow distribution; B, flow distribution packing of glass helices; C, activated carbon trickle bed; 1, mass flow controllers; 2, gas or liquid rotameters, 3, reactor (indicating point of gas phase introduction); 4, overflow tank for the liquid phase feed; 5, liquid phase hold-up tank; 6, absorber pump; 7, packed absorption column for saturation of the liquid phase; 8, gas–liquid disengager in the liquid phase saturation circuit. (Figure from Haure et al., 1989, with permission, © 1989 American Institute of Chemical Engineers.)

APPLICATION OF PERIODIC OPERATION TO SO₂ OXIDATION   251

Conversion and the mean rate of reaction were obtained by titrating the liquid recovered from the bed for total acidity before and after dissolved SO₂ and sulfurous acid were oxidized to sulfuric acid. The difference between these measurements allowed calculation of the SO₂ conversion. An extension of the Haure study (Metzinger *et al.*, 1992) added gas phase analysis and a check of the results through a sulfur balance. Computer control of flow interruption and acquisition of the SO₂ analyzer readings were also added.

Haure *et al.* (1989) conducted an exploratory investigation that considered the effect of cycle period, τ, cycle split, s, and time-average liquid superficial velocity, SV, on the spatial and time-average rate of oxidation. Cycle split, in this case, is the fraction of a period in which the liquid phase flows through the trickle bed. Figure 19 shows how these cycling variables affect the rate. Note that the period scale is logarithmic and data cover a range from 1 to 120 min. The figure also compares performance under periodic flow interruption with operation under steady state. The latter is given in the figure for the time-average liquid flow rate as the horizontal band in the center of the figure. These steady-state rates decrease as the time-average flow rate increases. The maximum steady state rate occurs

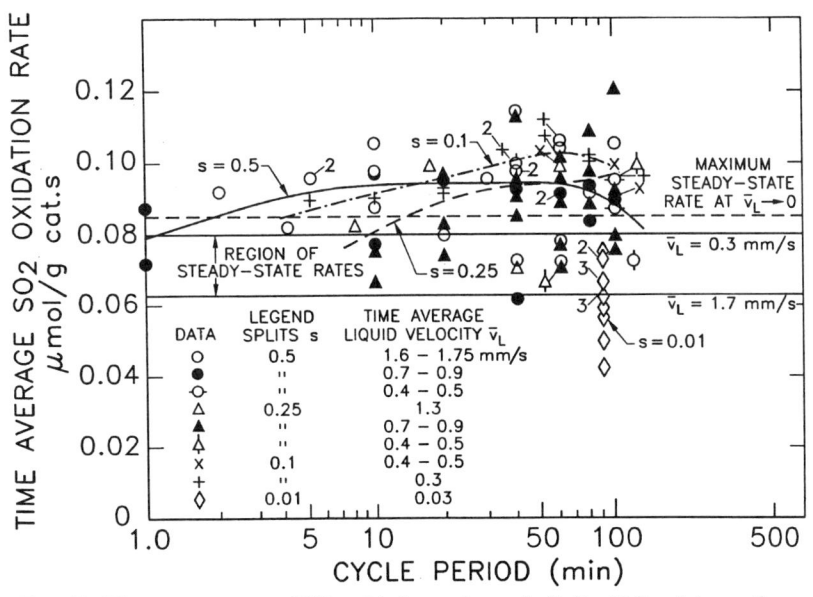

FIG. 19. Time-average rate of SO₂ oxidation under periodic liquid flow interruption as a function of cycle period, cycle split, and the time-average superficial liquid velocity at 26°C, 1 bar using a BPL activated carbon. (Figure from Haure *et al.*, 1990, with permission, © 1990 Elsevier Science Publishers.)

when the bed is loaded with liquid, but the phase is not flowing. The dashed horizontal line for a zero time-average liquid flow rate indicates this condition. Most of Haure's experimental data fall above the steady-state band and the maximum rate line. There is a separate symbol for each s and time-average liquid flow rate. The number beside a symbol indicates the number of data points the symbol represents. Curves shown in the figure have been drawn through data sets at $s = 0.5, 0.25,$ and 0.1, neglecting the variations of the time-average liquid flow rates.

An exception to improved performance under flow interruption is for a group of points at $\tau = 90$ min with $s = 0.01$. In these experiments water flowed through the trickle bed for less than 1 min during a cycle. The oxidation rate under steady state is close to the maximum steady-state rate. Even though the scatter of the experimental data at $s = 0.01$ is large, all the measurements under flow interruption lie below the line representing steady state.

The flow interruption data shown in the figure suggests that at any cycle split, s, there is a minimum cycle period needed for the time-average rate of SO_2 oxidation to exceed the maximum rate under steady-state operation. Once above this period, increasing τ exerts just a small effect on the rate until τ reaches about 120 min, when rate may begin to drop. The experimental data are inadequate to indicate an optimal period accurately. The cycle split, s, clearly affects the time-average oxidation rate. Notwithstanding the curves for different values of s shown in Fig. 19, the data do not indicate an optimal s. The time-average liquid flow rate also influences the time-average rate under flow interruption, and the data reflected in each of the three s curves in the figure cover a different range of time-average liquid flow rates. As may be seen from the caption in the figure, measurements made at $s = 0.1$ employed time-average liquid phase velocities between 0.3 and 0.5 mm/s, whereas those at $s = 0.5$ used velocities between 0.4 and 1.75 mm/s. Replotting the $s = 0.5$ data showed that at 1.65 mm/s the rate improvement is between 40 and 65%. At a time-average liquid velocity of 0.86 mm/s, the rate improvement drops to about 20%.

Haure et al. (1989) also undertook experiments in which the liquid flow rate was periodically reduced rather than interrupted. Switching between time-average liquid velocities of 4.0 and 1.2 mm/s at $s = 0.5$ resulted in about a 10% increase in the time-average rate of SO_2 oxidation over steady state. The rate improvement was independent of τ over the 2 to 60 min range explored. This is considerably less than the increase when flow interruption is utilized.

Part of the explanation for increased rates under flow interruption rests on the concept of segregation introduced earlier. At low liquid loading of a trickle bed, the external surface of a catalyst pellet is incompletely wetted. There is then direct access of the gas-phase reactants to the catalyst surface and, since both reactants are in the gas phase, higher oxidation rates are observed. Substantiation comes from steady-state measurements (Haure et al., 1992) that show a rise in SO_2 oxidation rate as the superficial liquid velocity, u, drops below 2 mm/s. This results from a decrease in wetting of the carbon catalyst as the liquid loading is reduced. There is also a rise in rate with increasing velocity above $u = 4$ mm/s caused by growing liquid-phase turbulence that augments the liquid-phase volumetric mass transfer coefficient. A consequence of these divergent trends is a minimum when the oxidation rate is plotted vs the liquid velocity. A minimum has been observed by others for aqueous trickle-bed systems (Ramachandran and Smith, 1979; Tan and Smith, 1980; Mata and Smith, 1981) and also for nonaqueous systems (Sedricks and Kenney, 1973; Satterfield and Ozel, 1973), so it appears to be a general phenomenon.

Periodic flow interruption changes liquid loading on a trickle bed from moderate or high to zero. Even at zero loading, the carbon bed is partially wetted through static hold-up. Because of the continued presence of water, oxidation proceeds at a high rate. If a low split is used, the bed operates for most of the cycle at the high oxidation rate. Indeed, performance under flow interruption or reduction can be predicted as illustrated in Fig. 20. The curve in this schematic shows the steady-state data mentioned above extrapolated to $u = 0$. The two operating states are SS_1 and ASS_2; the "A" signifies it is an apparent operating state because if the trickle bed is held at zero liquid loading for a long enough time, the liquid holdup eventually evaporates and the rate decays to zero as SO_3 poisons the catalyst surface. The zero rate condition, SS_2, is the extreme of other possible operating states, which must lie below ASS_2. A surprising result of the Haure study is that the trickle bed functions at the ASS_2 state for up to 120 min after liquid flow has been halted. Consequently, neglecting any dynamic behavior, the time-average rate of reaction will be the steady-state rate at the two states weighted by the relative duration the bed remains in each of the states. The latter depends on the cycle split. In Fig. 20, $s = 0.5$ and the virtual operating state (VSS in the schematic) is midway on the straight line joining ASS_2 and SS_1. The rate at VSS is clearly considerably higher than the steady-state rate at a time-average superficial liquid velocity (CSS in the figure). The figure also shows the oxidation rate at the quasi-steady-state operating point (QSS). This is a point on a straight line joining SS_1 and SS_2, and it represents the performance for a period of infinite duration. The lever rule can be applied to estimate the rates at VSS, CSS, and QSS

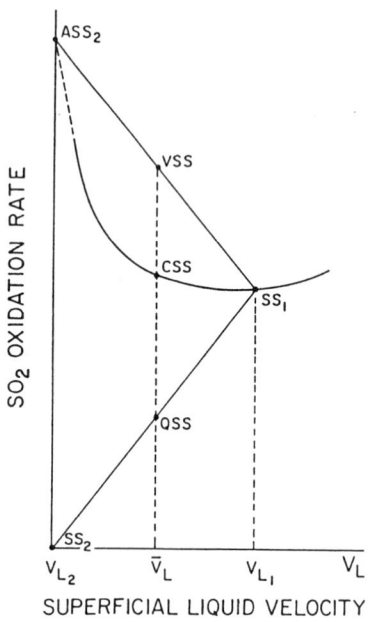

FIG. 20. Schematic diagram showing the estimation of the time-average rate of SO_2 oxidation under periodic flow interruption or reduction employing steady-state oxidation rate vs liquid loading data (Figure from Haure et al., 1989, with permission, © 1989, American Institute of Chemical Engineers.)

for s other than 0.5. Table XII shows that an estimate of the rate of SO_2 oxidation under periodic flow interruption can be obtained from steady-state rate data.

The role of water in SO_2 oxidation over activated carbon is to react with the SO_3 formed to yield sulfuric acid. This removes SO_3 from the catalyst

TABLE XII
PREDICTED[a] vs EXPERIMENTAL SO_2 OXIDATION RATES[b]

Cycle split	Mean superf. liquid, mm/s	CSS	Time avg. VSS	Exp. periodic rate	% diff.[c]
0.5	1.7	67	82	88.7	8
0.5	0.8	80	85	90.8	7
0.25	0.8	80	88	94	7
0.1	0.33	97	97	97.7	1

[a] VSS predicted from steady-state rates using construction shown in Fig. 20.
[b] Oxidation rates: μmol/kg · s.
[c] (Experimental − predicted)/predicted as %.

surface and prevents inhibition of the reaction. Consequently, in the absence of liquid flow as the reaction continues, the acid concentration builds up in the intraparticle liquid and interparticle hold-up. Once liquid flow is restarted, the initial concentration of acid flushed from the bed will be high and will then decay as acid is diluted by water added to the top of the trickle bed. Consequently, a low cycle split and a long period could be used to obtain a high acid concentration. Haure et al. varied s and τ and demonstrated that higher sulfuric acid concentrations were obtained. These researchers also examined what acid strength could be obtained starting with a water feed. For this purpose, they conducted a step change experiment and measured the strength of the acid leaving the bed in the first minute of liquid flow after steady-state liquid flow had been halted for a variable length of time. A nearly linear variation of acid molarity with the duration of flow interruption was observed. A molarity of about 1.2 was reached in their experiment. The acid concentration within the catalyst pellets was certainly much higher, possibly reaching 2 to 3 M.

The discrepancy between estimated and measured rates of SO_2 oxidation in Table XII arises, according to Haure et al. (1989), because the mean spatial temperature of the trickle bed depends on liquid loading. Flow interruption raises the time-average, space mean bed temperature. Figure 21 illustrates the effect. Time-average temperatures are shown for the three thermocouples located axially along the periphery of the carbon bed for periodic operation. Adjusting the steady-state data for the higher temperature using an activation energy of 10 kJ/mol for O_2 transport brings together the estimated and measured time-average rates of oxidation shown in Table XII.

Temperatures within the trickle bed during flow interruption vary with time in the form of temperature waves which move downward through the bed. Measured temperatures are shown in Fig. 22 as a function of cycle split for $\tau = 60$ min. Time-average superficial liquid velocity is constant for the three experiments at 1.65 mm/s. There are some interesting features of these waves: (1) The temperature of the top thermocouple, located some 30 mm below the top of the carbon bed, decreases prior to water addition; (2) temperature increases rather then decreases when the liquid flow begins if $s < 0.25$; and (3) the lags between the temperature maxima are reasonable for the $s = 0.5$ experiment, but are an order of magnitude longer than the plug flow filling times at the smaller cycle splits. The temperature rise at small s appears to be caused by hotter fluid being forced downward by the entering water after flow starts; heating through acid dilution is much too small at the concentrations encountered in the bed to account for the temperature rise. Cooling of the top thermocouple is caused by inhibition of the reaction as the acid concentration builds up and by evaporation as

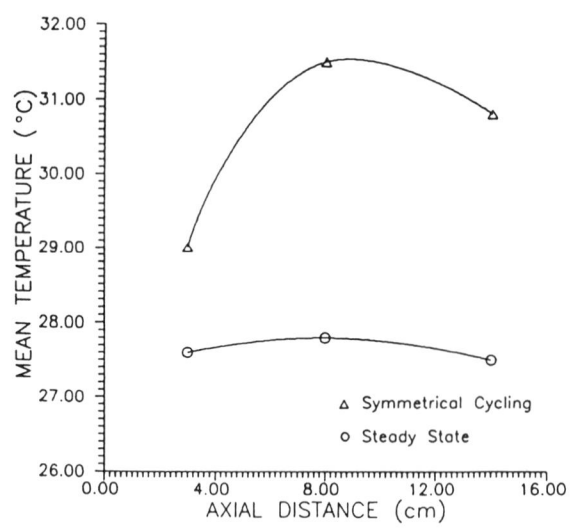

FIG. 21. Comparison of time-average temperature reading from thermocouples located at different axial points on the periphery of the carbon bed for periodic flow interruption and steady-state operation. Time-average $u = 1.65$ mm/s; for flow interruption, $\tau = 60$ min and $s = 0.5$. Gas and liquid inlet temperature is about 26.5°C. (Figure from Haure et al., 1989, with permission, © 1989 American Institute of Chemical Engineers.)

dry air enters the carbon bed. The time lags may be caused by poor flow distribution. Changes in peak shape with time suggest that some sort of flow distortion is occurring in the trickle bed.

B. Modeling and Simulation

The model presented in Table XIII was developed (Haure et al., 1990; Stegasov et al. 1992) from modeling of phase change in trickle beds (Kirillov and Stegasov, 1988). The model is dynamic, but it neglects the filling and draining steps. These are assumed to occur instantaneously. This assumption has been tested (Hasokowati et al., 1994). The filling step appears to occur as plug flow at the spatial mean liquid velocity in the packing (u_2/ε_{bed}). Even in a bed several meters deep, it is complete in minutes. Gravity drainage, which governs most of the bed, occurs almost as fast as the filling step. However, the bottom of the bed drains by a film thinning mechanism and is much slower, taking perhaps several minutes. Fortunately, the error introduced by neglecting the different static hold-up at the bottom of the trickle bed is small. Thus, the filling and draining steps are not important for beds of several meters in depth unless periods of less

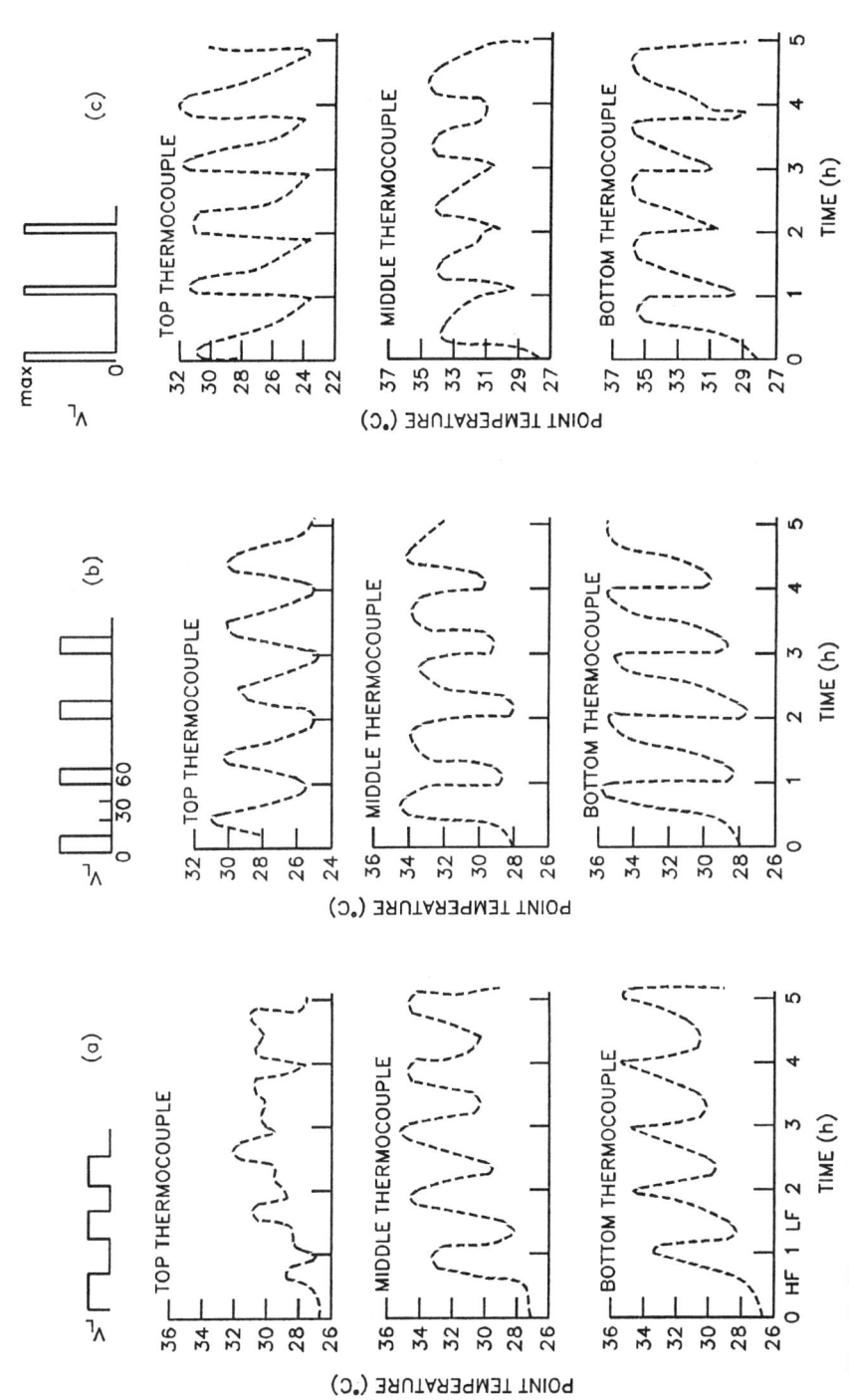

APPLICATION OF PERIODIC OPERATION TO SO$_2$ OXIDATION   257

FIG. 22. Thermocouple temperature readings in an experimental trickle bed operating under periodic flow interruption with $\tau = 60$ min and a time-average $u = 1.65$ mm/s: (a) $s = 0.5$, (b) $s = 0.25$, (c) $s = 0.1$. (Figure from Haure et al., 1990, with permission, © 1990 Elsevier Science Publishers.)

TABLE XIII
GENERAL MODEL FOR A PERIODICALLY OPERATED TRICKLE BED (STEGASOV ET AL., 1992)

Heat and material balances	Constraints and initial conditions

Material balances, Gas (vapor) phase:

$$\alpha_g \rho_g \left[\frac{\partial y_i}{\partial t} + u_g \frac{\partial y_i}{\partial z} \right] = R_i \quad (i = 1, 2) \quad (67)$$

$$\alpha_g \rho_g \left[\frac{\partial y_i}{\partial t} + u_g \frac{\partial y_i}{\partial z} \right] = (R_{evap})_i \quad (i = 3, 4) \quad (68)$$

$$\alpha_g \rho_g \left[\frac{\partial y_i}{\partial t} + u_g \frac{\partial y_i}{\partial z} \right] = 0 \quad (i = 5) \quad (69)$$

Liquid phase:

$$\alpha_l \rho_l \left[\frac{\partial C_i}{\partial t} + u_l \frac{\partial C_i}{\partial z} \right] = R_i + \nu_i R_p \quad (i = 1, 2) \quad (70)$$

$$\alpha_l \rho_l \left[\frac{\partial C_i}{\partial t} + u_l \frac{\partial C_i}{\partial z} \right] = R_{evap} + \nu_i R_p \quad (i = 3, 4) \quad (71)$$

Heat balance equations:

$$\varepsilon (\rho u C_p)_{mix} \frac{\partial T}{\partial z} = h_l(\theta - T) + \varepsilon \sum_{i=1}^{2} R_i H_i - \varepsilon \sum_{i=3}^{4} (R_{evap})_l (\Delta H_{evap})_i \quad (72)$$

$$(1 - \varepsilon) \rho_s (C_p)_s \frac{\partial \theta}{\partial t} = \lambda \frac{\partial^2 \theta}{\partial z^2} - h_l(\theta - T) + R_p \Delta H_R - \varepsilon \sum_{i=1}^{2} R_i H_i + \varepsilon \sum_{i=3}^{4} (R_{evap})_l (\Delta H_{evap})_i \quad (73)$$

Equation for variation of liquid phase bulk content:

$$\frac{\partial}{\partial t}(\alpha_l) + \frac{\partial}{\partial z}(\alpha_l u_l) = 0 \quad (74)$$

Equation of state:

$$P = \rho_g RT \quad (75)$$

Constraints:

$$\alpha_g + \alpha_l = 1, \quad u_g = \frac{Q_g}{A_x}, \quad u_l = \frac{Q_l}{A_x} \quad (76)$$

Initial and boundary conditions:

At $t = 0$: $\bar{u}(0, z) = \bar{u}^0(z), \quad \theta(0, z) = \theta^0(z)$:

$$\alpha_l(0, z) = \begin{cases} \alpha_g -, & \text{at the start of waterflow} \\ \alpha_g + \alpha_d, & \text{at the cut off of waterflow} \end{cases} \quad (77)$$

At $z = 0$; $\bar{u}(t, 0) = \bar{u}_0(t), \quad \frac{\partial \theta}{\partial z} = 0,$ (78)

$$\alpha_l(t, 0) = \begin{cases} \alpha_g + \alpha_d -, & \text{at the start of waterflow} \\ \alpha_g -, & \text{at the cut off of waterflow} \end{cases} \quad (79)$$

At $z = L$: $\frac{\partial \theta}{\partial z} = 0$ (80)

than 30 min are used. As may be seen in the table, the model is one-dimensional and thus assumes perfect liquid and gas distribution in the carbon bed and negligible radial heat loss. Plug flow of both phases is assumed. Interphase transport of all but sulfuric acid is allowed. Diffusion in and out of the carbon pellet is accounted for by an effectiveness factor.

A well-substantiated correlation for air–water systems taken from the trickle bed literature (Morsi and Charpentier, 1981) was used for the volumetric mass transfer coefficients in the R_i and $(R_{evap})_i$ terms in the model. The h_i term was taken from a correlation of Kirillov et al. (1983), while the liquid hold-up term α_l in Eqs. (70), (71), (74), (77), and (79) were estimated from a hold-up model of Specchia and Baldi (1977). All of these correlations require the pressure drop per unit bed length. The correlation of Rao and Drinkenburg (1985) was employed for this purpose. Liquid static hold-up was assumed invariate and a literature value was used. Gas hold-up was obtained by difference using the bed porosity.

Stegasov et al. (1992) expressed the kinetic rate of SO_2 oxidation as

$$R_p = \frac{v_4 k_2 \eta \eta_o C_1 C_2}{1 + AC_4}, \tag{81}$$

where C_1, C_2, and C_4 are the liquid-phase concentrations of SO_2, O_2 and H_2SO_4, respectively. The kinetic rate constant, k_2, the effectiveness factor, η, and the constant A in the denominator were taken from Komiyama and Smith (1975). Haure et al. (1989) were able to closely predict their steady-state measurements using these data. Stegasov et al. chose the wetting efficiency (η_o) function to fit the line representing Haure's steady-state data, but demonstrated that the function used agreed closely with wetting efficiency correlations proposed in the literature. Essentially, then, Table XIII is an apriori model.

For integration, Stegasov et al. approximate the equations in Table XIII by forward differences, which provides at each grid point a set of implicit algebraic equations that must be solved. The Runge–Kutta procedure was used for solution of temperature distribution in the bed and then an iterative procedure was used to find concentrations at the $n + 1$ time step. Time step was variable and was optimized continually during integration. A Richardson-type improvement was also used (Stegasov et al., 1992).

Figure 23 shows a simulation of Haure's periodic flow interruption data at time-average $u = 0.86$ mm/s (Curves 1 and 1') and $u = 1.65$ mm/s (Curves 2 and 2'). Data points and the fit (dashed line) are from Fig. 19. The simulation predictions are 28 to 35% too high, which is not bad considering the 25% variation in the experimental data shown in Fig. 19. The trends, however, are not properly represented. Figure 23 predicts enhancement factors declining with τ, whereas data indicate an increase at low τ and

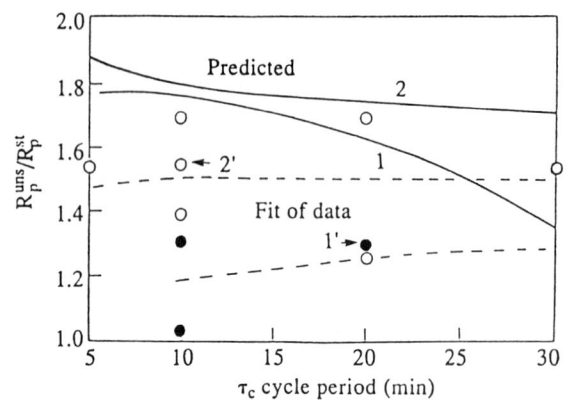

FIG. 23. Comparison of predicted and experimental enhancement factors for the experimental conditions of Fig. 19. Curves designated 1 and 1' are for $v_L = 0.86$ mm/s, while curves 2 and 2' are for $v_L = 1.65$ mm/s. The prime designates the data fits in Fig. 19. (Figure from Stegasov et al., 1992, with permission, © 1992 Elsevier Science Publishers.)

then no effect of period as τ augments. Improper estimate of the trend with period is traceable to two considerations: (1) At periods below 10 min the draining time of the bed cannot be neglected, and (2) the inhibition effect of acid buildup in the bed under zero water flow seems overpredicted.

The Stegasov et al. model does a much better job of predicting the effect of cycle split on time-average acid concentration leaving the trickle bed. Figure 24 shows that the model prediction for $v_L = 1.65$ mm/s and $\tau =$

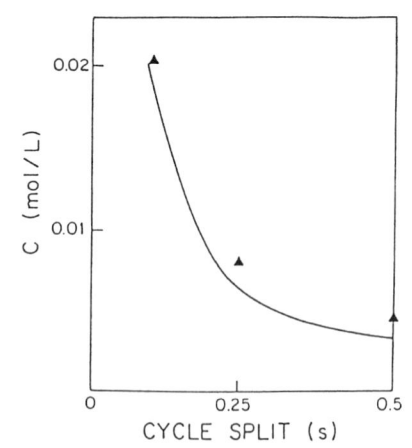

FIG. 24. Comparison of predicted and experimental time-average acid concentrations leaving the periodically operated trickle bed for the experimental conditions of Fig. 19 and $\tau = 10$ min. (Figure from Stegasov et al., 1992, with permission, © 1992 Elsevier Science Publishers.)

10 min agrees well with experimental data. However, prediction of the thermocouple data given in Fig. 22 was poor. Temperatures are overpredicted at $s = 0.25$ and particularly at $s = 0.1$, but the increasing temperature observed from the top to bottom thermocouples is properly represented. At $s = 0.5$, the temperatures and the increase from top to bottom seen in the data are well predicted. Indeed, the model predicts the constant temperature during the water flush, the temperature rise after water flow is discontinued, and the sharp drop when the water flow commences for the middle and bottom thermocouples at $s = 0.5$. The simulation poorly reproduces the behavior of the top thermocouple, but that temperature–time record is so irregular that the data are suspect. With respect to the data in Fig. 22 for $s = 0.1$, the simulation fails to predict the shape of the temperature–time record. It also fails to predict the drop in temperature prior to restart of liquid flow, and it predicts a short constant-temperature period when experimentally none is observed.

Some of the poor agreement is undoubtedly attributable to the data. Measurements were made at the periphery of the bed where radial heat loss becomes important. Certainly this contributes to the poor agreement at $s = 0.1$ when temperatures are highest in the bed. The Stegasov model assumes no radial gradients. If allowance is made for an a priori model with no data fitting, not even for reaction rate constants, then the performance is quite good. Certainly, tuning of the rate and mass transfer coefficients should improve the agreement of model and data in Fig. 23. Consequently, it is evident that periodic flow interruption can be modeled. The model proposed by Stegasov et al. (1992) will be useful for data interpretation and for scale-up of experimental results.

A useful application of the model is to examine the SO_2 and O_2 concentration profiles in the trickle bed. These are shown for the steady-state conditions used by Haure et al. (1989) in Fig. 25. The equilibrium SO_2 concentration drops through the bed, but the O_2 concentration is constant. In Haure's experiments O_2 partial pressure is 16 times the SO_2 partial pressure. At the catalyst particle surface, however, O_2 concentration is much smaller and is only about one-third of the SO_2 concentration. This explains why O_2 transport is rate limiting and why experimentally oxidation appears to be zero-order in SO_2.

C. APPLICATION TO STACK-GAS SCRUBBING

Several papers describe extension of the Haure work to the scrubbing of industrial stack gases as one step of a carbon based stack-gas cleanup process referred to as the RTI–Waterloo Process (Gangwal et al., 1992;

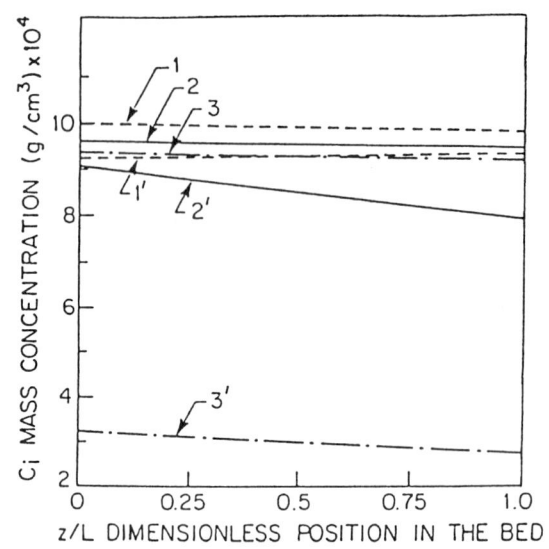

FIG. 25. Predicted mass concentration profiles for SO_2 and O_2 in the experimental trickle bed used by Haure *et al.* (1989) at $v_L = 1.65$ mm/s. (Figure from Stegasov *et al.*, 1992, with permission, © 1992 Elsevier Science Publishers.)

Metzinger *et al.*, 1992; 1994; Lee *et al.*, 1995, 1996a,b). Figure 26 provides a schematic of the process as conceived for stack gas from a coal burning power plant using an Illinois #6 coal containing 2.8 wt% sulfur. Hot gas leaving an electrostatic precipitator is cooled by gas–gas heat exchange to a temperature just above the boiling point of the product acid and enters a trickle bed. With a sufficient depth of the carbon bed, SO_2 removal and conversion to sulfuric acid is between 98 and 99%. By washing the carbon bed with product acid in place of water, a 5 to 10 N acid can be made, even with some water condensation. Acid leaves the bottom of the bed and the stream is split. Some flows to product storage, while the remainder is diluted with water and is recycled to flush the carbon bed. Although space velocities up to 10,000 h^{-1} seem feasible, several beds in parallel will be needed for satisfactory liquid distribution. Acid flow is continuous; the flow, however, is switched periodically from bed to bed. The wet but desulfurized gas is reheated in the gas heat exchanger before entering the SCR deNO$_x$ unit. With the very low level of SO_2 achieved in the scrubber, precipitation of ammonium sulfite/sulfate in the SCR can be avoided.

The attraction of periodic flow interruption for scrubbing is that the pressure drop is low relative to continuous, packed-bed scrubbers and a commercially useful acid can be produced. The catalyst is cheap, but that

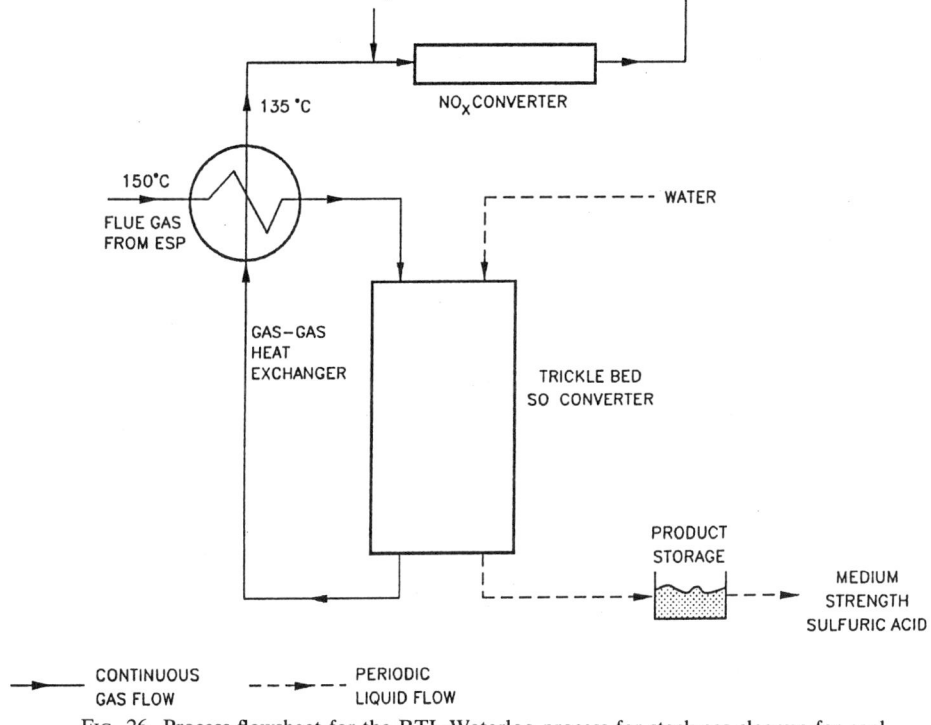

FIG. 26. Process flowsheet for the RTI–Waterloo process for stack-gas cleanup for coal burning power plants. (Figure taken from Gangwal *et al.*, 1992, with permission of the authors.)

is not a serious consideration, because laboratory experiments indicate that the carbon can be used for long periods.

The first set of experiments on SO_2 scrubbing used a series of step on and off experiments to test the performance of candidate carbons. Flooding the ca. 1-m deep bed of the test carbon with water or acid of different strength for an extended time period was followed by draining the bed and introducing simulated stack gas (600 to 2500 ppmv SO_2, 0 to 500 ppmv NO_x, 5 vol% O_2, 15 vol% CO_2, and 2 to 10 vol% H_2O). Gas flow was discontinued when breakthrough occurred and the cycle was repeated. Breakthrough was defined as the condition when SO_2 leaving the bed reached 5% of the inlet concentration. Different activated carbons displayed different breakthrough behavior, indicating that the choice of carbon is important. One carbon exhibited rate inhibition at 2500 ppmv SO_2 in the feed. All carbons showed a decrease in breakthrough time when 4.3 N acid replaced water as the flushing liquid. Adding a noble metal promoter to the carbon extended breakthrough time, while raising the operating temperature from 80 to 130°C sharply reduced the breakthrough time. Gangwal *et al.* (1992) give further experimental details.

Further experiments were carried out in the equipment shown in Fig. 18. Some important changes were made. Gas mixing was added to provide a simulated stack gas. Moisture level was adjusted either with a saturator or by adding live steam. The liquid side of the system was constructed entirely of glass so acid strengths up to 10 N could be used. Thermocouples were mounted radially as well as axially and could be inserted to variable depths into the bed. Finally, an SO_2 detector was placed in the off-gas line following the separator at the bottom of bed. Steady-state and periodic flow interruption measurements with the modified equipment are reported by Metzinger *et al.* (1994) for the BPL activated carbon and for a noble-metal-impregnated carbon. Since the bed depth of the experimental trickle bed was just 180 mm, breakthrough was not observed.

The dependence of SO_2 removal and SO_2 conversion to sulfuric acid on operating conditions is shown in Fig. 27 for experiments carried out at a bed temperature of 21°C using water as the flushing fluid. Steady-state measurements are also shown. Contrary to the results of Haure *et al.* (1989), periodic interruption of the liquid flow did not increase SO_2 removal at SV = 1000 h^{-1}; however, removal and conversion to acid at SV = 3000 h^{-1} was more than twice the steady-state values. As expected, gas space velocity is important. Removal drops as SV increases. The figure shows that washing the carbon bed with 2 N acid suppresses SO_2 removal, but has only a small effect on conversion to acid. A flush duration, D, of 3 min is significantly better than a 30-s duration and higher removals and conversion are found when $\tau = 5$ min rather than 30 min. Only about 50% of the SO_2 absorbed is converted to H_2SO_4 at high levels of removal; however, at higher SVs the percentage becomes about 70%. Consequently, unconverted SO_2 is present in the liquid phase as sulfurous acid or dissolved SO_2. These results came from a $2^5 - 1$ factorial experimental design that was used to identify system variables with the largest impact on SO_2 removal, conversion, and pressure drop over the bed. Measurements were made once flow interruption reached a stationary condition, that is, the variation with time of the SO_2 concentration leaving the bed during the zero liquid flow portion of the cycle was reproducible.

A less comprehensive factorial design was used at 80°C. Flush duration and τ were not considered. Very little sulfurous acid or dissolved SO_2 were found in the liquid collected, so conversion to acid is essentially the same as SO_2 removal. Figure 28 plots percent removal vs the normality of the wash liquid. Increasing normality lowers the percent SO_2 scrubbed from

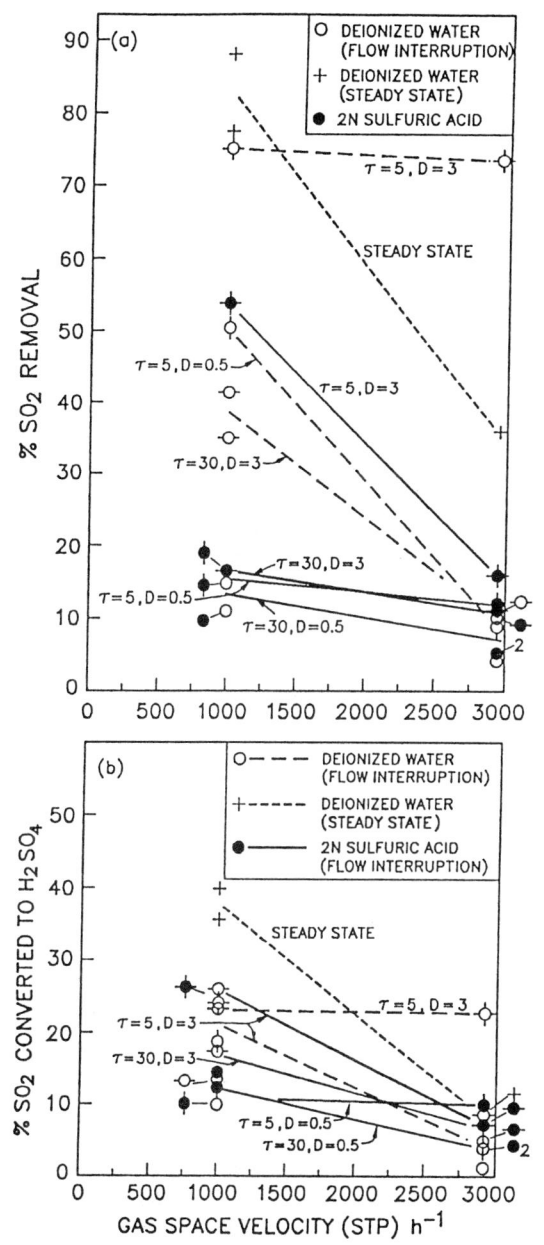

FIG. 27. Effect of space velocity, cycle period, normality of the flushing liquid, and flush duration on SO_2 scrubbing from a simulated stack gas at 21°C employing a bed of BPL carbon: (a) SO_2 removal, (b) SO_2 conversion to H_2SO_4. (Figure reproduced from Metzinger et al., 1994, with permission, © 1994 Elsevier Science Ltd.)

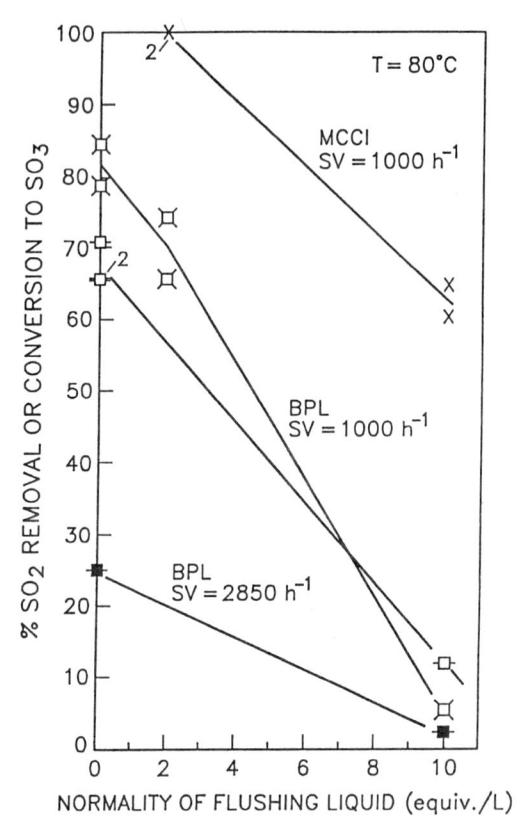

FIG. 28. Effect of acid strength, space velocity, and activated carbon on SO_2 removal and conversion to acid at 80°C in an intermittantly flushed trickle bed for $\tau = 60$ min and $s = 0.1$. (Figure reproduced from Metzinger *et al.*, 1994, with permission, © 1994 Elsevier Science Ltd.)

the stack gas. The MCCI carbon, impregnated with a noble metal, provides far better SO_2 removal than the BPL carbon. Indeed, it appears a bed of this activated carbon could produce a 10 N H_2SO_4 from 2500 ppm SO_2 in a typical stack gas from a high-sulfur coal. Space velocity continues to exert a strong influence on removal at 80°C.

Assuming plug flow of both phases in the trickle bed, a volumetric mass transfer coefficient, $k_L a$, was calculated from the measurements. The same plug flow model was then used to estimate bed depth necessary for 95% SO_2 removal from the simulated stack gas. Conversion to sulfuric acid was handled in the same way, by calculating an apparent first-order rate constant and then estimating conversion to acid at the bed depth needed for 95% SO_2 removal. Pressure drop was predicted for this bed depth by multiplying the depth by the experimental pressure drop per unit bed depth. Bed depths and anticipated pressure drops for a portion of the experiments are collected in Table XIV. The table shows that low pressure drops can be achieved at 95% SO$_2$ removal only at 80°C provided acid strength does not exceed 10 N and SV = 1000 h^{-1} is acceptable. Because of the large flows encountered in power plants, space velocity of 10,000 h^{-1} or more will be needed. In this case, pressure drops will be more than 100 times those given in the last column of Table XIV, as pressure drop increases with the square of the velocity. Pressure drops greater than 50 kPa will require placing a blower in the stack, greatly increasing the cost of SO$_2$ removal.

A systematic study of the cycling variables, cycle period (τ), flush duration (D) or cycle split (s), and the liquid loading (u_l) for the poorer performing BPL activated carbon is reported by Lee et al. (1995). All their measurements were undertaken at 80°C. Contrary to the influence of period seen in Fig. 27, Lee et al. observed that SO$_2$ removal and conversion to acid increase as τ augments while holding s constant. Results of this study are summarized in Fig. 29. The middle figure suggests there is an optimum

TABLE XIV
Estimated Carbon Bed Depth and Pressure Drop at 95% SO$_2$ Removal[a]

Catalyst temperature, flush liquid, space velocity	SO$_2$ removal in 190 mm deep bed (%)	Bed depth for 95% SO$_2$ removal (m)	Pressure drop at 95% SO$_2$ removal (kPa)
BPL—Activated Carbon			
21[a]/water			
SV = 1000 h^{-1}	38	1.19	0.84
= 2850	10	5.36	3.75
21[a]/2 N acid			
SV = 1000 h^{-1}	16	3.20	21
= 2850	11	4.98	68
80[a]/water			
SV = 1000 h^{-1}	67 (80)	0.270 (0.224)	0.71 (0.59)
= 2850	26	0.70	1.85
80[a]/2 N acid			
SV = 1000 h^{-1}	70	0.258	0.68
80[a]/10 N acid			
SV = 1000 h^{-1}	12 (6)	146 (3.0)	19 (24)
= 2850	2.4	7.68	101
MCCI—Impregn. Carbon			
Acid			
SV = 1000 h^{-1}	(100)	(0.18)	(1.7)
80[a]/10 N acid			
SV = 1000 h^{-1}	(60)	(0.30)	(4.0)

[a] Values in parentheses means τ = 60 min rather than 30 min.

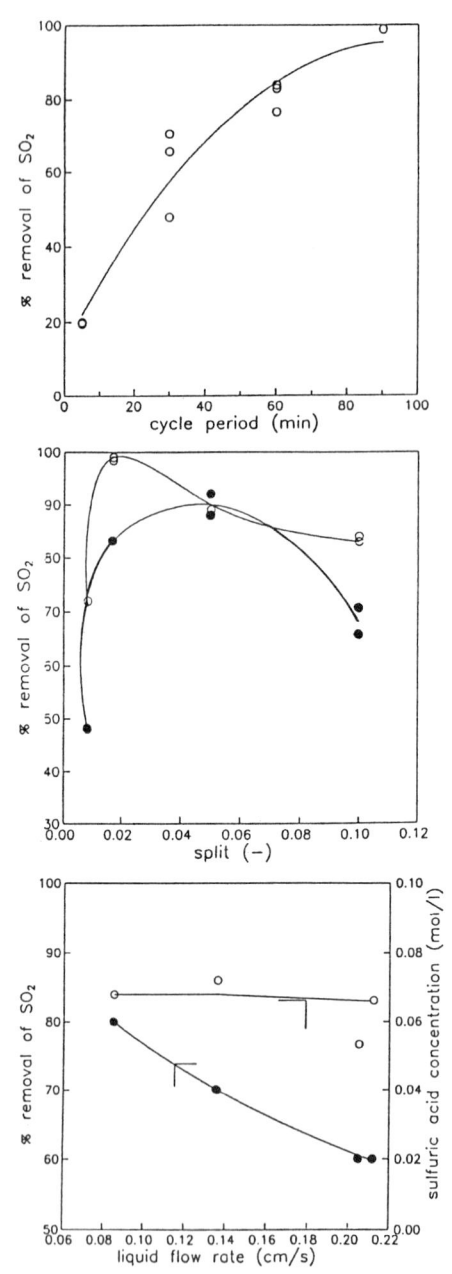

FIG. 29. Dependence of SO$_2$ removal on cycling variables using BPL activated carbon with gas and liquid feed temperatures at 80°C and a gas SV = 1000 h^{-1}: (a) influence of τ at $s = 0.1$ and u_l between 0.085 and 0.212 cm/s, (b) influence of s at τ (○ = 60 min, ● = 30 min) and u_l between 0.059 and 0.212 cm/s, (c) influence of u_l at $\tau = 60$ min and $s = 0.1$. (Figure reproduced from Lee et al., 1995, with permission, © 1995 Elsevier Science Ltd.)

cycle split, which depends on τ. Unfortunately, Fig. 29 does not isolate the dependence of scrubbing performance on flush duration. The duration, D, at any τ changes as s varies in (b). This is also true in (a): D increases in step with τ. The bottom figure shows that SO_2 removal is unaffected by u_l at $\tau = 60$ min and $s = 0.1$. Of course, increasing the liquid loading during washing of the carbon bed dilutes the acid formed.

In a follow up study, Lee et al. (1996a) returned to the choice of activated carbon for the scrubbing process. A new activated carbon, Centaur (Calgon Carbon Corp.), was tested under condition used for Lee's earlier work discussed in the previous paragraph. Levels of SO_2 removal approaching 99% were obtained with a 180-mm deep bed at 80°C and SV = 1000 h^{-1} with this carbon. Impregnating the carbon with Pt at a 0.1 wt% loading did not increase the SO_2 removed. Evidently, impregnation with noble metals is not needed to achieve high activity for SO_2 oxidation. Figure 30 shows percent SO_2 removal from stack gas as a function of cycle split and period for BPL and Centaur activated carbons. The figure suggests that for both carbons there is a minimum s or actually D for high SO_2 removal. Results for the Centaur carbon did not indicate an optimum s or D.

D. PHYSICAL EXPLANATION

The success of periodic flow interruption is due to the liquid static holdup within the porous catalyst pellets and the interstices of the catalyst bed.

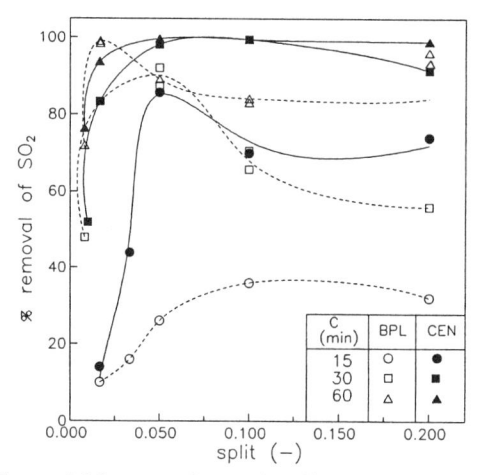

FIG. 30. Dependence of SO_2 removal on cycle split and period for a 178-g bed of either Centaur or BPL activated carbon with gas and liquid feed temperatures at 80°C, SV = 1000 h^{-1}, and $u_l = 0.2$ cm/s. (Figure reproduced from Lee et al., 1996a, with permission of the authors.)

Thus, the liquid continues to be present even though it is no longer flowing through the bed. It is the dynamic hold-up that varies through this form of periodic operation. When this hold-up diminishes, the mass transfer resistance for the gas phase reactant decreases and the rate of reaction rises. The role of the liquid flow portion of the cycle then is to recharge the internal particle and the interstitial static hold-up and to carry away the reaction products. It is the latter role that is important in SO_2 oxidation because the buildup of acid reduces gas solubility and inhibits the rate of oxidation. The negative effect of acid concentration on SO_2 oxidation, and thus removal from the stack gas, is illustrated in Fig. 27. The liquid phase also removes the heat released by the chemical reaction. The temperature rise with zero liquid flow contributes importantly to the higher rates measured by Haure *et al.* (1989) for SO_2 oxidation under periodic operation. It is less important in the stack gas application because the SO_2 concentration is much smaller. Stack gas is not saturated with water vapor, so a more important effect is cooling of the bed through evaporation.

In a trickle bed with intermittent liquid flow and significant static hold-up, cooling through evaporation, heating due to reaction, solubility changes with temperature and acid concentration, and mass transfer changes on flow interruption lead to complex and changing levels of SO_2 removal and bed temperatures within a cycle. These are illustrated in Fig. 31, which presents measurements for the Centaur carbon at 80°C. Narrow vertical bars in the figure show the duration of liquid flow through the carbon bed. The SO_2 detector is located downstream of the trickle bed and the plot does not allow for the time lag. Nevertheless, the rise in SO_2 concentration begins when water is reintroduced. This appears to be a mass transfer effect, because bed temperature increases at the same time. The increase in SO_2 leaving the bed continues after the liquid flow ceases because evaporation cools the bed and decreases the rate of conversion to acid. Early in the cycle, most of the SO_2 is extracted in the upper portion of the carbon bed. About halfway through the cycle, the top of the bed becomes saturated with H_2SO_4 and the reaction ceases. Lower sections of the bed begin to participate to a greater extent in SO_2 removal and oxidation. Temperatures are lower and solubilities are higher, so the reaction rate increases, causing SO_2 in the outlet to drop. Meanwhile, evaporation has halted at the top of the bed and hotter entering gas begins to heat the bed (Fig. 31b).

The pattern shown in Fig. 31 is similar to that observed with BPL carbon, except that the peak reached after water is introduced and then cut off is narrow and climbs to about half of the SO_2 concentration in the feed. If the flush duration is 30 s, bed temperatures tend to be constant and the lowest level of SO_2 in the outlet occurs before water is admitted to the bed. This results in two SO_2 peaks within a cycle.

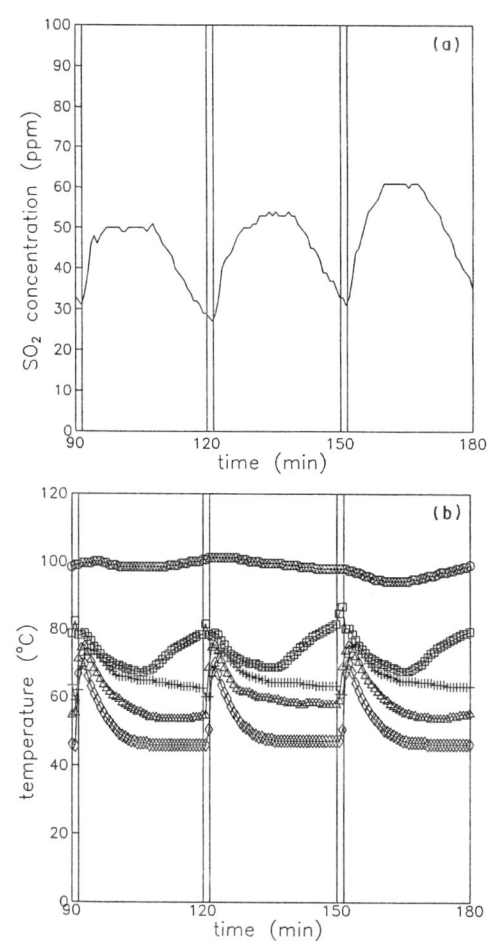

FIG. 31. Variation of outlet SO$_2$ concentration in ppm (vol) and bed temperatures at different locations within the carbon bed during a cycle. Measurements are for the Centaur carbon with gas and liquid feed temperatures = 80°C and gas SV = 1000 h^{-1}. Cycling variables were τ = 30 min, D = 1.5 min, and u_l = 0.17 cm/s. Symbols: ○ = top of bed, r = 0; □ = middle of bed, r = R; Δ = middle of bed, r = R/2; ◇ = middle of bed, r = 0; + = bottom of bed, r = 0. (Figure reproduced from Lee *et al.,* 1996b, with permission of the authors.)

The preceeding considerations suggest some requirements for rate or conversion improvement through periodic flow interruption in trickle beds: (1) wetting of the catalyst surface by the liquid phase, (2) significant liquid hold-up either interstitially or within the porous catalyst, (3) high liquid loading during the liquid flow portion of a cycle to ensure complete wetting of bed particles, (4) a sufficient duration of liquid flow to replenish internal and static hold-up, and (5) reaction and operating conditions such that mass transfer of a gas-phase reactant either is rate limiting or is one of the slower steps in the reaction-transport network. For the stack-gas scrubbing application, it is the duration of the liquid flush that is crucial to performance.

V. The Future of Periodic Operations for SO_2 Oxidation

The multistage catalytic converter used when sulfuric acid is produced by burning sulfur or from a hot smelter discharge with a moderately high SO_2 content represents fully developed, low-cost technology. It seems unlikely that a periodically operating process will displace this continuous process. Thus, the different ways of using periodic operation we have examined in this review seem relegated to special situations.

With respect to periodic reversal of flow direction, this seems to be competitive for low levels of smelter gas SO_2, probably under 4 vol%, at ambient temperature. If SO_2 is too high, thermal runaway can occur. A second, important application is to reduce SO_2 in the off-gas from a conventional acid plant. Use of flow reversal avoids the problem of reheating the gas leaving the first-stage absorber.

As for periodic air blowing of the final stage of a converter, this seems also to be technology for a special situation. It is a method of reducing SO_2 emissions to the environment, and thus it competes with periodic reversal of flow direction. It is not clear which of the two gives the better performance. Periodic flow reversal is a simple add-on at the tail end of a plant. Periodic air blowing, on the other hand, involves modification of a converter such as doubling the final stage and putting in piping to bring hot air from this stage to the primary and secondary air inlets of the sulfur burner. Modifications required can be seen from Fig. 4.

Periodic air blowing requires further development. For one, it has not been optimized with respect to cycle variables. The technology also needs to be tested on a pilot scale. Unlike periodic flow reversal, periodic air blowing requires a rapid-action three-way valve that will be exposed to SO_2 and SO_3 at temperatures between 350 and 400°C. Both flow reversal and air blowing subject the catalyst to significant periodic swings in temperature that may influence catalyst life. Data on catalyst life is not available. Nieken et al. (1994) and Budman et al. (1996) have pointed out that reactors incorporating periodic reversal of flow direction exhibit high parametric sensitivity, that is, an upward shift in SO_2 concentration could lead to thermal runaway, while a reduction in concentration may extinguish the reaction. Suitable control strategies for SO_2 oxidation need development.

Periodic interruption of liquid flow in a trickle-bed scrubber is another technology for special situations. It is directed at parts-per-million levels of SO_2 and probably cannot compete with conventional technology or other periodic operations, because it produces acid at a low concentration. Even though the carbon-based, periodic flow interruption process seems to be the most economic scrubbing process for 95% SO_2 removal (Gangwal *et al.,* 1992), that process, too, seems destined only for special applications. Fuel desulfurization is certainly the most economical method of reducing SO_2 emissions from power stations. Lime injection into the fire zone seems to be the favored route for coal-based plants, or, when higher levels of SO_2 removal are needed, scrubbing with a lime solution in a spray tower can be used. Thus, scrubbing seems limited to industrial discharges, such as emissions from acid plants where lime injection cannot be used. It may also have an application as a tail-end process for power-plant stack gases. Lime injection appears to be limited to 75 to 85% SO_2 removal at reasonable levels of lime addition, and scrubbing with a lime solution produces a solid waste.

Several uncertainties in this periodic process have not been resolved. Pressure drop is too high at SV = 10,000 h^{-1} when packed beds of carbon are used. Study of carbon-coated structured packing or of monoliths with activated carbon washcoats is needed to see if lower pressure drops at 95% SO_2 removal can be achieved. Stack gas from coal or heavy oil combustion contains parts-per-million or -per-billion quantities of toxic elements and compounds. Their removal in the periodically operated trickle bed must be examined, as well as the effect of these elements on acid quality. So far, laboratory experiments have been done to just 80°C; use of acid for flushing the carbon bed should permit operation at temperatures up to 150°C. Performance of periodic flow interruption at such temperatures needs to be determined. The heat exchange requirements for the RTI–Waterloo process shown in Fig. 26 depend on the temperature of SO_2 scrubbing. If operation at 150°C is possible, gas leaving the trickle bed can be passed directly to the $deNO_x$ step without reheating.

Nomenclature

ASS   apparent steady state
A_x cross-sectional area (m^2)
a volume specific interfacial area (m^{-1})

a_g	geometric specific surface area
B	dimensionless activation energy
CSS	steady state at the time-average liquid superficial velocity
C	concentration (mol/m^3)
C_i	concentration of the ith species
	concentration of the ith species in the liquid phase (Table XIII)
C_j	concentration of the jth species in the melt (Table XI)
C_v	concentration in the melt
C_o	maximum concentration of SO$_3$ in the melt phase
C^o	total concentration of pyrosulfate and sulfate anions
C_v	concentration of biatomic vanadium complexes
$C_3{}^L$	concentration of SO$_3$ in the melt phase
C_p	specific heat (kcal/kg °C)
C_r	ratio of solid to gas heat capacities [dimensionless—see just after Eq. (49)]
c_p	specific heat (kcal/kg °C)
D	dissymmetry factor (dimensionless)
	duration of flushing (min)
D_e	effective diffusion coefficient (m^2/s)
D_i	axial dispersion coefficient for species i (m^2/s)
	diffusion coefficient of species i mixture (m^2/s)
d	particle diameter (m)
d_e	equivalent particle diameter (m)
E_k	activation energy for the k reaction ($-k$ for the reverse reaction) (kcal/mol)
g	acceleration of gravity (m/s^2)
H_i	Henry's law constant for species i (mol/m^3 bar)
	enthalpy of component i (kcal/mol)
H_j	heat of the jth reaction (kcal/mol)
	Henry's law constant for species j (Table XI)
H_{in}	heat effect on active to inactive transformation (kcal/mol)
H^{dis}	heat of solution (kcal/mol)
ΔH	heat of reaction
ΔH_{evap}	heat of vaporization
h	heat transfer coefficient (kcal/m^2 °C s)
h_l	liquid particle heat transfer coefficient (kcal/m^2 °C s)
h_p	fluid–particle heat transfer coefficient (kcal/cm^3 °C s)
h_w	bed-to-wall heat transfer coefficient (kcal/cm^3 °C s)
K	equilibrium constant
K_{ad}	adsorption equilibrium constant
$(K_L)_k$	equilibrium constant in the melt for the kth reaction (various units)

K_H	solubility constant (mol/cm^3 bar)
k	rate constant
k_k	rate constant for the kth forward reaction
k_k	rate constant for the kth backward reaction
$k_L a$	volumetric mass transfer coefficient (s^{-1})
k_m	transport coefficient from pore volume to melt phase (mol/m^2 s)
k_p	fluid–particle mass transfer coefficient (m/s)
L	length of catalyst bed (cm, m)
n	index integer
P	pressure (bar, kPa)
P_i	partial pressure of species i (bar, kPa)
ΔP	pressure drop (bar, kPa)
Pe_H	Péclet number for heat
Pe_M	Péclet number for mass
Phg	modified Bodenstein number for gas phase (dimensionless—see just after Eqn. [49])
Phs	modified Bodenstein number for solid phase (dimensionless—see just after Eqn. [49])
Q	volumetric flow rate (m^3/s)
R	gas constant (various units)
R_i	rate of mass transfer of species i (kg/m^3 s)
R_p	rate of reaction (kg/m^3 s)
R_{evap}	evaporation rate
r	rate of reaction
r_j	rate of the jth reaction
r_{qss}	quasi-steady-state rate
QSS	quasi steady state
S	specific surface area (various units)
SS	steady state
SV	space velocity (s^{-1}, h^{-1})
STP	standard temperature and pressure
S_m	specific surface are of the melt phase on bed volume basis (m^{-1})
S_v	volumetric specific surface area (m^{-1})
S_{ext}	external surface
s	cycle split
s^v	complex density (mol/cm^3)
T	temperature (°C, K)
	gas temperature (Tables XI, XIII)
	dimensionless gas temperature
ΔT_D	model parameter [$H^{dis}/(\rho C_p)_g V_o$, Table XI]
ΔT_i	model parameter [$H_j/(\rho C_p)_g V_o$, Table XI]
ΔT_{ad}	adiabatic temperature rise (°C)

t	time (min)
t	space time (s)
t_c	cycle period (s, min Table XIII)
u	superficial velocity (m/s)
u_1	superficial velocity of the liquid (m/s, cm/s)
u_L	liquid interstitial velocity (m/s)
u_T	total interstitial velocity (m/s)
VSS	virtual steady state
V_o	molar gas volume (22.4 L/mol, Table XIII)
v_L	liquid superficial velocity (m/s)
v_{sv}	average velocity
X	conversion
X_1, X_2	conversion in the upflow(1) and downflow (2) directions in periodic flow reversal
x_i	mole fraction of i in the melt phase
x	conversion
y_i	gas phase mass fraction
y_j	mole or mass fraction of complex j in melt phase
y	adsorbate conversion
z	axial position (cm, m)

Greek

α	heat transfer parameter [dimensionless—see just after Eq. (49)] model parameter [$h_p S v \tau_s/(\rho C_p)_g$, Table XI] volume fraction or holdup expressed as a bed fraction (Table XIII)
β	equilibrium term defined by Eq. (58)
β_j	model parameter [$(k_p)_j S_v V_o P_{\pi s}/RT$, Table XI]
β_1	heat generation coefficient [dimensionless—see just after Eq. (49)]
β_2	generation term coefficient [dimensionless—see just after Eq. (49)]
γ	parameter [$\alpha(\rho C_p)_g$] dimensionless mass transfer coefficient—see just after Eq. (49)
γ_D	model parameter ($2\varepsilon_m \tau_s V_o/\tau$, Table XI)
γ_Θ	$(\rho C_p)_p (1_{-\varepsilon B}) \tau_s/(\rho C_p)_g \tau$
γ_v	model parameter ($2/\tau$, Table XI)
ε	void fraction
ε_B	void fraction in bed
ε_g	gas hold-up as fraction of bed volume ()
ε_p	void fraction in catalyst particle
ε_m	fraction occupied by the melt phase

APPLICATION OF PERIODIC OPERATION TO SO$_2$ OXIDATION   277

Θ	temperature of solid (K in Table XI)
θ	dimensionless solid temperature $(1-T_s/T_o)$
	solid temperature (Table XIII)
	dimensionless melt species concentration
$\Delta\theta_{ad}$	dimensionless adiabatic temperature rise
η	overall effectiveness factor
Λ	model parameter ($\varepsilon_m\tau_s C_v V_o$, Table XIII)
λ	conductivity
λ_1	axial conductivity (kcal/m$^{2°C}$ s)
μ_{jk}	stoichiometric coefficient for component j in the melt for the kth reaction
ν_{ik}	stoichiometric coefficient for species i in the kth reaction
ξ	dimensionless axial position
ρ	density (kg/m^3)
ρ_{cat}	catalyst density (kg/m^3)
ρ_g	gas density (kg/m^3)
τ	cycle period (s, min)
	dimensionless time (Table XI)
τ_s	space time (s, min)
τ'	dimensionless time (t/τ_s)
$\tau_{1/2}$	half cycle time (s, min)
Ψ	enhancement

Indices—Subscripts

d	dynamic
e	equivalent
g	gas
H	solubility
i	reactant species (generally 1 = SO$_2$, 2 = O$_2$ 3 = SO$_3$, 4 = H$_2$O, 5 = H$_2$SO$_4$) reaction index (Table XI)
j	melt component
k	reaction
	complex index (generally 1 = V$_2^{5+}$O$_2^{2-}$, 2 = V$_2^{5+}$O^{2-}, 3 = V$_2^{5+}$SO$_3^{2-}$, 4 = V$_2^{4+}$
L	liquid
l	liquid
m	melt species (Table XIII)
p	particle, constant pressure
R	reaction
s	space time
	solid or particle
	static

in	inactive
	inlet (Table XIII)
evap	evaporation
ext	external
max	maximum
o	time zero condition (Table XI)
3	SO_3

Indices-Superscripts

f	fluid or gas phase (Table XIII)
g	gas
p	particle
L	liquid or melt phase (Table XIII)
m	melt
in	inactive
	inlet (Table XIII)
°	initial or entrance state or condition
+	after time zero, maximum
−	before time zero, maximum
3	SO_3
‾	mean

REFERENCES

Balzhinimaev, B. S., Belyeava, N. P., and Ivanov, A. A., Kinetics of dissolution of inactive crystalline phase in vanadium catalysts for SO_2 oxidation. *React. Kinet. Catal. Letters* **29**, 465–472 (1985).

Balzhinimaev, B. S., Ivanov, A. A., Lapinaa, O. B., Mastikhin, M., and Zamaraev, K. I., The mechanism of SO_2 oxidation over supported vanadium catalysts. *Disc. Faraday Chem. Soc.* **87**(8), 1–15 (1989).

Berruti, F., Hudgins, R. R., Rhodes, E., and Sicardi, S., Oxidation of sulfur dioxide in a trickle-bed reactor. A study of reactor modelling. *Can. J. Chem. Eng.* **62**, 644–650 (1984).

Boreskov, G. K., and Matros, Yu. Sh., Unsteady-state performance of heterogeneous catalytic reactions. *Catal. Rev.—Eng. Sci.* **25**, 551–590 (1984).

Boreskov, G. K., Matros, Yu. Sh., Kiselov, O. V., and Bunimovich, G. A., Realization of heterogeneous catalytic processes under unsteady-state conditions. *Dokl. Acad. Nauk USSR* **237**, 160–163 (1977).

Boreskov, G. K., Bunimovich, G. A., Matros, Yu. Sh., and Ivanov, A. A., *Kinet. Catal.* **23**, 402 (1982).

Briggs, J. P., Hudgins, R. R., and Silveston, P. L., Composition cycling of an SO_2 oxidation reactor. *Chem. Eng. Sci.* **32**, 1087–1092 (1977).

Briggs, J. P., Kane, D. M., Hudgins, R. R., and Silveston, P. L., Reduction of SO_2 emissions from an SO_2 converter by means of feed composition cycling to the final stage, *in* "Proc.

1st Intern. Waste Treatment and Utilization Conf." (Moo Young, M. W., and G. Farquhar, Ed.), (Univ. of Waterloo, Waterloo, Ontaro) pp. 521–533, 1978.

Budman, H., Kryzonak, M., and Silveston, P. L., Control of a nonadiabatic packed bed reactor under periodic flow reversal. *Can. J. Chem. Eng.* **74,** 751–759 (1996).

Bunimovich, G. A., Matros, Yu. Sh., and Boreskov, G. K., Unsteady state performance of sulphur dioxide oxidation in production of sulfuric acid. *in* "Frontiers in Chemical Reaction Engineering," Vol. 2 (Doraiswarmy, L. K., and R. A. Mashelkar, Eds.). Wiley Eastern, New Delhi, 1984.

Bunimovich, G. A., Strots, V. O., and Goldman, O. V., Theory and industrial application of SO_2 oxidation reverse-process for sulphuric acid production, *in* "Unsteady-state Processes in Catalysis" (Matros, Yu. Sh., Ed.). VNU Science Press, Utrecht, 1990, pp. 7–24.

Bunimovich, G. A. Vernikovskaya, N. V., Strots, V. O., Balzhinimaev, B. S., and Matros, Yu. Sh., SO_2 oxidation in a reverse-flow reactor: influence of a vanadium catalyst dynamic properties. *Chem. Eng. Sci.* **50,** 565–580 (1995).

Carlsson, L., Sandgren, B., Simonsson, D., and Rihousky, M., Design and performance of a modular, multi-purpose electrochemical reactor. *J. Electrochem. Soc.* **130,** 342–350 (1983).

De Vos, R., Hatziantoniou, V., and Schöön, N.-H., The cross-flow catalyst reactor. An alternative for liquid phase hydrogenations. *Chem. Eng. Sci.* **37,** 1719–1726 (1982).

De Vos, R., Smedler, G., and Schöön, N.-H., Selectivity aspects of using the cross-flow catalyst reactor for liquid phase hydrogenations. *Ind. Eng. Chem. Process Des. Dev.* **25,** 197–202 (1986).

Eigenberger, G., and Nieken, U., Catalytic combustion with periodic flow reversal. *Chem. Eng. Sci.* **43,** 2109–2115 (1988).

Gangwal, S. K., McMichael, W. J., Howe, G. B., Spivey, J. J., and Silveston, P. L., Low-temperature carbon-based process for flue-gas cleanup. *IChemE Symp.* Series, No. 131 (1992).

Hasokowati, W., Hudgins, R. R., and Silveston, P. L., Loading, draining and hold-up in periodically operated trickle-bed reactors. *Can. J. Chem. Eng.* **72,** 405–410 (1994).

Hatziantoniou, V., Andersson, B., and Schöön, N.-H., Mass transfer and selectivity in liquid phase hydrogenation of nitro compounds in a monolithic catalyst reactor with segmented gas–liquid flow. *Ind. Eng. Chem. Process Des. Dev.* **25,** 964–970 (1986).

Haure, P. M., Hudgins, R. R., and Silveston, P. L., Periodic operation of a trickle bed reactor, *AIChEJ.* **35,** 1437–1444 (1989).

Haure, P. M., Bogdashev, S. M., Bunimovich, M., Stegasov, A. N., Hudgins, R. R., and Silveston, P. L., Thermal waves in the periodic operation of a trickle- bed reactor, *Chem. Eng. Sci.* **45,** 2255–2261 (1990).

Haure, P. M., Hudgins, R. R., and Silveston, P. L., Investigation of SO_2 oxidation rates in trickle-bed reactors operating at low liquid flow rates. *Can. J. Chem. Eng.* **70,** 600–603 (1992).

Isozaki, C., Katagiri, T., Nakamura, Y., Hirabayashi, S., Yabe S., and Yamaki, T., Demonstration of DC/DA sulphuric acid plant based on unsteady-state *in* "Unsteady-state Processes in Catalysis" (Matros, Yu. Sh., Ed.). VNU Science Press, Utrecht, 1990, pp. 637–642.

Ivanov, A. A., and Balzhinimaev, B. S., New data on kinetics and reaction mechanism for SO_2 oxidation over vanadium catalysts. *React. Kinet. Catal. Lett.* **35,** 413–424 (1987).

Kirillov, V. A., and Stegasov, A. N., On the simulation of heterogeneous catalytic processes with phase transitions in fixed catalyst bed (in Russian). *Teoreticheskiye osnoy khimicheskoi teknhnologii* **22,** 340–345 (1988).

Kirillov, V. A., Kuzmin, V. A., Pyanov, V. I., and Khanaev, V. M., Analysis of transfer process in a packed bed with concurrent down-flow of gas and liquid. *Hung. J. Ind. Chem.* **11,** 263–274 (1983).

Komiyama, H., and Smith J. M., Sulfur dioxide oxidation in slurries of activated carbon. *AIChEJ.* **21,** 664–676 (1975).
Lee, J.-K., Hudgins, R. R., and Silveston, P. L., A cycled trickle-bed reactor for SO₂ oxidation. *Chem. Eng. Sci.* **50,** 2523–2530 (1995).
Lee, J.-K., Hudgins, R. R., and Silveston, P. L., SO₂ oxidation in a periodically operated trickle-bed comparison of activated carbon catalysts. *Environ. Prog.* **15**(4), 239–244 (1996a).
Lee, J.-K., Ferrero, S., Hudgins, R. R., and Silveston, P. L., Catalytic SO₂ oxidation in a periodically operated trickle-bed: reactor. *Can. J. Chem. Eng.* **74,** 706–712 (1996b).
Mata, A. R., and Smith, J. M., Oxidation of sulfur dioxide in a trickle bed reactor. *Chem. Eng. J.* **22,** 229–235 (1981).
Matros, Yu. Sh., "Unsteady Processes in Catalytic Reactors." Elsevier Science Publishers, Amsterdam, 1985.
Matros, Yu. Sh., Unsteady-state oxidation of sulphur dioxide in sulphuric acid production. *Sulphur* **183,** 23 (1986).
Matros, Yu. Sh., "Catalytic Processes under Unsteady-State Conditions." Elsevier Science Publishers, Amsterdam, 1989.
Matros, Yu. Sh., and Bunimovich, G. A., Reverse-flow operation in fixed bed catalytic reactors. *Catal. Rev.—Sci. Eng.* **38**(1), 1–68 (1996).
Metzinger, J., Hudgins, R. R., Silveston, P. L., Gangwal, S. K., Application of a periodically operated trickle bed to sulfur removal from stack gas. *Chem. Eng. Sci.* **47,** 3723–3727 (1992).
Metzinger, J., Kühter, A., Silveston, P. L., and Gangwal, S. K., Novel periodic reactor for scrubbing SO₂ from industrial stack gases. *Chem. Eng. Sci.* **49,** 4533–4546 (1994).
Morsi, B. I., and Charpentier, J. C., On mass transfer with chemical reaction in multiphase systems. *NATO ASI CESME,* Izmir, Turkey, (1981).
Nieken, U., Kolios, G., and Eigenberger, G., Control of the ignited steady state in autothermal fixed-bed reactors for catalytic combustion. *Chem. Eng. Sci.* **49,** 5507–5518 (1994).
Purwono, S., Budman, H., Hudgins, R. R., Silveston, P. L., and Matros, Yu. Sh., Runaway in packed bed reactors operating with periodic flow reversal. *Chem. Eng. Sci.* **49,** 5473–5487 (1994).
Ramachandran, P. A., and Smith, J. M., Dynamic behavior of trickle bed reactors. *Chem. Eng. Sci.* **34,** 75–91 (1979).
Rao, V. G., and Drinkenburg, A. A. H., A model for pressure drop in two-phase gas-liquid down-flow through packed columns. *AIChEJ.* **31,** 1010–1018 (1985).
Rhee, H.-K., and Amundson, N. R., Equilibrium theory of creeping profiles in fixed-bed catalytic reactors. *Ind. Eng. Chem. Funda.* **13,** 1–4 (1974).
Sapundzhiev, H., Grozev, G., and Elenkov, D., Influence of geometric and thermophysical properties of reaction layer on sulphur dioxide oxidation in transient conditions. *Chem. Eng. Technol.* **13,** 131–135 (1990).
Satterfield, C. N., and Ozel, F., Direct solid-catalyzed reaction of a vapor in an apparently completely wetted trickle bed reactor. *AIChEJ.* **19,** 1259–1261 (1973).
Sedricks, W., and Kenney, C. N., Partial wetting in trickle bed reactors—the reduction of crotonaldehyde over a palladium catalyst. *Chem. Eng. Sci.* **28,** 558–1261 (1973).
Silveston, P. L., and Hudgins, R. R., Reduction of sulfur dioxide emissions from a sulfuric acid plant by means of feed modulation. *Environ. Sci. Technol.* **15,** 419–422 (1981).
Silveston, P. L., Hudgins, R. R., Bogdashev, S., Vernijakovskaja, N and Matros, Yu. Sh., Modelling of a periodically operated packed bed SO₂ oxidation reactor at high conversion in "Unsteady-state Processes in Catalysis (Matros, Yu. Sh., Ed.). VNU Science Press, Utrecht, 1990.
Silveston, P. L., Hudgins, R. R., Bogdashev, S., Vernijakovskaja, N., and Matros, Yu. Sh.,

Modelling of a periodically operated packed bed SO$_2$ oxidation reactor at high conversion. *Chem. Eng. Sci.* **49**(3), 335–341 (1994).

Snyder, J. D., and Subramaniam, B., Numerical simulation of a periodic flow reversal reactor for sulfur dioxide production. *Chem. Eng. Sci.* **48**, 4051–4064 (1993).

Specchia, V., and Baldi, G., Pressure drop and liquid hold-up for two-phase concurrent flow in packed beds. *Chem. Eng. Sci.* **32**, 515–523 (1977).

Stegasov, A. N., Kirillov, V. A., and Silveston, P. L., Modelling of catalytic SO$_2$ oxidation for continuous and periodic liquid flow through a trickle bed. *Chem. Eng. Sci.* **49**, 3699–3710 (1992).

Strots, V. O., Matros, Yu. Sh., Bunimovich, G. A., Periodically forced SO$_2$ oxidation in CSTR. *Chem. Eng. Sci.* **47**(9–11), 2701–2706 (1992).

Tan, C. S., and Smith, J. M. Catalyst particle effectiveness with unsymmetrical boundary conditions. *Chem. Eng. Sci.* **35**, 1601–1609 (1980).

Unni, M. P, Hudgins, R. R., Silveston, P. L., Influence of cycling on the rate of oxidation of SO$_2$ over a vanadia catalyst. *Can. J. Chem. Eng.* **51**, 623–629 (1973).

Van den Bussche, K. M., Neophytides, S. N., Zolotarskii, I. A., and Froment, G. F., Modelling and simulation of the reverse flow operation of a fixed bedreactor for methanol synthesis. *Chem. Eng. Sci.* **48**(12), 3335–3345 (1993).

Wu, H., Zhang, S., Li, C., and Fu, J., Modeling of unsteady-state oxidation process of SO$_2$ (I) axial heat transfer of fixed bed packed with vanadium catalyst. *J. Chem. Ind. Eng. (China)* **46**, 416–423 (1995).

Wu, H., Zhang, S., and Li, C., Study of unsteady-state catalytic oxidation of sulphur dioxide by periodic flow reversal. *Can. J. Chem. Eng.* **74**, 766–771 (1996).

Xiao, W.-D., and Yuan, W.-K., Modelling and simulation for adiabatic fixed-bed reactor with flow reversal. *Chem. Eng. Sci.* **49**(21), 3631–3641 (1994).

Xiao, W.-D., and Yuan, W.-K., Modelling and experimental studies on unsteady-state SO$_2$ converters. *Can. J. Chem. Eng.* **74**, 772–782 (1996).

Yang, M.-C., and Cussler, E. L., A hollow fiber trickle bed reactor. *AIChEJ.* **33**, 1754–1760 (1987).

Zhang, S., Wu, H., and Li, C.-Y., 'Modeling of unsteady-state oxidation process of SO$_2$ (II)—Model simulation and parameter analysis.' *J. Chem. Ind. Eng.(China)* **46**, 424–430 (1995).

INDEX

A

Adaptive Resonance Theory (ART)
  clustering in process data interpretation, 63–64, 73–77, 92–93
  input analysis of process data, 30–31
Advection problem, solution, 109
Affine deformation, drop breakup, 132–134
Agglomerate, *see also* Aggregation; Fragmentation
  definition, 159–160
  flow type effect on separation following rupture, 165–167
  particle interactions, 161–163
  size distributions, 160, 181, 183–184
  strength prediction, 164–165
Aggregation, *see also* Agglomerate
  bonding levels, 160–161
  coalescence relationship, 106
  diffusion-limited aggregation, 180–181
  flow-induced aggregation
    area-conserving cluster aggregation in two-dimensional chaotic flows, 189–190
    cluster–cluster aggregation, 186
    constant capture radius, aggregation in chaotic flows, 187, 189
    features, 180–181, 186
    fractal structure aggregation in chaotic flows, 191–192, 194
    particle–cluster aggregation, 186
  hierarchical cluster–cluster aggregation, 181
  linear trajectory aggregation, 180–181
  mass of clusters, 161
  mechanisms, 180
  polydispersity computations, 194
  reaction-limited aggregation, 180–181
  short-term behavior in well-mixed systems, 184–185
  sizes, 160
ART, *see* Adaptive Resonance Theory

B

Back propagation network (BPN)
  correspondence with linear discriminant pattern recognition, 53–55
  input–output analysis of process data, 38–39
Basis function
  adaptive-shape basis functions, 13
  fixed-shape basis functions, 12–13
BPN, *see* Back propagation network
Breakup
  affine deformation, 132–134
  capillary number, 130, 132
  critical capillary number, 132, 134
  drop size
    daughter droplets, 143, 145
    distribution from chaotic flows, 145, 147
    viscosity ratio effect on distribution, 147, 149
  flow reorientation importance, 134, 136
  fragmentation relationship, 106–107
  mechanisms
    capillary instability, 141–143, 145, 149, 151, 195
    combined mechanisms, 142
    end-pinching, 139, 149
    necking, 139, 143, 145, 149
    tipstreaming, 139
  modeling with coalescence, 155–159
  satellite formation in capillary breakup, 143
  small-scale mixing, 130–132
  stretching of low-viscosity-ratio elongated drops, 137–138
  thread breakup time during flow, 142–143
  velocity field equations for common flow types, 131

C

Capillary instability, *see* Breakup
Capillary number
  breakup problem, 130
  critical capillary number, 132, 134
  definition, 106, 128
CART, *see* Classification and regression tree
Catalytic converter, *see* Sulfur dioxide oxidation
Cavity flow, characteristics, 110, 112–113
Classification and regression tree (CART), input–output analysis of process data, 41–42
Clustering, process data interpretation
  Adaptive Resonance Theory, 63–64, 73–77, 92–93
  ellipsoidal basis function networks, 62–63
  fixed cluster approaches, 61–63
  generalization and memorization, 60
  k-nearest neighbor rule, 59
  kernel identification, 46
  1-nearest neighbor rule, 59
  performance, 61
  fixed cluster approaches, 61–63
  proximity indices, 59
  qualitative trend analysis, 63
  radial basis function networks, 61–62
  two-class problem example, 58–59
  variable cluster approaches, 63–64
Coalescence
  aggregation relationship, 106
  collisions
    flow type effects on shear-induced collisions, 151–153, 155
    frequency, 151, 155
    film drainage, 153, 155
    modeling with breakup, 155–159

D

Data analysis
  abnormal situation detection in continuous polymer process, 82–86
  batch operation variability, 86–90
  challenges with process data
    discriminant uncertainty, 8
    dynamic changes to process conditions, 8
    lack of abnormal situation exemplars, 8
    measurement uncertainty, 8
    scale and scope of process, 7–8
  feature mapping
    pattern exemplars, 4
    training set, 4
  input mapping, *see* Input analysis, process data
  input–output mapping, *see* Input–output analysis, process data
  numeric–numeric transformation, 3
  pattern recognition, overview, 2–3
  Q-statistics, 90
  selection of techniques, 9
Data interpretation
  backpropagation network, correspondence with linear discriminant pattern recognition, 53–55
  challenges with process data
    discriminant uncertainty, 8
    dynamic changes to process conditions, 8
    lack of abnormal situation exemplars, 8
    measurement uncertainty, 8
    scale and scope of process, 7–8
  diagnosis of batch polymer reactor operating problems, 90–96
  hierarchical modularization, 79–82
  hyperplane orientations, 49–50
  knowledge-based system approaches
    applications, 66–67
    decomposition of complex problems, 72, 93–94
    digraph, 70
    fault tree, 69–70
    feed injection system example, 65–66
    model-based methods, 68–69
    overview, 44, 64–65
    semantic networks, 67
    tables, 65, 67, 70–71
  labels
    context dependence, 6–7
    types, 5–6
  limit checking, 48–49
  local interpretation methods
    clustering

INDEX

Adaptive Resonance Theory, 63–64, 73–77, 92–93
ellipsoidal basis function networks, 62–63
fixed cluster approaches, 61–63
generalization and memorization, 60
k-nearest neighbor rule, 59
kernel identification, 46
1-nearest neighbor rule, 59
performance, 61
proximity indices, 59
qualitative trend analysis, 63
radial basis function networks, 61–62
two-class problem example, 58–59
variable cluster approaches, 63–64
probability density function approaches, 56–58
statistical measures for interpretation, 55–56
nonlocal interpretation methods
multivariate linear discriminant methods, 49–55
univariate methods, 47–49
numeric–symbolic mapping, 6, 43–55, 72–78
selection of techniques, 9
symbolic–symbolic mapping, 6, 64–72
Dispersion, *see* Powder dispersion in liquids
Duct flow
characteristics, 113–114
conversion into efficient mixing flows, 114, 116
velocity field, 113

E

EBFN, *see* Ellipsoidal basis function network
Ellipsoidal basis function network (EBFN)
clustering in process data interpretation, 62–63
input analysis of process data, 30
End-pinching, *see* Breakup
Erosion, *see* Fragmentation

F

FCCU, *see* Fluidized catalyst cracking unit
Flocculation, *see* Aggregation

Fluidized catalyst cracking unit (FCCU), input process data interpretation
Adaptive Resonance Theory degradation with complexity, 73–77
fast ramp training scenario, 73–74
hierarchical decomposition of malfunction groupings, 74–77
malfunction types, 73
slow ramp malfunction scenario, 73–74
Fragmentation, *see also* Agglomerate
breakup relationship, 106–107
definition, 160
deformation of agglomerates, 163
dispersion kinetics in two-zone model, 170–171, 173
distribution of fragments, 163
erosion
definition, 160, 179–180
kinetics, 169–170, 176–178
simultaneous rupture, 173, 178–179
fragmentation number, 106, 164
particle size distribution analysis, 173–176
rupture
critical fragmentation number, 167–169
definition, 160, 180
simultaneous erosion, 173, 178–179
scaling solutions of fragmentation equation, 107
shattering, 160

G

Gel permeation chromatography (GPC), diagnosis of batch polymer reactor operating problems, 91–95
GPC, *see* Gel permeation chromatography

H

Hierarchical classification, process data interpretation, 80
Horseshoe map, chaotic mixing, 110

I

Immiscible fluid mixing
breakup, *see* Breakup

286   INDEX

Immiscible fluid mixing (*Continued*)
  coalescence, *see* Coalescence
  dispersed fluid phases, role in flow structure, 129–130
  initial stages, 128–129
  large versus small scales, 125–128
  physical picture, 124–125
  polymer processing example, 129
  velocity field, 128
  viscous immiscible liquid mixing model, 156–159
Input analysis, process data
  adaptive resonance theory, 30–31
  basis function, 14
  classification of methods, 11, 14
  cluster *see*king, 28–29
  definition, 4, 13
  ellipsoidal basis function network, 30
  latent variable representation, 10, 14
  linear projection methods, 14–27
  multiscale filtering
    finite impulse response median hybrid filter, 19–20
    steps in filtering, 22
    temporal resolution, 19
    threshold for filtering, 23
    wavelet, 21–22
    wavelet packet, 23–24
  multivariate methods, 4–5, 10, 13, 24–27
  nonlinear principle component analysis, 28
  nonlinear projection methods, 27–32
  principle component analysis
    eigenvectors, 26–27
    hierarchical blocking, 81
    loadings, 24–25
    multidimensional analysis, 27, 85
    scores, 25
    steps, 25
    two principle component model example, 88–89
  probabilistic methods, 4
  process considerations in determining filter characteristics, 14
  radial basis function network, 29–30
  single-scale filtering
    basis functions, 15
    filter gain, 16
    filter matching to response times, 17–18
    finite impulse response, 15–16, 18–19
    infinite impulse response, 15–16, 18–19
    linear phase filters, 19
    phase shift, 16
    univariate methods, 4, 13–19
Input–output analysis, process data
  back propagation networks, 38–39
  basis functions
    adaptive-shape basis functions, 13
    fixed-shape basis functions, 12–13
  classification and regression tree, 41–42
  classification of methods, 11–12
  linear projection methods, 33–40
  multivariate adaptive regression splines, 42–43
  nonlinear partial least squares, 37–38, 81
  nonlinear principle component regression, 37
  nonlinear projection methods, 40–41
  ordinary least squares regression, 33–35, 52
  overview, 4–5, 32–33
  partial least squares, 36–37, 52, 85
  partition-based methods, 41–43
  principle component regression, 35–36
  process output variable representation, 11
  projection pursuit regression, 39–40
  radial basis function network, 40–41

K

KAM tube, *see* Kolmogorov–Arnold–Moser tube
KBS, *see* Knowledge-based system
Kenics static mixer, velocity field, 122
Knowledge-based system (KBS)
  applications, 66–67
  decomposition of complex problems, 72, 93–94
  digraph, 70
  fault tree, 69–70
  feed injection system example, 65–66
  model-based methods, 68–69
  overview, 44, 64–65
  semantic networks, 67
  tables, 65, 67, 70–71
Kolmogorov–Arnold–Moser (KAM) tube, features, 116

INDEX   287

M

Multiscale filtering, input analysis of process data
  finite impulse response median hybrid filter, 19–20
  steps in filtering, 22
  temporal resolution, 19
  threshold for filtering, 23
  wavelet, 21–22
  wavelet packet, 23–24
Multivariate adaptive regression splines (MARS), input–output analysis of process data, 42–43

N

Necking, *see* Breakup

O

OLS, *see* Ordinary least squares
Ordinary least squares (OLS), input–output analysis of process data, 33–35, 52

P

Partial least squares (PLS)
  input–output analysis of process data, 36–37, 52, 85
  nonlinear partial least squares, 37–38, 81
Partitioned-pipe mixer (PPM), features, 114, 116
Pattern recognition, data analysis and interpretation, 2–3
PCA, *see* Principle component analysis
PCR, *see* Principle component regression
Periodic operation, sulfur dioxide oxidation
  applications, 206–207, 272–273
  heat removal, 206
  periodic air blowing of final stage of multistage catalytic converter
    composition cycling results, 211–212
    composition forcing, 216–217, 223
    geographic distribution, 206–207
    isothermal back-mixed reactor application, 217, 219–223

  mechanistic model, 215–216
  performance, 209–211, 272
  rate enhancement, 208–209
  temperature changes, 211–212, 272–273
  unsymmetrical cycling, 213–214
periodic flow interruption in trickle-bed catalytic reactors
  cycle period effects, 251–252
  cycle split effects, 251–252
  estimated oxidation rates, 255
  geographic distribution, 206–207
  intracycle gas removal, 270
  liquid flow rate effects, 252–253
  liquid loading, 253–254
  modeling and simulation
    accuracy of models, 259–261
    cycle split effects, 260
    equations, 258
    filling and draining steps, 256, 259
    gas concentration estimation, 261
    kinetics, 259
  physical mechanism, 269–272
  reactor design, 250
  segregation processes, 248–249
  space velocity, 251
  stack gas scrubbing
    activated carbon performances, 263–264, 267, 269–270
    advantages, 262–263
    application, 273
    bed depth estimation, 266–267
    operating conditions, 264
    pressure drop estimation, 267, 273
    recommendations for improvement, 271–272
    RTI–Waterloo process, 262, 273
    wash liquid normality effects on removal, 264, 266
  temperature, 255–256, 270
  water role, 255
periodic reversal of flow direction in single-stage converter
  advantages, 247, 272
  bench-scale unit design, 230, 232, 234
  catalyst beds, 224–225
  double contacting-double absorption process, 229–230
  geographic distribution, 206–207, 223
  modeling and simulations

288 INDEX

Periodic operation (*Continued*)
  accuracy of models, 238–239
  boundary conditions, 238
  equations, 234–236
  heat recuperator effects, 242–243
  inlet sulfur dioxide concentration effects, 240, 242
  nonisothermal tubular reactor, 244–246
  operating problem identification, 243–244
  pseudo-homogeneous model, 236–237
  rate model, 238
  simplification, 237
  startup temperature jump, 243
  temperature, 239–240, 247
  performance of sulfuric acid plants, 225–227
  pilot-scale reactors, 227–229
  recuperator beds, 225
  sulfur dioxide concentration variation in feed, 225, 227
  temperature, 225
PLS, *see* Partial least squares
Polymer blend, advantages, 124
Powder dispersion in liquids
  aggregation, *see* Aggregation
  fragmentation, *see* Fragmentation
  mixing optimization, 195, 198
  physical picture, 159–161
  stages, 160
PPM, *see* Partitioned-pipe mixer
Principle component analysis (PCA), input analysis of process data
  eigenvectors, 26–27
  hierarchical blocking, 81
  loadings, 24–25
  multidimensional analysis, 27, 85
  scores, 25
  steps, 25
  two principle component model example, 88–89
Principle component regression (PCR)
  input–output analysis of process data, 35–36
  nonlinear principle component regression, 37
Probability density function
  approaches in process data interpretation, 56–58
  chaotic flow, 112–113
Process data, *see* Data analysis; Data interpretation

Q

Q-statistics, process data analysis, 90

R

Radial basis function network (RBFN)
  clustering in process data interpretation, 61–62
  input analysis of process data, 29–30
  input–output analysis of process data, 40–41
RBFN, *see* Radial basis function network
Rupture, *see* Fragmentation

S

Shattering, *see* Fragmentation
Single-scale filtering, input analysis of process data
  basis functions, 15
  filter gain, 16
  filter matching to response times, 17–18
  finite impulse response, 15–16, 18–19
  infinite impulse response, 15–16, 18–19
  linear phase filters, 19
  phase shift, 16
Single screw extruder, improvement of mixing, 116
Smoluchowski's equation, scaling solutions, 107, 183
Stack scrubber, *see* Trickle-bed catalytic reactor
Stokes equation, interface modeling for Newtonian fluids, 126–127
Strain
  per period optimization in shear flows with periodic reorientation, 120–122
  shear zone dependence, 125
Stretching
  chaotic flow islands, 113, 124
  distribution of values, 118, 120

equations, 109–110
flow classification, 131–132
low-viscosity-ratio elongated drops, 137–138
preconditions, 125
quantitative estimation, 122
viscous immiscible liquid mixing model, 157
Sulfur dioxide oxidation
  applications, 206–207, 272–273
  heat removal, 206
  periodic air blowing of final stage of multistage catalytic converter
    composition cycling results, 211–212
    composition forcing, 216–217, 223
    geographic distribution, 206–207
    isothermal back-mixed reactor application, 217, 219–223
    mechanistic model, 215–216
    performance, 209–211, 272
    rate enhancement, 208–209
    temperature changes, 211–212, 272–273
    unsymmetrical cycling, 213–214
  periodic flow interruption in trickle-bed catalytic reactors
    cycle period effects, 251–252
    cycle split effects, 251–252
    estimated oxidation rates, 255
    geographic distribution, 206–207
    intracycle gas removal, 270
    liquid flow rate effects, 252–253
    liquid loading, 253–254
    modeling and simulation
      accuracy of models, 259–261
      cycle split effects, 260
      equations, 258
      filling and draining steps, 256, 259
      gas concentration estimation, 261
      kinetics, 259
    physical mechanism, 269–272
    reactor design, 250
    segregation processes, 248–249
    space velocity, 251
    stack gas scrubbing
      activated carbon performances, 263–264, 267, 269–270
      advantages, 262–263
      application, 273
    bed depth estimation, 266–267
    operating conditions, 264
    pressure drop estimation, 267, 273
    recommendations for improvement, 271–272
    RTI–Waterloo process, 262, 273
    wash liquid normality effects on removal, 264, 266
    temperature, 255–256, 270
    water role, 255
  periodic reversal of flow direction in single-stage converter
    advantages, 247, 272
    bench-scale unit design, 230, 232, 234
    catalyst beds, 224–225
    double contacting-double absorption process, 229–230
    geographic distribution, 206–207, 223
    modeling and simulations
      accuracy of models, 238–239
      boundary conditions, 238
      equations, 234–236
      heat recuperator effects, 242–243
      inlet sulfur dioxide concentration effects, 240, 242
      nonisothermal tubular reactor, 244–246
      operating problem identification, 243–244
      pseudo-homogeneous model, 236–237
      rate model, 238
      simplification, 237
      startup temperature jump, 243
      temperature, 239–240, 247
    performance of sulfuric acid plants, 225–227
    pilot-scale reactors, 227–229
    recuperator beds, 225
    sulfur dioxide concentration variation in feed, 225, 227
    temperature, 225

T

Tipstreaming, *see* Breakup
Trickle-bed catalytic reactor, periodic

Trickle-bed catalytic reactor (*Continued*)
flow interruption in sulfur dioxide oxidation
  cycle period effects, 251–252
  cycle split effects, 251–252
  estimated oxidation rates, 255
  geographic distribution, 206–207
  heat removal, 206
  intracycle gas removal, 270
  liquid flow rate effects, 252–253
  liquid loading, 253–254
  modeling and simulation
    accuracy of models, 259–261
    cycle split effects, 260
    equations, 258
    filling and draining steps, 256, 259
    gas concentration estimation, 261
    kinetics, 259
  physical mechanism, 269–272
  reactor design, 250
  segregation processes, 248–249
  space velocity, 251
  stack gas scrubbing
    activated carbon performances, 263–264, 267, 269–270
    advantages, 262–263
    application, 273
    bed depth estimation, 266–267
    operating conditions, 264
    pressure drop estimation, 267, 273
    recommendations for improvement, 271–272
    RTI–Waterloo process, 262, 273
    wash liquid normality effects on removal, 264, 266
  temperature, 255–256, 270
  water role, 255

V

VILM, *see* Viscous immiscible liquid mixing model
Viscoelastic fluid, mixing, 116–118, 124
Viscous immiscible liquid mixing model (VILM)
  drop size distribution, 157–158
  physical aspects, 156
  stretching distributions, 157
  viscosity ratio, dispersed phase size dependence, 158–159
Viscous mixing, overview, 107–108
Vortex mixing flow, characteristics, 110

W

Wavelet, *see* Multiscale filtering

CONTENTS OF VOLUMES IN THIS SERIAL

Volume 1

J. W. Westwater, *Boiling of Liquids*
A. B. Metzner, *Non-Newtonian Technology: Fluid Mechanics, Mixing, and Heat Transfer*
R. Byron Bird, *Theory of Diffusion*
J. B. Opfell and B. H. Sage, *Turbulence in Thermal and Material Transport*
Robert E. Treybal, *Mechanically Aided Liquid Extraction*
Robert W. Schrage, *The Automatic Computer in the Control and Planning of Manufacturing Operations*
Ernest J. Henley and Nathaniel F. Barr, *Ionizing Radiation Applied to Chemical Processes and to Food and Drug Processing*

Volume 2

J. W. Westwater, *Boiling of Liquids*
Ernest F. Johnson, *Automatic Process Control*
Bernard Manowitz, *Treatment and Disposal of Wastes in Nuclear Chemical Technology*
George A. Sofer and Harold C. Weingartner, *High Vacuum Technology*
Theodore Vermeulen, *Separation by Adsorption Methods*
Sherman S. Weidenbaum, *Mixing of Solids*

Volume 3

C. S. Grove, Jr., Robert V. Jelinek, and Herbert M. Schoen, *Crystallization from Solution*
F. Alan Ferguson and Russell C. Phillips, *High Temperature Technology*
Daniel Hyman, *Mixing and Agitation*
John Beck, *Design of Packed Catalytic Reactors*
Douglass J. Wilde, *Optimization Methods*

Volume 4

J. T. Davies, *Mass-Transfer and Interfacial Phenomena*
R. C. Kintner, *Drop Phenomena Affecting Liquid Extraction*
Octave Levenspiel and Kenneth B. Bischoff, *Patterns of Flow in Chemical Process Vessels*
Donald S. Scott, *Properties of Concurrent Gas–Liquid Flow*
D. N. Hanson and G. F. Somerville, *A General Program for Computing Multistage Vapor–Liquid Processes*

Volume 5

J. F. Wehner, *Flame Processes–Theoretical and Experimental*
J. H. Sinfelt, *Bifunctional Catalysts*
S. G. Bankoff, *Heat Conduction or Diffusion with Change of Phase*
George D. Fulford, *The Flow of Liquids in Thin Films*
K. Rietema, *Segregation in Liquid–Liquid Dispersions and Its Effect on Chemical Reactions*

Volume 6

S. G. Bankoff, *Diffusion-Controlled Bubble Growth*
John C. Berg, Andreas Acrivos, and Michel Boudart, *Evaporation Convection*
H. M. Tsuchiya, A. G. Fredrickson, and R. Aris, *Dynamics of Microbial Cell Populations*
Samuel Sideman, *Direct Contact Heat Transfer between Immiscible Liquids*
Howard Brenner, *Hydrodynamic Resistance of Particles at Small Reynolds Numbers*

Volume 7

Robert S. Brown, Ralph Anderson, and Larry J. Shannon, *Ignition and Combustion of Solid Rocket Propellants*
Knud Østergaard, *Gas–Liquid–Particle Operations in Chemical Reaction Engineering*
J. M. Prausnitz, *Thermodynamics of Fluid–Phase Equilibria at High Pressures*
Robert V. Macbeth, *The Burn-Out Phenomenon in Forced-Convection Boiling*
William Resnick and Benjamin Gal-Or, *Gas–Liquid Dispersions*

Volume 8

C. E. Lapple, *Electrostatic Phenomena with Particulates*
J. R. Kittrell, *Mathematical Modeling of Chemical Reactions*
W. P. Ledet and D. M. Himmelblau, *Decomposition Procedures for the Solving of Large Scale Systems*
R. Kumar and N. R. Kuloor, *The Formation of Bubbles and Drops*

Volume 9

Renato G. Bautista, *Hydrometallurgy*
Kishan B. Mathur and Norman Epstein, *Dynamics of Spouted Beds*
W. C. Reynolds, *Recent Advances in the Computation of Turbulent Flows*
R. E. Peck and D. T. Wasan, *Drying of Solid Particles and Sheets*

Volume 10

G. E. O'Connor and T. W. F. Russell, *Heat Transfer in Tubular Fluid–Fluid Systems*
P. C. Kapur, *Balling and Granulation*
Richard S. H. Mah and Mordechai Shacham, *Pipeline Network Design and Synthesis*
J. Robert Selman and Charles W. Tobias, *Mass-Transfer Measurements by the Limiting-Current Technique*

Volume 11

Jean-Claude Charpentier, *Mass-Transfer Rates in Gas–Liquid Absorbers and Reactors*
Dee H. Barker and C. R. Mitra, *The Indian Chemical Industry–Its Development and Needs*
Lawrence L. Tavlarides and Michael Stamatoudis, *The Analysis of Interphase Reactions and Mass Transfer in Liquid–Liqid Dispersions*
Terukatsu Miyauchi, Shintaro Furusaki, Shigeharu Morooka, and Yoneichi Ikeda, *Transport Phenomena and Reaction in Fluidized Catalyst Beds*

Volume 12

C. D. Prater, J. Wei, V. W. Weekman, Jr., and B. Gross, *A Reaction Engineering Case History: Coke Burning in Thermofor Catalytic Cracking Regenerators*

Costel D. Denson, *Stripping Operations in Polymer Processing*
Robert C. Reid, *Rapid Phase Transitions from Liquid to Vapor*
John H. Seinfeld, *Atmospheric Diffusion Theory*

Volume 13

Edward G. Jefferson, *Future Opportunities in Chemical Engineering*
Eli Ruckenstein, *Analysis of Transport Phenomena Using Scaling and Physical Models*
Rohit Khanna and John H. Seinfeld, *Mathematical Modeling of Packed Bed Reactors: Numerical Solutions and Control Model Development*
Michael P. Ramage, Kenneth R. Graziano, Paul H. Schipper, Frederick J. Krambeck, and Byung C. Choi, *KINPTR (Mobil's Kinetic Reforming Model): A Review of Mobil's Industrial Process Modeling Philosophy*

Volume 14

Richard D. Colberg and Manfred Morari, *Analysis and Synthesis of Resilient Heat Exchanger Networks*
Richard J. Quann, Robert A. Ware, Chi-Wen Hung, and James Wei, *Catalytic Hydrometallation of Petroleum*
Kent David, *The Safety Matrix: People Applying Technology to Yield Safe Chemical Plants and Products*

Volume 15

Pierre M. Adler, Ali Nadim, and Howard Brenner, *Rheological Models of Suspensions*
Stanley M. Englund, *Opportunities in the Design of Inherently Safer Chemical Plants*
H. J. Ploehn and W. B. Russel, *Interations between Colloidal Particles and Soluble Polymers*

Volume 16

Perspectives in Chemical Engineering: Research and Education

Clark K. Colton, *Editor*

Historical Perspective and Overview

L. E. Scriven, *On the Emergence and Evolution of Chemical Engineering*
Ralph Landau, *Academic–Industrial Interaction in the Early Development of Chemical Engineering*
James Wei, *Future Directions of Chemical Engineering*

Fluid Mechanics and Transport

L. G. Leal, *Challenges and Opportunities in Fluid Mechanics and Transport Phenomena*
William B. Russel, *Fluid Mechanics and Transport Research in Chemical Engineering*
J. R. A. Pearson, *Fluid Mechanics and Transport Phenomena*

Thermodynamics

Keith E. Gubbins, *Thermodynamics*
J. M. Prausnitz, *Chemical Engineering Thermodynamics: Continuity and Expanding Frontiers*
H. Ted Davis, *Future Opportunities in Thermodynamics*

Kinetics, Catalysis, and Reactor Engineering

Alexis T. Bell, *Reflections on the Current Status and Future Directions of Chemical Reaction Engineering*
James R. Katzer and S. S. Wong, *Frontiers in Chemical Reaction Engineering*
L. Louis Hegedus, *Catalyst Design*

Environmental Protection and Energy

John H. Seinfeld, *Environmental Chemical Engineering*
T. W. F. Russell, *Energy and Environmental Concerns*
Janos M. Beer, Jack B. Howard, John P. Longwell, and Adel F. Sarofim, *The Role of Chemical Engineering in Fuel Manufacture and Use of Fuels*

Polymers

Matthew Tirrell, *Polymer Science in Chemical Engineering*
Richard A. Register and Stuart L. Cooper, *Chemical Engineers in Polymer Science: The Need for an Interdisciplinary Approach*

Microelectronic and Optical Materials

Larry F. Thompson, *Chemical Engineering Research Opportunities in Electronic and Optical Materials Research*
Klavs F. Jensen, *Chemical Engineering in the Processing of Electronic and Optical Materials: A Discussion*

Bioengineering

James E. Bailey, *Bioprocess Engineering*
Arthur E. Humphrey, *Some Unsolved Problems of Biotechnology*
Channing Robertson, *Chemical Engineering: Its Role in the Medical and Health Sciences*

Process Engineering

Arthur W. Westerberg, *Process Engineering*
Manfred Morari, *Process Control Theory: Reflections on the Past Decade and Goals for the Next*
James M. Douglas, *The Paradigm After Next*
George Stephanopoulos, *Symbolic Computing and Artificial Intelligence in Chemical Engineering: A New Challenge*

The Identity of Our Profession

Morton M. Denn, *The Identity of Our Profession*

Volume 17

Y. T. Shah, *Design Parameters for Mechanically Agitated Reactors*
Mooson Kwauk, *Particulate Fluidization: An Overview*

Volume 18

E. James Davis, *Microchemical Engineering: The Physics and Chemistry of the Microparticle*
Selim M. Senkan, *Detailed Chemical Kinetic Modeling: Chemical Reaction Engineering of the Future*
Lorenz T. Biegler, *Optimization Strategies for Complex Process Models*

Volume 19

Robert Langer, *Polymer Systems for Controlled Release of Macromolecules, Immobilized Enzyme Medical Bioreactors, and Tissue Engineering*
J. J. Linderman, P. A. Mahama, K. E. Forsten, and D. A. Lauffenburger, *Diffusion and Probability in Receptor Binding and Signaling*
Rakesh K. Jain, *Transport Phenomena in Tumors*
R. Krishna, *A Systems Approach to Multiphase Reactor Selection*
David T. Allen, *Pollution Prevention: Engineering Design at Macro-, Meso-, and Microscales*
John H. Seinfeld, Jean. M. Andino, Frank M. Bowman, Hali J. L. Forstner, and Spyros Pandis, *Tropospheric Chemistry*

Volume 20

Arthur M. Squires, *Origins of the Fast Fluid Bed*
Yu Zhiqing, *Application Collocation*
Youchu Li, *Hydrodynamics*
Li Jinghai, *Modeling*
Yu Zhiqing and Jin Yong, *Heat and Mass Transfer*
Mooson Kwauk, *Powder Assessment*
Li Hongzhong, *Hardware Development*
Youchu Li and Xuyi Zhang, *Circulating Fluidized Bed Combustion*
Chen Junwu, Cao Hanchang, and Liu Taiji, *Catalyst Regeneration in Fluid Catalytic Cracking*

Volume 21

Christopher J. Nagel, Chonghun Han, and George Stephanopoulos, *Modeling Languages: Declarative and Imperative Descriptions of Chemical Reactions and Processing Systems*
Chonghun Han, George Stephanopoulos, and James M. Douglas, *Automation in Design: The Conceptual Synthesis of Chemical Processing Schemes*
Michael L. Mavrovouniotis, *Symbolic and Quantitative Reasoning: Design of Reaction Pathways through Recursive Satisfaction of Constraints*
Christopher Nagel and George Stephanopoulos, *Inductive and Deductive Reasoning: The Case of Identifying Potential Hazards in Chemical Processes*
Keven G. Joback and George Stephanopoulos, *Searching Spaces of Discrete Solutions: The Design of Molecules Processing Desired Physical Properties*

Volume 22

Chonghun Han, Ramachandran Lakshmanan, Bhavik Bakshi, and George Stephanopoulos, *Nonmonotonic Reasoning: The Synthesis of Operating Procedures in Chemical Plants*
Pedro M. Saraiva, *Inductive and Analogical Learning: Data-Driven Improvement of Process Operations*
Alexandros Koulouris, Bhavik R. Bakshi and George Stephanopoulos, *Empirical Learning through Neural Networks: The Wave-Net Solution*
Bhavik R. Bakshi and George Stephanopoulos, *Reasoning in Time: Modeling, Analysis, and Pattern Recognition of Temporal Process Trends*
Matthew J. Realff, *Intelligence in Numerical Computing: Improving Batch Scheduling Algorithms through Explanation-Based Learning*

Volume 23

Jeffrey J. Siirola, *Industrial Applications of Chemical Process Synthesis*
Arthur W. Westerberg and Oliver Wahnschafft, *The Synthesis of Distillation-Based Separation Systems*
Ignacio E. Grossmann, *Mixed-Integer Optimization Techniques for Algorithmic Process Synthesis*
Subash Balakrishna and Lorenz T. Biegler, *Chemical Reactor Network Targeting and Integration: An Optimization Approach*
Steve Walsh and John Perkins, *Operability and Control in Process Synthesis and Design*

Volume 24

Raffaella Ocone and Gianni Astarita, *Kinetics and Thermodynamics in Multicomponent Mixtures*
Arvind Varma, Alexander S. Rogachev, Alexandra S. Mukasyan, and Stephen Hwang, *Combustion Synthesis of Advanced Materials: Principles and Applications*
J. A. M. Kuipers and W. P. M. van Swaaij, *Computational Fluid Dynamics Applied to Chemical Reaction Engineering*
Ronald E. Schmitt, Howard Klee, Debora M. Sparks, and Mahesh K. Podar, *Using Relative Risk Analysis to Set Priorities for Pollution Prevention at a Petroleum Refinery*

Volume 25

J. F. Davis, M. J. Piovoso, K. A. Hoo, and B. R. Bakshi, *Process Data Analysis and Interpretation*
J. M. Ottino, P. DeRoussel, S. Hansen, and D. V. Khakhar, *Mixing and Dispersion of Viscous Liquids and Powdered Solids*
Peter L. Silveston, Li Chengyue, Yuan Wei-Kang, *Application of Periodic Operation to Sulfur Dioxide Oxidation*

ISBN 0-12-008525-9